AF356606

Circular Economy Strategies for Sustainable Development: Applications and Impacts

Circular Economy Strategies for Sustainable Development: Applications and Impacts

Editor

Ana Ramos

Basel • Beijing • Wuhan • Barcelona • Belgrade • Novi Sad • Cluj • Manchester

Ana Ramos
Institute of Science and
Innovation in Mechanical and
Industrial Engineering
Porto
Portugal

Editorial Office
MDPI AG
Grosspeteranlage 5
4052 Basel, Switzerland

This is a reprint of the Special Issue, published open access by the journal *Sustainability* (ISSN 2071-1050), freely accessible at: www.mdpi.com/journal/sustainability/special_issues/circular_economy_strategies_for_sustainable_development.

For citation purposes, cite each article independently as indicated on the article page online and using the guide below:

Lastname, A.A.; Lastname, B.B. Article Title. *Journal Name* **Year**, *Volume Number*, Page Range.

ISBN 978-3-7258-1662-0 (Hbk)
ISBN 978-3-7258-1661-3 (PDF)
https://doi.org/10.3390/books978-3-7258-1661-3

Cover image courtesy of Ana Ramos

© 2024 by the authors. Articles in this book are Open Access and distributed under the Creative Commons Attribution (CC BY) license. The book as a whole is distributed by MDPI under the terms and conditions of the Creative Commons Attribution-NonCommercial-NoDerivs (CC BY-NC-ND) license (https://creativecommons.org/licenses/by-nc-nd/4.0/).

Contents

About the Editor

Ana Ramos

Ana Ramos is a researcher at INEGI - Institute of Science and Innovation in Mechanical and Industrial Engineering focused on developing waste-to-energy strategies for the sustainable production of renewable energy.

She holds a Chemistry degree, an M.Sc. in Environmental Analytical Chemistry, and a Ph.D. in Environmental Engineering from the University of Porto. After having worked and developed studies in the fields of circular economy and sustainability, Ana has recently earned an individual research project to conduct the work "Closing the loop on solid residues: co-gasification strategies for the sustainable production of renewable energy". In recent years, Ana has enriched her personal skills in project management, life cycle assessment, circular economy and sustainability, science communication, and environmental sustainability and is an active member of several COST actions and other international consortia. Her most recent research covers the assessment of waste-to-energy schemes, evaluating the environmental and socio-economic aspects of the techniques, namely for the thermal conversion of biomass and municipal solid wastes through gasification-assisted solutions.

So far, circa fifty scientific publications in top peer-reviewed journals have been achieved throughout her career, as well as two guest-edited special issues and two book chapters. Ana has been supervising M.Sc. and Ph.D. students in various fields of Engineering and Environmental Science, as well as coordinating and lecturing advanced training courses, mainly related to the circular economy and sustainability, to life cycle and social assessment, and to industry decarbonization themes.

Preface

The chapter "Circular Economy Strategies for Sustainable Development: Applications and Impacts" explores the potential of the circular economy as a strategic framework to foster sustainable development. By redefining traditional linear economic models that emphasize take–make–dispose patterns, this chapter shows how circular economy principles can mitigate environmental degradation, enhance resource efficiency, and promote social equity, providing a comprehensive overview of various circular economy approaches. It aims to offer a holistic understanding of how these strategies can be implemented across different sectors, such as manufacturing, fuels, and chemistry, and at organizational/business level. The purpose is to disclose the benefits of circular practices, with case studies and empirical evidence that underscore their impact on environmental sustainability, economic resilience, and societal well-being.

The motivation behind this scientific piece stems from the urgent need to address global challenges such as climate change, resource scarcity, and waste management, proposing innovative solutions that can harmonize economic growth with ecological preservation. This chapter seeks to contribute to the body of knowledge on sustainable development, showcasing circular economy strategies as viable pathways to a more sustainable future.

This chapter is directed to a diverse audience including researchers, policymakers, industry practitioners, and students. Researchers will find detailed analyses and frameworks that can inform further academic inquiry. Policymakers will gain insights into the regulatory and institutional mechanisms necessary to facilitate circular economy transitions. Industry practitioners can explore practical applications and best practices to integrate circular economy principles into their business models. Students and educators will benefit from the conceptual clarity and empirical data that support the integration of these concepts into educational curricula.

This is a collaborative effort by a team of interdisciplinary scholars and experts, each bringing unique perspectives and expertise to the subject. Together, they aim to present a balanced and comprehensive exploration of circular economy strategies for sustainable development.

The authors would like to express their deepest gratitude to all those who have contributed to the development of this chapter. Special thanks are due to the editorial team for their invaluable support and to all peer reviewers, whose constructive feedback has significantly enhanced the quality of this work. We extend heartfelt appreciation to our families and colleagues for their encouragement throughout the process.

We hope this chapter inspires and empowers stakeholders across various sectors to embrace circular economy strategies, ultimately contributing to a more sustainable and resilient global society.

Ana Ramos
Editor

Editorial

Editorial for the Special Issue "Circular Economy Strategies for Sustainable Development: Applications and Impacts"

Ana Ramos

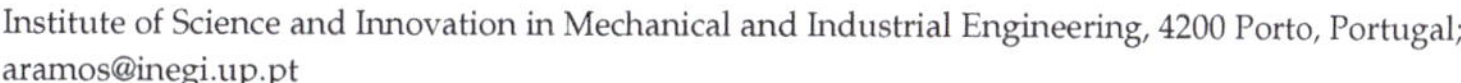

Institute of Science and Innovation in Mechanical and Industrial Engineering, 4200 Porto, Portugal; aramos@inegi.up.pt

Citation: Ramos, A. Editorial for the Special Issue "Circular Economy Strategies for Sustainable Development: Applications and Impacts". *Sustainability* **2022**, *14*, 12831. https://doi.org/10.3390/su141912831

Received: 28 September 2022
Accepted: 30 September 2022
Published: 8 October 2022

Publisher's Note: MDPI stays neutral with regard to jurisdictional claims in published maps and institutional affiliations.

Copyright: © 2022 by the author. Licensee MDPI, Basel, Switzerland. This article is an open access article distributed under the terms and conditions of the Creative Commons Attribution (CC BY) license (https://creativecommons.org/licenses/by/4.0/).

1. Introduction

The severe extraction of fossil resources and the extreme degradation of natural capital to attend to the increasing demands of production and consumption has generated a surplus of waste and emissions, the final destination of which is commonly landfills. Currently, the average temperature on Earth is already about 1.1 °C higher than it was in the late 1800s, and emissions continue to rise. To keep global warming lower than 1.5 °C (as requested by the Paris Agreement) emissions need to be reduced by 45% by 2030 and reach net zero by 2050 [1]. Nevertheless, the latest trends show that the actual commitments made by governments fall far short of what is expected. The combined current national climate plans for all 193 Parties to the Paris Agreement would lead to a sizable increase of almost 14% in global greenhouse gas emissions by 2030, compared to 2010 levels [2].

Apart from all of the environmental impacts associated with these events, they also represent a major bottleneck in the production chain and create new challenges for sustainable development and society in general. Circular economy has been proposed by the European Commission and other regulatory agencies and leading entities as a tool to enhance sustainability and its ambitions, including the attainment of carbon neutrality and promoting the efficient use of resources while keeping them in the economy at their highest value for the longest amount of time [3–6]. Within the myriad of possible alternatives to approach these issues, all of the domains concerning society as a whole should be considered, namely the environmental, economic, social, and technical spheres [4].

This Special Issue aims to show different aspects promoting the transition from the traditional linear economy to a circular ecosystem. Within this view, the economy is decarbonized, biodiversity is maintained, and natural resources are restored, while a more resilient and balanced society is enforced.

2. Motivation

The continuous and excessive production of goods and the expanded services offered to attend to society's requirements have led to an impasse in the environmental reservoirs and in the availability of non-renewable resources [5,6]. Mass consumption with no restrictions to the end use has had devastating effects, as reflected by the waste arriving along our shores. Indeed, the pollution generated while seeking to satisfy today's needs is another factor contributing to an announced destruction scenario, where biodiversity is endangered and social dissimilarities are enlarged. Some of the ongoing changes in the Earth's climate systems are already considered irreparable, implying a state of planetary emergency [7].

Nine planetary boundaries have been defined as safe operating spaces for humanity, supporting the long-lasting and thriving development for generations to come. Currently, more than half of these boundaries have already been crossed, compromising this safety zone and the overall wellbeing of the planet: two of these planetary boundaries are situated in an "uncertainty zone", while three are considered as having a high risk level, reinforcing the threat for large-scale abrupt and irreversible changes [8]. A catastrophic situation may escalate from a cascade of tipping points, with multiple earth systems reaching a point of no return. The loss of the West Antarctic ice sheet and the losses in the Amazon rainforest as well as the extensive melting of permafrost and other key components of the climate system are close to cross critical thresholds that will lead to steep and irrevocable changes, with the main responsibility being attributed to climate change.

These are the main challenges that our biosphere and communities face in the upcoming decades, and immediate action needs to be taken to reverse their progression. Hence, it is imperative to act fast, with unparalleled changes being taken to limit the climate change catastrophes and guaranteeing sustainable development from now on.

3. Background and Contents

Circular economy appears as a response to the need to create a cohesive and flourishing society. It guarantees sustainable development and promotes the regeneration of resources in a safe and clean environment [9]. Specific circular strategies include designing longer-lasting products, enhanced services, and business models (among other approaches); calling upon alternative reusable, repairable, and recyclable options to lower fossil-based resource consumption; and using bio-based materials and more energy-efficient processes to emit less or no CO_2 emissions. These contribute to limiting the pollution of natural systems and restoring biodiversity, as resource recirculation in value cascades or repurpose routes are applied instead of end-of-life scenarios [10]. Decarbonizing the economy and decoupling growth from intensive consumption are the ultimate goals to achieve sustainability and a more resilient society [9].

This Special Issue is dedicated to original full-length works, reviews, and case study applications describing circular economy schemes as well as their impacts in the environment and in society as enablers of a healthier environment, a balanced humankind, and the successful achievement of sustainable development.

Funding: This work is funded by national funds through the FCT—Fundação para a Ciência e a Tecnologia, I.P. and, when eligible, by COMPETE 2020 FEDER funds, under the Scientific Employment Stimulus—Individual Call (CEEC Individual)—2021.03036.CEECIND/CP1680/CT0003.

Conflicts of Interest: The author declares no conflict of interest.

References

1. United Nations. Paris Agreement. In Proceedings of the Conference of the Parties to the United Nations Framework Convention on Climate Change (21st Session), Paris, France, 12 December 2015.
2. United Nations. *Climate Action—For a Livable Climate: Net-Zero Commitments Must Be Backed by Credible Action*; United Nations: New York, NY, USA, 2022.
3. European Commission. *COM(2020) 98 Final, A New Circular Economy Action Plan for a Cleaner and More Competitive Europe*; European Commission: Brussels, Belgium, 2020.
4. Ekins, P.; Gupta, J.; Boileau, P. *Global Environment Outlook—GEO-6: Healthy Planet, Healthy People*; Cambridge University Press: Cambridge, UK, 2019; p. 745.
5. OECD. *Global Material Resources Outlook to 2060 Economic Drivers and Environmental Consequences*; OECD Publishing: Paris, France, 2019.
6. International Resource Panel, United Nations Environment Programme. *Global Resources Outlook: 2019*; International Resource Panel, United Nations Environment Programme: Paris, France, 2019.
7. Circle Economy. *The Circularity Gap Report 2021*; Circle Economy: Amsterdam, The Netherlands, 2021; p. 37.
8. Stockholm Resilience Center. *The Nine Planetary Boundaries*; Stockholm Resilience Center: Stockholm, Sweden, 2015. Available online: https://www.stockholmresilience.org/research/planetary-boundaries/planetary-boundaries/about-the-research/the-nine-planetary-boundaries.html (accessed on 10 April 2021).
9. Schulze, G. *Growth within: A Circular Economy Vision for a Competitive Europe*; Ellen MacArthur Foundation, the McKinsey Center for Business Environment: Cowes, UK, 2016; pp. 1–22.
10. European Commission. *COM(2018) 773 Final, A Clean Planet for All. A European Strategic Long-Term Vision for a Prosperous, Modern, Competitive and Climate Neutral Economy*; European Commission: Brussels, Belgium, 2018.

MDPI

Review

A Systematic Review on Seaweed Functionality: A Sustainable Bio-Based Material

Pranav Nakhate and Yvonne van der Meer *

Aachen Maastricht Institute for Biobased Materials, Faculty of Science and Engineering, Maastricht University, Urmonderbaan 22, 6167 RD Geleen, The Netherlands; n.nakhate@maastrichtuniversity.nl
* Correspondence: yvonne.vandermeer@maastrichtuniversity.nl; Tel.: +31-6-46705568

Abstract: Sustainable development is an integrated approach to tackle ongoing global challenges such as resource depletion, environmental degradation, and climate change. However, a paradigm shift from a fossil-based economy to a bio-based economy must accomplish the circularity principles in order to be sustainable as a solution. The exploration of new feedstock possibilities has potential to unlock the bio-based economy's true potential, wherein a cascading approach would maximize value creation. Seaweed has distinctive chemical properties, a fast growth rate, and other promising benefits beyond its application as food, making it a suitable candidate to substitute fossil-based products. Economic and environmental aspects can make seaweed a lucrative business; however, seasonal variation, cultivation, harvesting, and product development challenges have yet not been considered. Therefore, a clear forward path is needed to consider all aspects, which would lead to the commercialization of financially viable seaweed-based bioproducts. In this article, seaweed's capability and probable functionality to aid the bio-based economy are systematically discussed. The possible biorefinery approaches, along with its environmental and economic aspects of sustainability, are also dealt with. Ultimately, the developmental process, by-product promotion, financial assistance, and social acceptance approach are summarized, which is essential when considering seaweed-based products' feasibility. Besides keeping feedstock and innovative technologies at the center of bio-economy transformation, it is imperative to follow sustainable-led management practices to meet sustainable development goals.

Keywords: sustainable development goals; bio-based economy; seaweed functionality

Citation: Nakhate, P.; van der Meer, Y. A Systematic Review on Seaweed Functionality: A Sustainable Bio-Based Material. *Sustainability* **2021**, *13*, 6174. https://doi.org/10.3390/su13116174

Academic Editor: Ana Ramos

Received: 27 April 2021
Accepted: 24 May 2021
Published: 31 May 2021

Publisher's Note: MDPI stays neutral with regard to jurisdictional claims in published maps and institutional affiliations.

Copyright: © 2021 by the authors. Licensee MDPI, Basel, Switzerland. This article is an open access article distributed under the terms and conditions of the Creative Commons Attribution (CC BY) license (https://creativecommons.org/licenses/by/4.0/).

1. Highlights

- A systematic review on seaweed functionality;
- Seaweed characteristics and potential value chains;
- The exploitation of quantitative and qualitative analysis seaweed value chain benefits;
- Summary of the "PBFS" approach, essential for considering seaweed-based products' feasibility.

2. Introduction

'Sustainability' is a central topic in today's research, and is a crucial pillar in the socio-political landscape. The world's population is expected to reach 9 billion by 2050, putting prodigious pressure on environmental resources [1]. Climate change, resource depletion, and toxicity potentials are the concerning threats to the society for which a new solution is urgently needed. Focusing on reducing our dependency on the fossil-based economy and shifting toward a bio-based economy could help in tackling these situations, as well as for achieving sustainable development goals (SDG's) [2]. The accomplishment of a bio-based economy is entirely dependent on the utilization of renewable resources, especially biomass, to produce multi-functional applications, including food, animal feed, bio-based materials, energy, and pharmaceuticals. Recently, much focus has been given to either producing novel bio-based materials, or replacing the existing fossil-based products by the scientific community and start-up industries [3]. Moreover, the prevailing question arises: can sustainability and economic growth synchronize to lead the way forward? Policymakers and researchers must answer this question and set up a framework within the boundaries of the circular economy.

The circular economy is the most discussed and promoted concept, wherein the life cycle of materials, products, and resources are extended as much as possible to extract their economic benefits [1]. The bio-based economy can manage resources efficiently, thereby improving the life cycle of the system at every stage by minimizing waste and improving economic benefits [4]. The European Union (EU) has made considerable efforts in making the bio-economy sustainable by establishing a European bio-based industries-joint undertaking (BBI-JU) fund worth 3.7 billion euros via public-private partnership (PPP) [5]. Moreover, the EU has established initiatives to encourage and help the bio-based industry to interlink within new value chains. It is estimated [6] that the bio-based industry generated nearly 2.3 trillion euros worth of turnover in 2015, wherein, 50% was contributed by food and beverage related industries, ~17% was accompanied by the agricultural industry ~8% by the paper industry [7]. The bio-based economy is at its early stage, albeit having created more than 22 million job opportunities in 2012, with the numbers increasing yearly [8]. The United States (US) has also made a significant effort to boost the bio-based economy, as 4.2 million new employees were created with nearly a USD 400 billion contribution to the US economy in 2014 [9].

Camia et al., 2018 [10] reported that nearly 1400 Mt dry matter is produced in Europe, out of which approximately 950 Mt comes from the agricultural sector and 150 Mt from the forestry sector. It was also estimated that out of 1 billion tons of available biomass, the EU utilizes roughly 60% of it in the food sector, 20% in the bio-based energy sector, and 19% in the novel bio-based material sector [7]. The EU chemical sector used 77.7 Mt of organic raw material, with 10% of its share coming from renewable material [10]. The prospects for the bio-based economy look promising; however, the cost associated with the development of bio-based materials is significantly dependent upon the availability and efficacy of feedstock [1].

In the past few decades, the environmental outlook has changed from loss of biodiversity to resource depletion and climate change. As a result, seaweed has been getting exponential interest as an innovative feedstock for bio-based materials from different sectors as a sustainability target. From an industrial point of view, such biomass has potential application in various fields, including pharmaceutical, food, feed, cosmetics, bioenergy, etc. A sufficient equilibrium between social, environmental, and economic performances can set up a benchmark for future sustainable development, and newly established industries can benefit from these astonishing consequences. However, most of such industries are inceptive, and even institutional research is rudimentary. Therefore, it is valuable to cumulate the recent progress, coherent potential value chains, and subsequent sustainability impacts in a comprehensive overview.

According to Scopus, nearly 7000 articles related to seaweed applications were published in 2020 alone, wherein, a significant focus was placed upon agricultural and medicinal applications. Subsequently, >5000 reviews in total have been published so far; however, ~70% of such articles discuss a seaweed strain. The remaining review articles have focused on domain-specific applications. It is essential to understand that seaweed's application suitability has not been evaluated in every case. Moreover, the economic, social, and environmental aspects of seaweed's value chains are discussed collectively in just one article [11]; despite this, the entire value chain's inclusiveness is lacking. The approach to gathering the information presented in this review is shown in Figure 1 below.

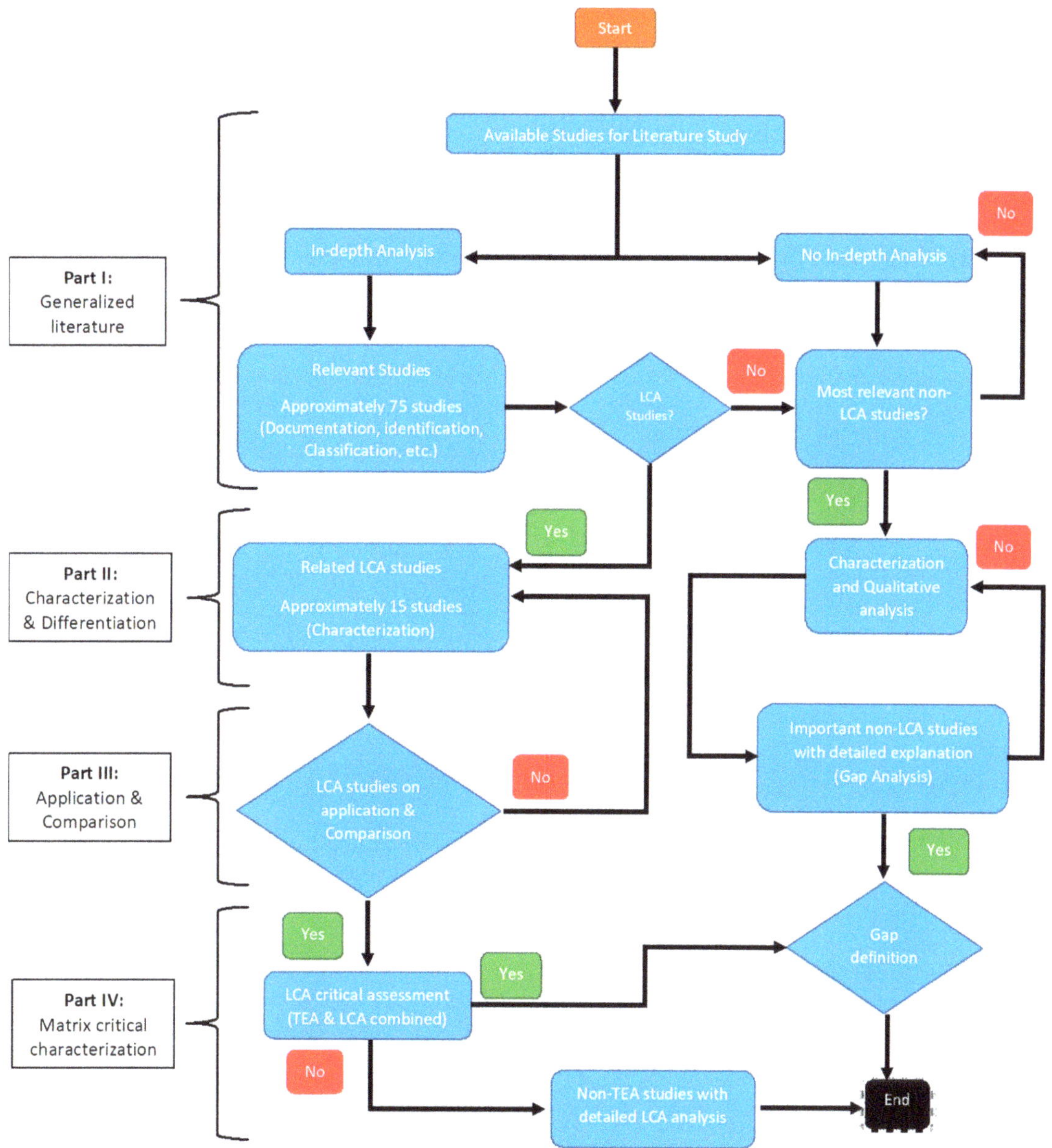

Figure 1. Literature survey strategy.

The prerequisite of the seaweed value chain can provide a valuable addition to the actual delivery of the product through nutrient uptake and carbon sequestration. The primary aim of the present manuscript is to assess whether seaweed biomass can deal with ongoing sustainability issues, and to identify routes of potential improvement. First, the manuscript discusses the current seaweed market and seaweed characteristics. Next, it deals with the various types of seaweed cultivation and harvesting, as well as seaweed functionality and its application in various fields, in order to give a comprehensive overview to the instigators. Later, this manuscript discusses the aspects surrounding sustainability, including environmental and economic perspectives. Finally, a PBFS approach is discussed to develop a trajectory for future scenarios. In this review, the latest information on seaweed functionality and its sustainability proceedings is collected, which may help policymakers, industries, and researchers to further develop a bio-based economy.

3. What Is Seaweed?

Seaweed is a macroscopic alga, with its usual habitat at the bottom of shallow coastal waters. It typically grows at a depth of 180 m, and can be found on rocks, pebbles, shells, black water, and seawater plants. They are classified into green algae (Chlorophyceae), red algae (Rhodophyceae), and brown algae (Phaeophyceae) based on pigmentation. Approximately 624 algal species have been identified in India alone, wherein green seaweed contributes nearly 72%, red algae 27%, and brown algae nearly 1% [12]. It is essential to understand that seaweeds are the only species responsible for producing several phytochemicals, including agar-agar, align, and carrageen which has extensive additional applications [13]. Various consumptive and non-consumptive applications such as, fertilizers, animal feed, medicines, building materials, soups, sushi, salads, and other snacks are mentioned in the literature [14]. Seaweeds have been prominently used as a feedstock for thousands of years, and are mentioned in Greek texts extensively. It is also reported that during the scarcity of fodder, seaweeds were dried and fed to horses, sheep, and cattle until the early 1900s [15].

Recently, the EU and various countries' interest have increased for seaweed's cultivation and application due to its multi-dimensional functionality. However, historical aspects of seaweed functionality must be considered in order to gain a better understanding. The first commercial use of seaweed was reported in the 17th century, where it was used to replace wood ash in glass production, especially in France and Norway [14]. Norway followed a similar path; seaweed burning to produce potash was one of the significant incomes during the 18th century. It was reported that during 1913, 150,000 tons of Kelp (a brown seaweed) was dried and burned for the production and export of 6000 tons of potash [14]. In France, seaweed was also an essential ingredient in the production of iodine in 1823. The production of seaweed meals was first industrialized in 1937, produced in nine factories, all of which are still functional [14]. Traditionally, algae cultivation was a household activity with a focus on agricultural purposes only. However, the first commercial algal harvesting plant was established in 1947 in the western part of Ireland to harvest enough algae for feed and food applications. Though seaweed has comprehensive functionality, it was advised to focus on the techno-economic improvements in seaweed farming and cultivation, as knowledge and skills are not yet advanced enough to encourage stakeholders to invest.

4. Seaweed Market

The Food and Agricultural Organization (FAO) of the USA estimated that nearly 30 Mt of seaweed was utilized in 2014, wherein most of it was produced via aquaculture and nearly 6% was harvested through wild species. Seaweed usage has been increased by nearly 176% since 1995 due to the scientific and technical enhancement in harvesting practices worldwide [16]. The world seaweed product market is dominated by the Philippines, Indonesia, and China, whereas European countries and the USA are the latest emerging players. It is estimated that if a special economic zone can be implemented for the marine aquaculture program globally, the seaweed yield could be increased exponentially to 300–1120 Mt [17,18]. The conversion of at least half of this available seaweed biomass into valuable products such as biogas would neutralize the natural gas import in the USA. As seaweed's product demand, production, and domestication increase, an improvement in seaweed biotechnology is ultimately encouraged [19].

Subsequently, research and innovation have increased exponentially in the past decade, resulting in many research publications and patents [20]. Although publications and patents do not necessarily indicate the market overview, it certainly shows a potential way forward [21]. A total of 15 countries have filed patents on seaweed and its application. Nearly 37% of the total scientific publications are within the food sector, followed by agricultural (19%) and human health applications (13%). Out of 795 seaweed species registered, 31 species were subjected to a patent application, and 138 species were discussed in publications [19]. The seaweed patent registration scenario in the EU is presented in Figure 2.

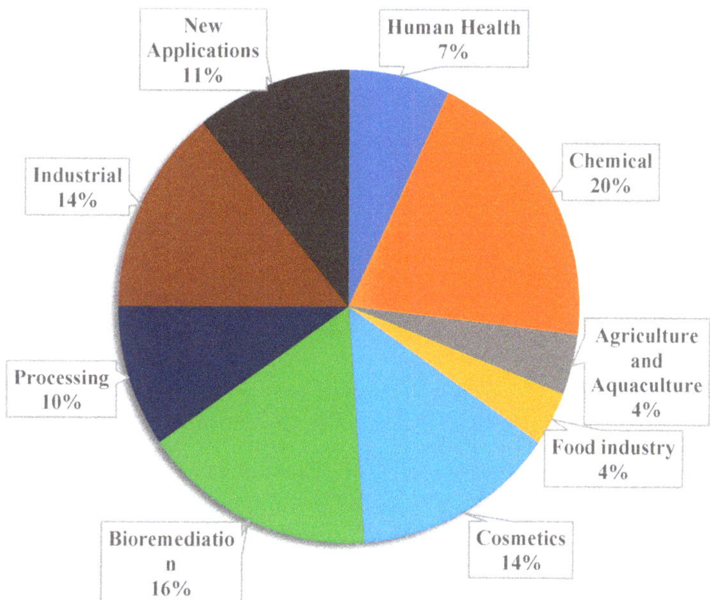

Figure 2. Seaweed patent registration in Europe per application.

5. Seaweed Characteristics

The most significant advantage with seaweed is that it does not contain distinct parts usually found in other terrestrial plants; therefore, the whole seaweed can be used as biomass [22]. Most of the green seaweeds, including *Monostroma, Caulerpa, and Ulva* species, are discussed in the literature for their capability to produce *Ulva. Ulva is* a complex acidic polysaccharide, which has cosmetic and medical applications [22,23]. In red seaweed, *Palmaria palmata, Kapapaphysus, Pyropia Yezoensis,* species are widely discussed to produce a large quantity of carrageenan, agar with applications in pharmaceutical, textile, paint, antibiotic, food, and biotechnology sectors [24]. Brown seaweed, *Ascophyllum nodosum,* and *Laminaria* are known for their capacity to produce various polysaccharides (alginate) with the primary application as a thickener, stabilizer, and gelling agent [22,25,26]. Typical seaweed characterization is presented in Table 1 below. Seaweed has a relatively similar composition to other bioenergy crops. Seaweed has excellent nutritional value and has a higher polysaccharide content, making it suitable for dried and fresh vegetables, as an ingredient for the commercial production of phycocolloids [27], and a perfect candidate in the changing fuel market [28].

Seaweeds are usually characterized by high ash content (13–30%), protein content (5–43%), fibers (5–53%), and a low amount of fats. Few macroalgae have high lipid content in their biomass, whereas some other macroalgae are considered to have high protein content. Their composition may vary according to the type, geographic condition, seasons, and cultivation method. Nevertheless, it is documented that green and red algal species have high carbohydrate content; moreover, brown algal species have a high protein and carbohydrate content [29].

Table 1. Typical Seaweed Characterization.

Constituents	Brown Seaweed (Phaeophyta)	Red Seaweed (Rhodophyta)	Green Seaweed (Chlorophyta)
Most discussed species	• *Ascophyllum nodosum,* • *Laminaria digitata,* • *Alaria esculanta*	• *Palmaria palmata,* • *Kapapaphysus,* • *Pyropia Yezoensis*	• *Monostroma,* • *Caulerpa,* • *Ulva* species
Water content	50–75%	60–88%	60–80%
Ash content	15–25%	15–30%	8–20%
Carbohydrate content	Alginate 15–30% Mannitol 5–10%	30–45%	30–45%
Protein	4–10%	8–40%	15–25%
Fats	1–4.5%	0.3–2%	0.6–0.7%
Potassium	3–4%	5–7%	0.7–1%
Magnesium	0.5–0.9%	0.4–0.5%	0.2–0.5%
Sodium	1–3%	1–3%	2–4%
Other	Fucoidan 4–10%	Xylan 20–40%	Xylan 30–40%

5.1. Lipids

As noted previously, lipid content is relatively lower in seaweed than in other plants, including sunflower and soy. Moreover, its lipid quality is of great interest, as it contains essential fatty acids, including omega-3 and other fat-soluble vitamins [30]. It was observed that the lipid content in brown seaweed varies between 1 to 4.5 g/100 g of dry seaweed biomass; however, there have been many discussions in the literature about its specific lipid content [31]. Few publications discuss the phospholipid as a significant lipid constituent in seaweed [32], while others argue that glycolipids are the prominent constituents [33,34]. The seasonal variation has a crucial impact on lipid content as well as on seaweed growth. In one study, it was observed that *Undaria*, a green seaweed, reaches a maximum height of 4 m by the spring; however, it degenerates in summer [35]. The lipid content was observed to increase during winter, and this also declined in summer. The same observation was noted by Nelson et al., 2002 of *Egregia menziesii,* a brown seaweed, wherein the highest lipid content of 13.3 mg/g was observed in spring, and the lowest lipid content of 6.3 mg/g was observed in summer [36].

5.2. Carbohydrates

Carbohydrates are the primary constituent of seaweed, with these accounting for at least 40% of its mass. The majority (approximately 45%) of the ash-free volatile solids found in *Sargassum*, an all-season brown seaweed, are complex carbohydrates. Fucoidan, a sulfated polysaccharide, has been extracted from brown seaweed in recent years [37]. *Ulva*, the most discussed green seaweed, is observed to have a high content of monosaccharides, including rhamnose, glucose, uronic acid, etc., [38]. In the case of red seaweed, *Gracilaria*, galactose and glucose are found to have a dominant presence. The variation in carbohydrate content also originates from the season, temperature and is species-dependent [39].

5.3. Proteins

Protein content is an essential aspect of seaweed functionality. However, it varies according to species, geographical region, growth environment, and season. It has been observed that brown seaweed contains a low protein concentration (4–10%), and green seaweed contains a moderately high protein concentration (15–25%). The highest protein concentration was observed in red seaweed (8–40%), which has a high commercial importance [40]. In the case of the *Laminaria digitata* species of brown seaweed, a protein content of nearly 15% has been reported, whereas in the case of *Undaria pinnatifida,* a species of brown seaweed, a protein content of up to 24% has been reported [41]. In green seaweed, seasonal changes have been found to affect the protein content, as evidenced by the fact that the protein content of *Ulva armoricana*, a species of green seaweed was found to be 18% in October and 24% in February [42], respectively. In the case of *Porphyra tenera*, a species of red seaweed, protein content was reported to reach up to 47% during ideal circumstances [43]. It was hypothesized that the seasonal variation in seaweed protein content is catalyzed by the nitrogen nutrients found in the cultivation water. Salinity was also found to impact seaweed growth directly, as it was observed that the algal growth slows down in the rainy season due to a decrease in salinity, resulting in a decreased protein concentration [44]. The light intensity was also found to impact protein content significantly, as light-harvesting pigments bound by phycobiliprotein were produced by red seaweed [45].

5.4. Pigments

Seaweeds are also reported to be a source of pigments that are of commercial importance. As mentioned previously, phycobiliproteins are the primary light-harvesting pigments observed in red seaweed, which are the only water-soluble pigments [46] and contribute nearly 50% of its total protein content. It is commercially significant in various applications, including dairy products, wasabi, and gums. Brown seaweed dominates the Fucoxanthin, a Xanthophyll pigment responsible for the brown coloration. It has been reported that the Fucoxanthin is the most stable compound, which can get through the drying and storage process at room temperature [47]. Moreover, it has been reported that chlorophyll-a contributes nearly 1.5% of seaweed organic content [48].

5.5. Other Constituents

The algal phenolic compounds are called Phlorotannins and are derived from the Phloroglucinol unit, i.e., 1–3-trihydroxybenzene. The brown seaweed is found to have a rich concentration of Phlorotannins. Researchers have found immense interest in replacing synthetic antioxidants with natural seaweed-derived antioxidants in the last decade [47,48]. Apart from antioxidants, seaweeds are a rich source of vitamins, including A, B-complex, C, D, niacin, folic acid, etc. The brown seaweed named *Ascophyllum* and *Fucus* has a higher vitamin E concentration than any other seaweed species. As the seaweed is grown in marine environments, it has a high concentration of minerals, including Mg, Ca, Cu, Fe, as well as some rare minerals [49]. In recent studies, seaweed was also found to have bioactive compounds that can be used as secondary metabolites.

6. Seaweed Cultivation

Seaweed cultivation is not frequently discussed in the literature, despite the fact that seaweed has a rich history of applications. Globally, however, few seaweed species are commercially cultivated. Cultivation methods are traditional in most cases, and the cultivation areas are significantly smaller in size than other vegetables. In the last decade, the focus has been on seaweed cultivation at the molecular level after realizing that seaweed has substantial economic potential. The commercial cultivation of seaweed is carried out at a sizable scale throughout the world. The whole process is mainly distinguished into juvenile plant (seed stock) cultivation, adult plant cultivation, harvesting, and final processing [50]. Seed stock cultivation is the primary step that determines the success of the whole seaweed cultivation process. It depends upon the quality of the crop and seasonal variations. The actual cultivation process starts with adult plant cultivation, wherein the seaweed grows to the full extent and is divided into microscopic and macroscopic processes. The macroscopic process starts with algal fragments, and only two life cycle stages are under control during the cultivation process. The microscopic cultivation process is a relatively new method that starts with the seaweed's microscopic pores, and the entire life cycle of seaweed is under control during the cultivation process. Based on the cultivation environment, the cultivation process is performed in five different setups:

1. Land-based—indoor cultivation;
2. Land-based—outdoor pond, open-air cultivation;
3. Shallow sea—semi-floating or pole system;
4. Shallow sea—U-type floating system (in a subtidal zone);
5. Shallow sea fixed support system (in an intertidal zone).

Harvesting is the ultimate purpose of seaweed cultivation, and the seaweed is harvested using different advanced techniques mentioned elsewhere [51]. However, the processing is the final, partially dependent stage in seaweed cultivation, carried out to produce the intended product [50].

7. Sustainability in Seaweed Cultivation

As seaweed cultivation and its application are still at the development stage, environmental sustainability may encourage future establishments. Environmental sustainability can be assessed using life cycle assessment (LCA), which evaluates both the benefits and burdens associated with the whole life cycle of seaweed, from seaweed production to the application and the end-of-life stages. In general, LCA appears to be one of the essential techniques to quantify various environmental impacts associated with a product, process, system, or service from cradle-to-grave, based on the ISO 14040 series guidelines [52,53]. The LCA methodology considers the life cycle starting from extraction of raw materials, manufacturing a product, transport, distribution, use, and end-of-life, including waste collection, segregation, treatment, recycling, and disposal [54]. However, very few studies are available in the literature that solely focuses on the seaweed cultivation's life cycle assessment. Most of the available literature focuses on the application and uses of seaweed. The seaweed applications in the different sectors and the related LCA studies are discussed in the latter part of this paper; however, the use of algal biomass in biorefinery as third-generation feedstock has gained a lot of interest worldwide. The value chain of seaweed cultivation is presented in Figure 3 below.

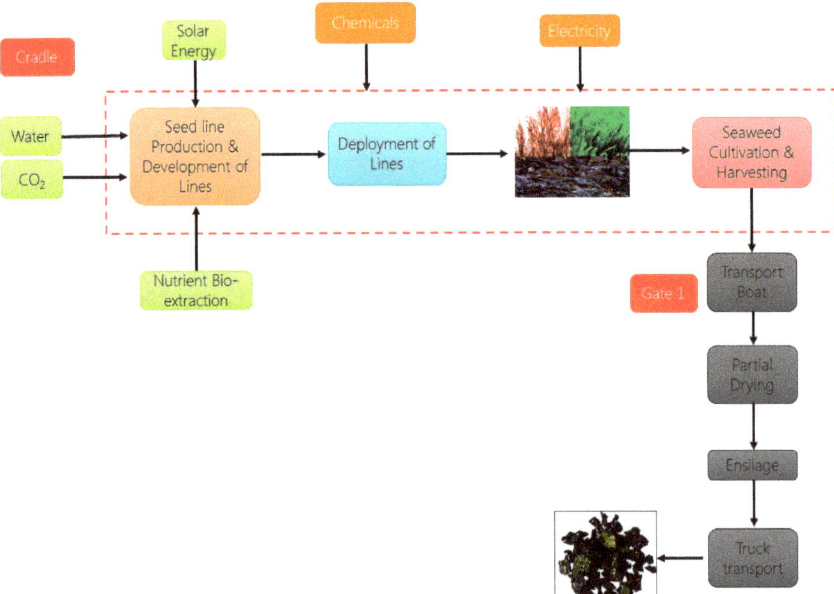

Figure 3. Seaweed Cultivation & Harvesting System Boundary.

The life cycle for seaweed cultivation started with the seed line production and development of lines, which is the juvenile plant cultivation process mentioned previously. The development of lines and seaweed harvesting is considered inside the system boundary by most available studies [54,55]. The inventories required for developing the LCA model of seaweed cultivation processes include electricity, water, seashore land required, nutrients, such as phosphorus (P), magnesium (Mg), zinc (Zn), nitrogen (N). Researchers have found that carbon dioxide is required to grow juvenile seaweed, which means that seaweed utilizes the oceanic carbon from the water column directly for their growth, and reduces oceanic carbon content [55].

At present, most seaweed cultivation activities are carried out in limited coastal locations; therefore, it is estimated that a maximum of 2.48 million tons of carbon have been extracted from the ocean, which is nearly 0.4% of the total expected oceanic carbon. The agricultural sector is expected to produce nearly 30% of the total global warming gases; seaweed has a great potential to reduce these emissions due to its efficient carbon sequestration. After the cultivation and harvesting process, seaweed is partially dried and transported to the next facility to produce intermediates and products. The seaweed cultivation output usually contains emissions such as liquid waste, solid waste, and seaweed as a product itself. The starch and lipid content of the seaweed can also be considered in the output based on the intended application of the sustainability studies [56]. The available literature focusing on the sustainability assessment of seaweed cultivation is presented in Table 2.

Table 2. Life Cycle Assessment of Seaweed Cultivation.

Sr. No.	LCA Study	FU	System Boundary	Impacts Studied *	Hotspot for Environmental Impact	Ref.
1	Seaweed as a Feedstock for Energy and Feed	1 ha of offshore cultivation area	Cradle-to-Gate 1	(ReCiPe Midpoint) CC, CED-T, ME FWE, HT-C, HT-NC	Design of Seeded line, Low protein output, Land occupation impacts	[57]
2	LCA for system design of Seaweed cultivation and drying	1 ton protein production from dried seaweed	Cradle-to-Gate 1	(CML 2001) OLD, HTP, FAETP, MAETP, TETP, AD, EP, CC	Design of dried Seaweed production system, Protein production, the drying process	[58]
3	LCA of microalgae cultivation	1 ha area of cultivation	Cradle-to-Gate 2	(CML-2001) GWP, AP, ODP, EP, HTP	Diesel consumption for harvesting, Hatchery, Cultivation process	[59]
4	Comparative LCA of two Seaweed cultivation systems	18 ha of floating longlines, and 0.6 ha of raft systems	Cradle-to-Gate 1	Exergetic life cycle Assessment expressed in $MJ_{ex} \cdot MJ_{ex}^{-1}$	Distance between cultural ropes, Biomass Yield, Hatchery, the distance between facilities	[55]
5	LCA of Biomethane from offshore cultivated Seaweed	1 km trip with a gas-powered car	Cradle-to-Grave	CC, ODP, FEW, TA, ME, FD, WD	Electricity consumption during the drying process, Drying process	[60]
6	LCA of Seaweed-based Biostimulant production	Production of 1 kg of Extract	Cradle-to-Gate 2	(ReCiPe Midpoint)ME, FWE, FD, HTP, IR, WD, TA	Electricity, Cultivation Process	[61]
7	LCA of Seaweed-based Biostimulant production	Production of 1 kL of Extract	Cradle-to-Gate 2	(ReCiPe Midpoint)ME, FWE, FD, HTP, IR, WD, TA	Electricity, Cultivation Process	[62]

* The abbreviations are presented at the end of the manuscript.

Few LCA studies were published on the seaweed's cultivation (Cradle-to-Gate-1) to its application (Cradle-to-Gate-2). Most LCA studies have focused on cultivation and energy application. Several flaws were identified using the life cycle assessment model after analyzing the seaweed cultivation system listed in Table 2. It has been consistently observed that the electricity required throughout the cultivation system has contributed to significant impacts in all the impact categories and resulted in resource depletion, global warming potential, and increased toxicity potential. The use of renewable energy to replace traditional energy may improve the overall environmental performance of the system. The design of seaweed lines has caused an increase in many of the impacts. Interestingly, it has been reported that as the distance between the rope increases from 2 m to 5 m, the resource demand was found to be increased, resulting in a nearly 4% increase in its environmental impacts [55]. Similarly, the alternate light floating material for the raft system was discussed in the literature, as the high-density polyethylene (HDPE) increases the environmental impacts [63,64]. The distance between the cultivation area and processing unit has also been found to have a notable impact from an environmental point of view. The cultivation of only economically feasible seaweed species, such as *Palmaria palmate*, requires fewer nutrients and a high carbon sequestration rate. This approach may also improve the environmental performance at the later stages, and is therefore encouraged in some techno-economic analysis studies [65–68]. The impact of raw materials consumed, including chemicals and minerals, is not discussed extensively in the literature; however, the carbon absorption has reportedly affected the environment by reducing the overall impacts, especially in terms of global warming. Moreover, the literature suggested that the traditional seaweed cultivation system needs to be technologically upgraded, which can mitigate the shortcomings related to large-scale handling [59].

8. Seaweed Applications

Seaweed has wide applications in biofuels, food and animal feed, energy, pharmacy, cosmetic, and other chemical industries. Interestingly, all the seaweed species have little to no lignin content, making them directly suitable for producing third-generation biofuels. After harvesting, seaweed traditionally goes through enzymatic hydrolysis and/or saccharification to separate the polysaccharides. The extracted C-5 or C-6 sugar is then used for the fermentation process to obtain various biofuels. In the past few decades, seaweed has only been considered for producing a single product, like alginate, or bioethanol, which stretched the economic, social, and environmental considerations. The cascading approach has been discussed in the literature for the past few years [69] as a means to fully explore seaweed's functionality. The cascading approach promotes the production/application of high-valued material first, followed by the next high-valued materials in descending order.

Cascading the seaweed valorization has a great potential in yielding various by-products and other chemicals, which could ultimately reduce its environmental burden and improve the socio-economic potential. Provided that food and energy are the primary concerns in the 21st century, a strategy can be made to process the seaweed to utilize food, animal feed, and energy. In contrast, the technology can be improvised to use the leftover biomass [70] for other applications. In this functional order, the seaweed biorefinery concept has recently been upgraded to utilize the excess seaweed biomass to produce biofuels and biogas. In this manner, the seaweed biomass value can be increased to multiple folds, which would benefit the supply chain, and promote the circular economy.

8.1. Seaweed in Food Applications

Food safety is of prime concern all over the world. The United Nation's SDGs prioritize food safety and encourage stakeholders to develop the policy around it. According to estimates, only half of the total 9 billion people would have access to nutritional food by 2050 [69]. Seaweed is well-known for its nutritional value to humans and animals, as it can provide minerals, vitamins, calories, and essential antioxidants [71,72]. The most significant benefit of seaweed cultivation is that it does not compete with other terrestrial plants. The total consumption of dried seaweed is estimated to be 2,000,000 tons, including sea vegetables, as a direct source and other applications, like phycocolloids. Asian countries, including China, Korea, Japan, Vietnam, the Philippines, and Thailand, are intensively involved in seaweed production. Moreover, some European countries, including France, Norway, Ireland, United Kingdom, Spain, and other countries, including Canada, New Zealand, and Chile, are active in seaweed production. Seaweeds are varied according to geographical location; however, *L. japonica, P. yezeonsis, U. pinnatifida* are the most consumed seaweeds. Seaweed is usually dried before any further processing, followed by salting, mixing, rolling, and slicing according to requirement. The link between the seaweed species and preparing seaweed food must be highlighted to get the customer's approval [73]. Moreover, the prominent challenge in using seaweed as food is that it has to be accepted by a wider global community.

Apart from traditional use, seaweed has other commercial importance as well. The alginic acid obtained from the seaweed has stabilizing and thickening properties, making it suitable for syrups, ice-creams, sauces, juices, shakes, desserts, and bakery products [74]. There are seven different forms of carrageenas observed in seaweed with the sulfated galactose unit, out of which three forms of carrageenas are commercially used. The carrageenas is a roughly thickening agent and is preferred to prepare pizzas, desserts, gels, canned foods, etc. It is also reportedly used as a preservation or additive agent, and the carrageenas additives are usually marked as E407 nomenclature. Seaweed is composed mainly of agaropectin and agarose, which are the polymers of galactopyranose. These can be used as a gelling agent, primarily in canned products, desserts, pie fillings, etc., under the nomenclature of E406 in the food industry [75].

Seaweeds are nutritious, with many health benefits associated; however, they are prone to accumulate undesirable heavy metals. Therefore, it is always advised to check the organic and inorganic components present in the seaweed before any further processing. Seaweed's daily consumption has not been prescribed as of now; moreover, Asian diets contain nearly 5–8 g of seaweed per day; therefore, it is safe to consider 5–8 g of seaweed (nearly 30 g of wet seaweed) consumption daily [49,75].

8.2. Seaweed into Furfural

Furfural is becoming an essential chemical that can be obtained after biomass conversion. The US Department of Energy (USDOE) has listed furfural as a platform chemical, and it is expected to increase its market potential [73] steadily. Furfural can be used as a base chemical to produce different derivatives, including various polymer units, liquid fuels, and organic solvents [76–78]. Furfural is produced from the extracted sugars of lignocellulosic biomass waste after catalytic conversion. The economic conversion of extracted biomass into furfural depends on various factors, including the availability of diversified feedstock, type, nature of the catalyst, and reacting conditions [79]. Park et al., 2016 have extensively researched the production of furfural from seaweed. The commercial H_2SO_4, $H_2PW_{12}O_{40}$, and Ambelyst15 catalysts were tested for their efficiency, and it was observed that the $H_2PW_{12}O_{40}$ catalyst exhibited significant catalytic activity [80]. Similarly, the tetrahydrofuran was an efficient reactive medium compared to water in the conversion of seaweed-derived alginate into the furfural. The seaweed extract was used to produce the alginic acid first, followed by alginic acid hydrolysis to the monomers. Later, the monomer was converted into the furfural through the series of decarboxylation-dehydration reactions.

8.3. Seaweed into Hydrogen as a Fuel Application

The monomers obtained from the biomass's carbohydrate content play an essential role in producing fuels [81,82]. It has been discussed in the literature that hydrogen produced from the fermentation of simple carbohydrates, such as glucose and xylose, has great potential to dominate the fuel sector [83–85]. Mannitol is a simple carbohydrate with high water solubility, which is observed at nearly 20–30% in the dried brown seaweed [86].

It has been reported the fermentative production of hydrogen yields 5 moles of hydrogen per mole of mannitol consumed for the acetic acid pathway, 1 mole of hydrogen per mole of mannitol consumed for the ethanol or lactic acid pathway, and 3 moles of hydrogen per mole of mannitol consumed for the butyric acid pathway [87]. The heat pre-treatment, mannitol concentration, and pH were reported to be the rate-determining factor in hydrogen production. The anaerobic digestion process yields nearly 1.8 moles of hydrogen per mole of mannitol used; however, maintaining the fermentation conditions is a challenging task [88].

8.4. Seaweed into Biofuel

Production of bioethanol from seaweed can reduce agricultural land, freshwater consumption, and chemical fertilizer usage, ultimately reducing the environmental impacts throughout the lifecycle. The pulp, which remains after seaweed processing, creates an unnecessary burden if not utilized. The pulp contains many carbohydrates, which can be used for bioethanol production, and the leftovers from the fermentative bioethanol production can be used as a fertilizer [89]. Statistically speaking, Kumar S. et al., 2013 [90] have reported that nearly 25% (*w/w*) of the pulp can be obtained after the seaweed biomass processing, out of which the cellulosic material constitutes 40%, and the hemicellulosic material constitutes nearly 20% of the overall leftover pulp (*w/w*). On the successive enzymatic hydrolysis process, almost 88% of the cellulosic material is converted into simple sugars and utilized further in the fermentation process with 86% of bioethanol yield efficiency. Moreover, it has been assumed that the CO_2 absorption in the algal biomass is nearly seven times higher than that of terrestrial woody biomass [91]. Therefore, the ultimate carbon sequestration can further reduce the environmental impacts associated with the lifecycle of bioethanol production. Several researchers have extensively worked on bioethanol production from the seaweed pulp, and several seaweed species have been reported in the literature in the same context. Yeon et al., 2016 have reported getting 0.386 g of bioethanol per g of *Sargassum sagamianum* seaweed pulp [92]. In contrast, Yanagisawa et al., 2011 have used a cluster of species, including *U. pertussa, G. elegans*, and *A. crassifolia*, to get the yield of 0.381 g, 0.376 g, and 0.281 g of bioethanol, respectively [93]. Kumar S. et al., 2013 have extensively collected the literature from the past decades to compare the bioethanol yield from various biomass [90], and it was reported that the highest bioethanol yield of 0.46 g of bioethanol was obtained per g pulp of *G. verrucosa* seaweed. Kim et al. (2015) have reported that red seaweed, especially *Gelidium amansii*, is a great potential source for bioethanol [94]. This red seaweed species has a high carbohydrate content, which can easily be converted into simple sugars such as galactose or glucose via hydrolysis, and produce bioethanol after fermentation. Approximately 11 million tons of red seaweed were produced in 2011, while there has been a constant increment reported in the production [94].

The production of bioethanol from seaweed has many technological barriers. The processes involved in bioethanol production, including seaweed harvesting, pulp pre-treatment, enzymatic hydrolysis, and the fermentation process, require techno-economic upgrades for achieving a higher yield. The adaptation of the new saccharification methods, i.e., simultaneous and combined saccharification process, and enzymatic hydrolysis, have reportedly increased the bioethanol yield by the 20%. The method was effective in most cases with low or single polysaccharide content in the seaweed [67].

8.5. Seaweed as an Animal Feed

The traditional raw materials used for animal feed include oats, soybean, wheat, barley, and sorghum. The critical setback in using the traditional feedstock is that, apart from being a healthy source of the human diet, they are seasonal crops, and they require a considerable amount of time to grow. The use of such resources throughout the year can create a feud among the supply value chain and lead to inflation in the food prices [95]. Seaweed possesses many advantages over seasonal crops as they can be harvested in any season, proliferates, and do not require terrestrial agricultural land [96]. Seaweed contains several bioactive compounds and healthy nutrients, which can serve the purpose and provide dietary benefits. The seaweed extract contains soluble fibers, including fucoidan, alginate, and ulvan, which have good antimicrobial properties. Gardiner et al., 2008 and Reilly et al., 2008 have mentioned the use of seaweed extract in pig diets as some feed antibiotics [97,98]. Kolb et al., 2004 mentioned using seaweed extract in the cattle's diet as a dietary supplement [99]. They also mentioned that the addition of seaweed extract improves animal health and increases milk quality. It has been reported that seaweed can replace nearly 15–20% of the traditional poultry diets based on the type of seaweed and the poultry animal [100].

8.6. Seaweed in Pharmaceuticals

In the past few decades, extensive research has been carried out on seaweed extracts in various medical and pharmaceutical products. Cancer prevention, tumor-suppressing, and health recovery treatments using seaweed extract are documented in the literature. Funahashi et al., 1999 documented a series of seaweed types and their effectiveness in preventing and suppressing cancer. The administration of seaweed extract at 1.6 g/kg body weight of the patient for 28 days reportedly inhibited 46–70% of the Ehrlich carcinoma [101]. The brown seaweeds such as *Scytosiphon lomentaria*, *Laminaria japonica*, *Lessonia nigrescens*, and *Sargassum ringgoldianum* used in the study showed 70%, 58%, 60%, and 46% of inhibition, respectively. Apart from brown seaweed, some of the red seaweeds, like *Eucheuma geltinae*, *Porphyra yezoensis*, and green seaweeds, such as *Enteromorpha prolifera*, have shown nearly 52% of inhibition. Various biomolecules obtained from seaweeds such as fucoids, glycolipids, and phospholipids have shown great potential against the MethA fibrosarcoma and have an antimetastatic effect on A549 lung cancer [102].

Some of the biomolecules extracted from the brown seaweed have excellent dietary fiber content, which plays a significant role in cancer prevention. Brown seaweeds such as *Undaria pinnatifida* and *Hijikia fusiformis* were found to have a high concentration of Fucoxanthin, and are frequently used as a daily dietary supplement in Asian countries. Other brown seaweeds such as *Ecklonia kurome* and *Laminaria japonica* were found to have laminarin as an active ingredient. Laminarin is a water-soluble polysaccharide with β-(1–3)-glucan linkages [103]. Moreover, brown seaweeds are bountiful with Carrageenans and Phlorotannins, linear galactans with antioxidative properties [104]. Some of the red seaweeds, including *Gracilaria arcuate* and *Actinotrichia fragilis*, have 3-amino-1-propanesulfonic acid, which is commercially known as alzhemed. The alzhemed drug treatment at stage III clinical trials is effective in Alzheimer's disease treatment [105]. The clinical trials have reported that the absorption of 500 mg of brown seaweed like *Fucus vesiculosus* can lead to an 8% increment in insulin sensitivity index than that of placebo in diabetic patients [106]. Alginic acid is one of the important biomolecules obtained from brown seaweeds, and has wide biomaterial applications to prepare nanoparticles or gel for targeted drug delivery [107].

8.7. Seaweed in Cosmetics

The cosmetic industry has been booming for the past few decades with approximately EUR 400 billion trade worldwide. Europe is the leading cosmetic market with a EUR 72 billion business, followed by the USA (~EUR 38 billion) and Japan (~EUR 30 billion). France has a substantial cosmetic market with nearly EUR 25 billion trade with roughly EUR 8 billion trade surplus [108]. The cosmetic industry is always hunting for new constituents for economic and social purposes. Bio-based materials have significant demand in the cosmetic industry, and the marine algal source has been considered extensively in cosmetics product development [109]. Seaweeds are the apparent option to consider in the cosmetic industry, as they contain various biomolecules, named phycocosmetics [108]. The brown seaweed *Macrocystis pyrifera* belongs to the *laminariaceae* family, which contains hyaluronic acid. Hyaluronic acid is the prominent constituent of the extracellular matrix and hence used as a treatment aid for burn victims since 1968 [110]. In many other cases, plastic surgeons use hyaluronic acid for treating volume loss and face wrinkles [108]. Fucoxanthin and Astaxanthin are major biomolecules constituted in the brown seaweeds. These biomolecules have excellent antioxidant properties, and therefore, have applications as an anti-aging ingredient in the cosmetic industry [111]. Micosporine-like amino acids (MAAs), such as shinorine and usujirene, are found in most seaweed algae. These molecules are responsible for absorbing most of the UV radiation between 310–360 nm and hence protect the seaweed algae. These MAAs have been exploited on an industrial scale for their application in sunscreen lotion [112]. Many brands such as Daniel Jouvance (LA Gacilly, France), Science & Mer (Paris, France), Phytomer (Brittany, France), Algotherm (Brittany, France), and Gelyma (Marseille, France) are working on commercializing the phycocosmetics, and many others are exploiting the future possibilities.

9. Sustainable Seaweed Applications

Researchers are exploiting the various ways through which seaweed can be utilized maximally. However, even though world seaweed production has been increased three times in the past 50 years, the sustainability of seaweed functionality is still a challenging concern [113]. Sustainability is a relative concept, especially in seaweed cultivation, and depends on the production region. Many reports on seaweed functionality have reported that the time of harvest, as discussed earlier, has a significant effect on its applicability. For example, Adam et al., 2011 reported a 30% higher biomethane yield when the *Laminaria digitate* seaweed is cultivated and harvested in June, compared to winter or spring [114]. Similarly, Jard et al., 2013 observed a 25% higher biomethane yield when *Saccharina latissimi* seaweed was cultivated and harvested in August, compared to the winter or summer [26]. Therefore, seaweed sustainability must be discussed and assessed to identify the sustainability benefits and drawbacks in relation to seaweed functionality.

Conversion of seaweed biomass has typically been presented as having strong potential for biorefinement in the present literature, wherein its composition, treatment technologies, and value chains are discussed, similar to that of microalgae. Few published reports have focused on the characteristics of seaweed and its application in bioenergy. The rest of the published research has noted its carbon capture capacity and viewed it as a solution to the 'food vs. fuel' debate. Moreover, the sustainability and thorough analysis of a value chain in the biorefinery process is still lacking in the literature. Sustainability is often a vaguely used term in literature on seaweed cultivation, focusing on ecology and the environment. Nonetheless, with an increase in the number of reports on seaweed functionality, it is necessary to implement sustainability assessments comprehensively, which could also support finding improved seaweed economics and social acceptance. In the present manuscript, available literature on environmental and techno-economical aspects of seaweed's value chain has been collected and described. The value chain of seaweed is presented in Figure 4 below.

In 2007, the Biorefinery International Energy Agency (BIEA) identified different tasks that need to be achieved in order to develop a low-carbon sustainable economy. Task 42 aimed to implement sustainable biorefineries with a zero-waste value chain and the production of both bio-based food and non-food-based value chains [115]. The biomass from terrestrial origins is still the predominant feedstock; however, aquatic biomass is considered for the biorefinery approach in recent years. A seaweed biorefinery process could provide an essential boost to the seaweed market, thereby closing the loops in the circular economy. With additional economic value given to seaweed processing, several technologies have increased along with interest in this field, as reported by Laurens et al., 2017 [116]. Since seaweed is still an expensive feedstock (~USD 50–80/ton), the biorefinery approach allows full use of available biomass, including high-value applications, giving more socio-economic and environmental benefits. Figure 4 depicts the potential biorefinery approaches in seaweed processing through different systems. System 1 illustrates the traditional methpd of utilizing seaweed for the production of furfural. System 1 is the most discussed pathway in the literature, wherein seaweed is processed via physical and thermochemical technologies to extract the complex carbohydrate content (such as xylose, alginate, levulinate, etc.), thereby converting it into furfural. This system's implementation is mostly at an early stage of development and at pilot scale in some cases. Furfural is a starting material for various products and a key intermediate in many bio-based value chains. Furfural derived from bio-based sources can be converted to solvents and chemicals used in the polymer and pharmaceutical industry, as projected in Systems 5, 6, and 7.

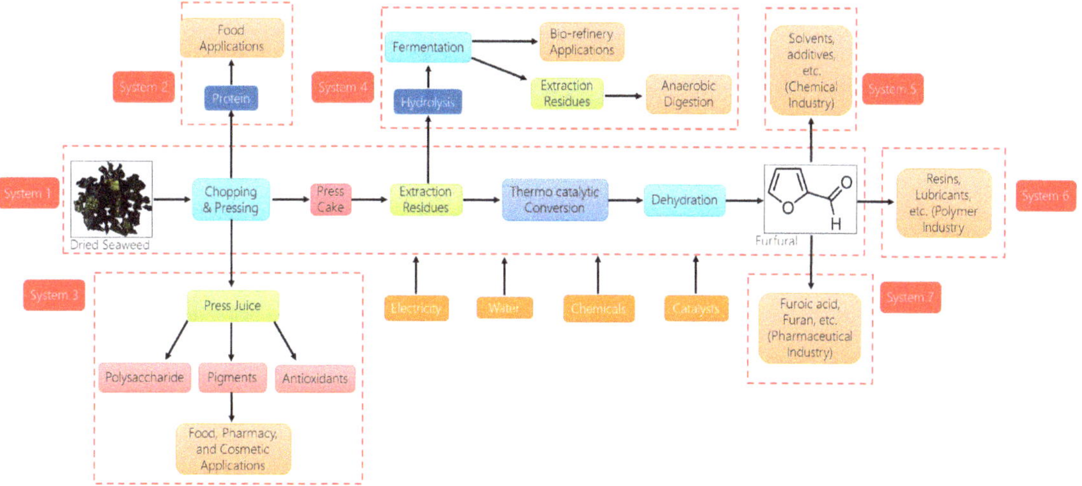

Figure 4. Seaweed biorefinery.

System 2 is the obvious choice for seaweed processing, wherein the proteins and other parts are separated for a variety of food applications. This system has been established in Asian countries, where seaweed is consumed as a food in various cuisines, including Furikake, Jerky, Sea-chi (kimchi), pickle, salsa, tea, etc. Seaweed-based food companies such as Cargill, Acadian seaplants, DuPont de Nemours, Irish seaweed, Mara seaweed, and Beijing Leili had nearly USD 5 billion/year collective trade in the last decade [117]. System 3 depicts another biorefinery route for seaweed processing, wherein the seaweed extracts are used as active biological ingredients in food, pharma, and cosmetics industries, as discussed previously. Carrageenan, alginate, and agar were found to have antiviral, anti-inflammatory, antibacterial, and antioxidant properties, attracting interest from various pharmaceutical companies [118]. Institutes such as IOTA pharmaceuticals, Oncology SME, University of York, and the University of Cambridge have collaborated to develop a seaweed-based drug system to fight cancer. According to a recent public report, scientists at Rensselaer Polytechnic Institute have claimed that the seaweed extract (RPI-27) could outperform the well-known COVID-19 drug Remdesivir to trap the virus before it can infect the human body. In addition, several cosmetic industries, such as, Earthrise Nutritionals (California, USA), Cyanotech Corporation (Kailua-Kona, Hawaii), BlueBioTech Int. GmbH (Kaltenkirchen, Germany), Bluetec Naturals Co., Ltd. (Erdos, Inner Mongolia, China), and Tianjin Norland Biotech Co., Ltd. (Tianjin, China), etc., have been developing an entire range of cosmetic products based on seaweed extracts. System 4 presents one possible biofuel concept from seaweed extract, which is widely discussed in the literature [93]. There are also potential economic benefits in producing other value-added products, such as bio-butanol and biogas, for implementing this route. First- and second-generation biomass, used for fuel production, has limitations in terms of its usability and sustainability. However, macroalgae's current developments can overcome these hurdles and provide a sustainable low-carbon economy by exploiting the output of value-added products, apart from bioenergy [119].

The LCA's quantitative and qualitative analysis can exploit the benefits of the seaweed value chain. The literature on the LCA of the seaweed value chain is limited, with most papers focusing on biofuels. The European Directives embraced the LCA methodology in 2009 to evaluate the environmental impacts generated by biofuels during their entire life cycle, and created an objective to reduce the GHG emissions by 50% in the next decade [120]. The available literature on LCA of seaweed applications is compiled in Table 3 below.

Table 3. Life Cycle Assessment of Seaweed Algae Applications.

Sr. No.	LCA Study	FU	System Boundary	Impacts Studied *	Hotspots	Ref.
1	LCA of 1,3-Propanediol production from fossil and Biomass comparison	Production of 1 kg 1,3-Propanediol produced from Biomass	Cradle-to-gate	GWP, EP, FFC	Land use, water, Soil erosion, Nutrient run-off, Ecosystem vulnerability	[121]
2	LCA of Algal Biodiesel and co-products	Production of 1 kg Biodiesel (Dry weight basis)	Cradle-to-Grave	Land use, GHG, FD	Value chain of co-products	[122]
3	LCA of Bio-based Adipic acid from Lignin	Production of 1 kg polymer grade Adipic acid	Cradle-to-Gate 2	(TRACI Midpoint) OD, GHG, Smog, AP, FD, HT-NC, HT-C	Valorization potential of feedstock, Nitrous oxide emission, NaOH utilization, and heating	[123]
4	LCA of Bio-based (Corn and Starch based) wood flooring coating	1 m^2 of coating	Cradle-to-Gate 2	(TRACI 2.1) GWP, OD, AP, EP,	Diacrylate monomer, Itaconic acid, Epoxy resins	[124]
5	LCA of Bio-based and Fossil-based Succinic acid	Production of 1 kg Succinic acid	Cradle-to-Gate 2	(ReCiPe 1.0) CC, OD, HT, IR, TA, FEW, ME, WD, FD	Ecosystem impacts are high, Dextrose production	[125]
6	LCA of Industrial production of Algal Biodiesel	Production of 10 GJ Biodiesel	Cradle-to-Gate 2	GWP, CED	Solar energy and temperature affect algal growth, Land use	[126]
7	LCA of Vetiver based Biorefinery	Biofuel production from 1 kg Vetiver Biomass	Cradle-to-Gate 2	(ReCiPe) FD, Carbon Dioxide emission	The energy required for Furfural production, Impact of Enzyme production not included	[127]
8	LCA of Biogas production from Marin algae	1.1 TJ of energy produced in 1 Year	Cradle-to-Grave	(Impact 2000+) CC, RD	Anaerobic Fermentation Unit, Energy input	[128]
9	LCA of Biofuel production from Brown seaweed	Cultivation and Processing of 1 t of dry biomass produced in Denmark for Biofuel production	Cradle-to-Grave	GWP, AP, TETP	Cultivation process, Energy consumption, Anaerobic digestion	[129]
10	LCA of Biogas production from Marine macroalgae	1 kg feedstock mixture fed to the digester, and 1 MJ energy produced from biogas	Cradle-to-Grave	(ReCiPe Midpoint 1.06) GWP, AP, EP	Restricted consideration of Pre-treatment methods, Feedstock variation, Toxicity of Phenols	[130]

Table 3. *Cont.*

Sr. No.	LCA Study	FU	System Boundary	Impacts Studied *	Hotspots	Ref.
11	LCA of Macroalgal Biorefinery for production of Bioethanol, Protein, and Fertilizer	1 ha of the sea under cultivation	Cradle-to-Grave	(ReCiPe Midpoint 1.06) CC, CED-T, HT-C, HT-NC, ME,	Energy consumption for the drying process, Biofertilizers,	[131]
12	LCA of Biogas production from co-digestion of macroalgae	Production of 1 m^3 of biogas	Cradle-to-Grave	CC, ME, FD, FWE, Land use	Eco-design phase, Energy	[132]
13	LCA of Seaweed into Biomethane	Production of 1 MJ of compressed Biomethane	Cradle-to-Gate 2	GWP, AP, FAETP, TETP, MAETP	Digestion process, Field application, Electricity consumption	[133]
14	LCA of Biogas production from Marine macroalgae	Production of 1 GJ of Biogas from macroalgae	Cradle-to-Grave	CC, Land use, AP, EP, OLD, FAETP, MAETP	Electricity, Process material	[134]
15	LCA of Seaweed into Fuel and Energy	1 ha of offshore cultivation area	Cradle-to-Grave	CC, Land use, CED, HT,	Seeded line, Energy, Sugar-to-Protein conversion,	[57]
16	LCA of Macroalgae derived single cell oil	Production of 1 t single cell oil	Cradle-to-Gate 2	CC, FAETP, MAETP, HT, TETP, WD	Fermentation, Acid pre-treatment, Enzymatic hydrolysis, Energy demand	[135]
17	LCA of Valorization strategies of macroalgae	1 kg of valorized biomass	Cradle-to-Gate 2	AP, EP, GWP, HTP, FAETP, MAETP, TETP, ADP	Electricity, Extraction process, Organic solvents, Pre-treatment	[136]

After evaluating available LCA studies, it was observed that most of the studies focus on the cradle-to-gate approach. Discussion around the application or the end-of-life section is missing from most of these studies; wherein, the end-of-life section is only considered in biofuel applications (combustion of the fuel). Energy consumption and seaweed cultivation were major impact contributors in fuel applications. In the non-fuel applications of seaweed, LCA studies indicate that seaweed cultivation plays a significant role in imposing environmental impacts throughout the cradle-to-gate scenario. Traditionally, infrastructure is excluded from LCA studies; however, seaweed cultivation requires specific infrastructure, different from terrestrial ones, to facilitate higher growth. Therefore, impacts associated with infrastructure development have to be considered to get a realistic overview; however, this aspect has been neglected in available studies. Technology usage was different in every study, starting from the seaweed cultivation until the intended application production. Therefore, it is not easy to compare all studies on the potential seaweed-based product and their commercially available counterpart. Due to the presence of reactive nitrogen in nutrients and the anoxic conditions that occurred during seaweed cultivation, nitrogen emission (N_2O, NH_3, etc.) most likely occurs, which leads to acidification and GHG emissions [137,138]. Provided that N_2O has a GWP of 250 times the GWP of CO_2, more emphasis must be put on such emissions, and the emission factors have to be considered while interpreting the environmental impacts. Similarly, the CO_2 fixation at the cultivation step and CO_2 emission during the user phase (especially combustion in biofuel applications) need to be considered. Most studies have omitted this part and considered the entire process as a CO_2-neutral process. This approach may significantly affect the

GWP impact, and therefore, the greenhouse gas flow must be considered. Additionally, these studies have considered various functional units and impact assessment methodologies, making it challenging to compare seaweed-based products on an environmental or economic basis. As mentioned previously, the LCA studies have been conducted for the high-end seaweed-based products only, whereas most low-end products and their feasibility are not discussed. Moreover, the discussion about co-products or by-products generated during the seaweed's processing is missing, which would improve the seaweed value chain. Feasibility studies, including techno-socioeconomic assessment, should be carried out to promote the seaweed-based products and support the selection of economically viable and environmentally sustainable value chains.

10. Techno-Economic Assessment

Apart from technological glitches, seaweed is at the center of attention for its potential to substitute fossil-based products and positive environmental impacts, such as nutrients recovery. However, seaweed farming has to be compatible and sustainable to accomplish either of these needs. Presently, seaweed is primarily produced for consumption purposes, whereas its production for high/low value-added products such as fertilizers, bioenergy, and biomaterials is still at a developmental stage. Scarce information is available in the literature for techno-economic assessment of seaweed-based value-added products; however, published literature implies that seaweed cultivation and harvesting is the decisive factor in estimating the cost of bio-based products [66,139]. It was reported that the seaweed harvesting cost would vary from USD 200–900/ton of dry mass, based on the type of seaweed cultivation [65]. Moreover, the use of seaweed as food and chemical seems a worthwhile option. Clarens et al., 2010 suggested the optimization of seaweed farming by expanding the current production line and producing value-added products or selling wet seaweed at a higher price (USD 2/kg) to have profitable farming [140]. Electricity production from seaweed has often been discussed in the literature, and a break-even selling price of USD 154/MWh was estimated. This break-even price of seaweed-derived electricity is comparable with the existing renewable sources like solar (USD 150 $\pm$ 10 MWh) and thermal (USD 250 $\pm$ 10 MWh). Jorquera O., 2010 [141] assumed that nearly 16 MMT of dry seaweed would be required to sell the electricity to the industrial sector at the rate of USD 70/MWh. However, it was also assumed that if the government subsidizes 20% of the production cost, then the required seaweed production would go down to nearly 11 MMT. The UK government has taken an initiative to subsidize the electricity production from seaweed anaerobic digestion, by which the failure in time price of the electricity has been set to 147 GBP/MWh for small-scaled units [66]. Dave et al., 2013 have reported the break-even selling price of 178 GBP/MWh at an internal rate of return of 8% while considering 86.4 tons/day dry seaweed consumption, 64% conversion rate at anaerobic digestion, 25% capital fees, and 4% operational cost [66]. Nevertheless, seaweed farming must develop business cases for future value chains and optimize strategies. In order to compete with already established products or chemicals, the risk factors, such as seaweed growth variation, nutrient recovery, CO_2 fixation, market volatility, CapEx, OpEx, and supply-demand value chains must be analyzed. Government viable funding, subsidies, and seed funding are needed to develop a viable seaweed market; otherwise, the seaweed business will not be sustainable.

11. A Way Forward

It is understood that seaweed has a great potential to go beyond its application as food and support the transition to a bio-based economy. However, specific shortfalls in knowledge have been identified based on the listed scenarios. It is unknown from the available literature how seaweed can mitigate climate change's ongoing problems and support the concept of a circular economy. The available literature does not discuss viable business models for seaweed apart from its application as food. The social aspects are still missing from the available documents, which play a significant role in product establishment. Producing seaweed-based products at a laboratory scale and comparing them with already established benchmarks will not elucidate the barriers to commercialization. Therefore, a clear path forward is needed to consider the aspects which would lead to the commercialization of financially viable seaweed-based bioproducts. Based on the present understanding, we have summarized the "PBFS" approach, which is essential while considering seaweed-based products' feasibility. The PBFS based product feasibility approach is presented in Figure 5 below. This approach considers four significant aspects of commercialization, including process development (P), by-product promotion (B), financial assistance (F), and social acceptance (S).

Figure 5. Seaweed product feasibility approach.

The first and foremost step is to achieve technological enhancement. In the process' economics, it is understood that nearly 30% of the production cost is accompanied by feedstock processing and handling. Technological improvements will help in increasing product yield, thereby reducing the feedstock processing cost. It may include better storage facilities, separation processes, enzymatic hydrolysis, pre-treatments, fermentation, etc. Emerging technologies and R&D in the field of biotechnology and biochemical processing can improve seaweed-based product economics. Apart from this, skilled labor would play a significant role in R&D, thereby developing novel technologies and improving the overall process.

The second part of the feasibility approach is promoting co-products and by-products developed during seaweed processing. It is understood that the bio-based products are not economically viable compared to their chemical counterparts if produced in a stand-alone facility. It is evident from the overall scheme in Figure 4 that several by-products and co-products are generated while processing seaweed, including proteins, polysaccharides, pigments, extraction residues, etc., have a commercial value in the market. If the production line were strategized so that co-products/by-products are separated efficiently, and their respective processing is considered within the system boundary, then the seaweed-based product would become more economically viable.

The third approach is assisting these projects financially. Stand-alone facilities with the production of seaweed-based products will not survive on their own. For promoting the bio-based economy, the EU and other countries must develop a policy bubble for producing bio-based products. These stand-alone companies should be provided with seed funding for start-up and R&D at the initial development stage. To compensate for the production cost, viability gap funding should be provided until the selling price becomes profitable.

Other financial initiatives such as collective tax reduction, incentivizing bio-based products, and defining minimum seaweed purchase price are also encouraged.

The fourth important aspect of achieving the feasibility of seaweed-based products is to improve social acceptance. End users are a significant part of the value chain, since it is they who utilize or consume the product. However, the end-user is always willing to get a premium experience at a low cost, and more often than not, is uninterested in changing the conventional route. It is essential to convey the value-added benefits and competitiveness of a bio-based product to the end-user. Target-specific information would help customers to understand the merits and usefulness of seaweed-based products. The EU may play a crucial role in developing a policy, awareness, and advertising benefits to attract potential customers. Apart from this, evaluating socio-economic aspects and developing benchmarks will also assist in decision-making and social acceptance.

12. Conclusions

The bio-based economy's prospects seem compelling, given that it is an innovation-based sustainable system, fulfilling the circular economy's goals. The latest understanding, research, modifications, and competitiveness of bio-based products have propelled unprecedented growth in the last couple of decades. However, some market-driven aspects such as commercial success, scale-up, and economic viability are severe obstacles to progress. Seaweed is a significant feedstock for various bio-based products, which is still in its infancy, and has excellent development potential. Seaweed cultivation has been restricted to food applications so far, and its multi-dimensional usage has not yet been explored commercially. Seaweed cultivation ensures several benefits, including its ability to trap sea nutrients, limit eutrophication, increase biodiversity, and achieve carbon sequestration. Seaweed processing requires extra steps, which could, holistically speaking, counteract environmental sustainability. However, full exploitation of seaweed feedstock with targeted biorefineries approaches and novel value chains is expected to bring environmental and socio-economic benefits.

Author Contributions: Conceptualization, formal analysis, resources and data curation, and original draft preparation, P.N.; funding acquisition, reviewing, supervision, and editing, Y.v.d.M. All authors have read and agreed to the published version of the manuscript.

Funding: This research and the APC were funded by the European Regional Development Fund, the Dutch provinces of Noord-Brabant, Zeeland and Limburg in the context of OP Zuid, grant number PROJ-02125, and Maastricht University.

Institutional Review Board Statement: Not applicable.

Informed Consent Statement: Not applicable.

Data Availability Statement: Not applicable.

Acknowledgments: The authors gratefully acknowledge the support from the ZCORE (Zeewier naar COating en Resin) consortium partners.

Conflicts of Interest: The authors declare no conflict of interest.

Abbreviations

CC	Climate Change
CED-T	Cumulative Energy Demand, Total
ME	Marine Eutrophication
FWE	Fresh Water Eutrophication
HT-C	Human Toxicity, Cancer
HT-NC	Human Toxicity, Non-Cancer
OLD	Ozone Layer Depletion

HTP	Human Toxicity Potential
FAETP	Fresh Aquatic Eco-toxicity Potential
MAETP	Marine Aquatic Eco-toxicity Potential
TETP	Terrestrial Eco-toxicity Potential
AD	Abiotic Depletion
EP	Eutrophication Potential
GWP	Global Warming Potential
AP	Acidification Potential
ODP	Ozone Depletion Potential
TA	Terrestrial Acidification
FD	Fossil Depletion
WD	Water Depletion
IR	Ionizing Radiation
GHG	Green House Gas
5-HMF	5-Hydroxymethylfurfural
FFC	Fossil fuel consumption
RD	Resource Depletion
NMVOC	Non-Methane Volatile Organic Compound
PM	Particulate Matter
NREU	Non-Renewable Energy Use
PBFS	Process development, By-product promotion, Financial assistance, and Social acceptance

References

1. Dupont-inglis, J.; Borg, A. Destination bioeconomy—The path towards a smarter, more sustainable future. *New Biotechnol.* **2017**, *6784*, 30041–30049. [CrossRef] [PubMed]
2. Mengal, P.; Wubbolts, M.; Zika, E.; Ruiz, A.; Brigitta, D.; Pieniadz, A.; Black, S. Bio-based Industries Joint Undertaking: The catalyst for sustainable bio-based economic growth in Europe. *New Biotechnol.* **2018**, *40*, 31–39. [CrossRef] [PubMed]
3. Schütte, G. What kind of innovation policy does the bioeconomy need? *New Biotechnol.* **2017**, *40*, 82–86. [CrossRef] [PubMed]
4. Bell, J.; Paula, L.; Dodd, T.; Németh, S.; Nanou, C.; Mega, V.; Campos, P. EU ambition to build the world's leading bioeconomy— Uncertain times demand innovative and sustainable solutions. *New Biotechnol.* **2017**, *S1871-6784*, 30022–30025. [CrossRef] [PubMed]
5. Dietz, T.; Börner, J.; Förster, J.J.; Von Braun, J. Governance of the Bioeconomy: A Global Comparative Study of National Bioeconomy Strategies. *Sustainability* **2018**, *10*, 3190. [CrossRef]
6. European Commission. *A Sustainable Bioeconomy for Europe: Strengthening the Connection between Economy, Society and the Environment*; European Commission: Luxembourg, 2018; pp. 1–107.
7. Moreno, A.D.; Susmozas, A.; Oliva, J.M.; Negro, M.J. Overview of bio-based industries. In *Biobased Products and Industries*; Elsevier: Amsterdam, The Netherlands, 2020; pp. 1–40.
8. Krüger, A.; Schäfers, C.; Schröder, C.; Antranikian, G. Towards a sustainable biobased industry—Highlighting the impact of extremophiles. *New Biotechnol.* **2018**, *40*, 144–153. [CrossRef]
9. United States Department of Agriculture. *An Economic Impact Analysis of the U.S. Biobased Products Industry*; United States Department of Agriculture: Washington, DC, USA, 2018; pp. 1–108.
10. Andrea, C.; Nicolas, R.; Klas, J.; Roberto, P.; Sara, G.-C.; Raul, L.-L.; van der Marijn, V.; Tevecia, R.; Patricia, G.; Robert, M.; et al. Biomass production, supply, uses and flows in the European Union. *Environ. Impacts Bioenergy* **2018**, 1–126. [CrossRef]
11. Resurreccion, E.P.; Colosi, L.M.; White, M.A.; Clarens, A.F. Comparison of algae cultivation methods for bioenergy production using a combined life cycle assessment and life cycle costing approach. *Bioresour. Technol.* **2012**, *126*, 298–306. [CrossRef] [PubMed]
12. Saravanan, K.R.; Ilangovan, K.; Khan, A.B. Floristic and macro faunal diversity of Pondicherry mangroves, South India. *Trop. Ecol.* **2008**, *49*, 91–94.
13. Ramani, G.; Tulasi, M.S.; Bhai, V.A. Seaweed: A Novel Biomaterial. *Int. J. Pharm. Pharm. Sci.* **2013**, *5*, 40–44.
14. Delaney, A.; Frangoudes, K.; Li, S. Society and Seaweed: Understanding the Past and Present. *Seaweed Health Dis. Prev.* **2016**, *2*, 7–40.
15. Evans, F.D.; Critchley, A.T. Seaweeds for animal production use. *J. Appl. Phycol.* **2014**, *26*, 891–899. [CrossRef]
16. White, W.L.; Wilson, P. World seaweed utilization. In *Seaweed Sustainability*; Elsevier Inc.: Amsterdam, The Netherlands, 2015; pp. 7–26.
17. Buschmann, A.H.; Camus, C.; Infante, J.; Neori, A.; Israel, A.; Hernández-gonzález, M.C.; Pereda, S.V.; Gomez-, J.L.; Golberg, A.; Tadmor-shalev, N.; et al. Seaweed production: Overview of the global state of exploitation, farming and emerging research activity. *Eur. J. Phycol.* **2017**, *52*, 391–406. [CrossRef]
18. Bruhn, A.; Dahl, J.; Nielsen, H.B.; Nikolaisen, L.; Rasmussen, M.B.; Markager, S.; Olesen, B.; Arias, C.; Jensen, P.D. Bioenergy potential of Ulva lactuca: Biomass yield, methane production and combustion. *Bioresour. Technol.* **2011**, *102*, 2595–2604. [CrossRef]

19. Mazarrasa, I.; Olsen, Y.S.; Mayol, E.; Marbà, N.; Duarte, C.M. Global unbalance in seaweed production, research effort and biotechnology markets. *Biotechnol. Adv.* **2014**, *32*, 1028–1036. [CrossRef]
20. Mazarrasa, I.; Olsen, Y.S.; Mayol, E.; Marbà, N.; Duarte, C.M. Rapid growth of seaweed biotechnology provides opportunities for developing nations. *Nat. Publ. Gr.* **2013**, *31*, 591–592. [CrossRef]
21. Arnaud-haond, S.; Arrieta, J.M.; Duarte, C.M. Marine Biodiversity and Gene Patents. *Sci. Glob. Genet. Resour.* **2011**, *331*, 1521–1522. [CrossRef] [PubMed]
22. Usman, A.; Khalid, S.; Usman, A.; Hussain, Z.; Wang, Y. Algal Polysaccharides, Novel Application, and Outlook. In *Algae Based Polymers, Blends, and Composites*; Elsevier Inc.: Amsterdam, The Netherlands, 2017; pp. 115–153.
23. Venkatesan, J.; Lowe, B.; Anil, S.; Manivasagan, P.; Kheraif, A.A.; Kang, K.-H.; Kim, S.-K. Seaweed polysaccharides and their potential biomedical applications. *Starch* **2015**, *66*, 1–10. [CrossRef]
24. Werner, A.; Clarke, D.; Kraan, S. *Strategic Review of the Feasibility of Seaweed*; National Development Plan: Galway, Ireland, 2004; pp. 1–123.
25. Lunardello, K.A.; Yamashita, F.; Benassi Toledo, M.; Rensis Barros, C.; Maciel, V. The physicochemical characteristics of nonfat set yoghurt containing some hydrocolloids. *Int. J. Dairy Technol.* **2011**, *65*, 260–267. [CrossRef]
26. Jard, G.; Marfaing, H.; Carrère, H.; Delgenes, J.P.; Steyer, J.P.; Dumas, C. French Brittany macroalgae screening: Composition and methane potential for potential alternative sources of energy and products. *Bioresour. Technol.* **2013**, *144*, 492–498. [CrossRef] [PubMed]
27. Hamid, N.; Ma, Q.; Boulom, S.; Liu, T.; Zheng, Z.; Balbas, J.; Robertson, J. Seaweed minor constituents. In *Seaweed Sustainability*; Elsevier Inc.: Amsterdam, The Netherlands, 2015; pp. 193–242.
28. Lang, I.; Hodac, L.; Friedl, T.; Feussner, I. Fatty acid profiles and their distribution patterns in microalgae: A comprehensive analysis of more than 2000 strains from the SAG culture collection. *BMC Plant Biol.* **2011**, *11*, 1–16. [CrossRef] [PubMed]
29. Colin, B.; Fereidoon, S. *Marine Netraceuticals and Functional Foods*; CRC Press: Boca Raton, FL, USA, 2008; pp. 1–512.
30. Laurens, L.M.L.; Lane, M.; Nelson, R.S. Sustainable Seaweed Biotechnology Solutions for Carbon Capture, Composition, and Deconstruction. *Trends Biotechnol.* **2020**, *38*, 1232–1244. [CrossRef] [PubMed]
31. Dawczynski, C.; Schubert, R.; Jahreis, G. Amino acids, fatty acids, and dietary fibre in edible seaweed products. *Food Chem.* **2007**, *103*, 891–899. [CrossRef]
32. Masakazu, M.; Nakazoe, J. Production and Use of Marine Algae in Japan. *Jpn. Agric. Res. Q. JARQ* **2001**, *35*, 281–290.
33. Narayan, B.; Miyashita, K.; Hosakawa, M. Comparative Evaluation of Fatty Acid Composition of Different Sargassum (Fucales, Phaeophyta) Species Harvested from Temperate and Tropical Waters. *J. Aquat. Food Prod. Technol.* **2008**, *13*, 53–70. [CrossRef]
34. Khotimchenko, S.V. Lipids from the Marine Alga Gracilaria verrucosa. *Chem. Nat. Compd.* **2005**, *41*, 230–232. [CrossRef]
35. Hay, C.H.; Villouta, E. Seasonality of the Adventive Asian Kelp Undaria pinnatifida in New Zealand. *Bot. Mar.* **1993**, *36*, 461–476. [CrossRef]
36. Nelson, M.M.; Nichols, P.D.; Scientific, T.C. Seasonal Lipid Composition in Macroalgae of the Northeastern Pacific Ocean Seasonal Lipid. *Bot. Mar.* **2002**, *45*, 58–65. [CrossRef]
37. Kim, K.; Lee, O.; Lee, B. Genotoxicity studies on fucoidan from Sporophyll of Undaria pinnatifida. *Food Chem. Toxicol.* **2010**, *48*, 1101–1104. [CrossRef] [PubMed]
38. Popper, Z.A.; Michel, G.; Domozych, D.S.; Willats, W.G.T.; Tuohy, M.G.; Kloareg, B.; Stengel, D.B. Evolution and Diversity of Plant Cell Walls: From Algae to Flowering Plants. *Annu. Rev. Plant Biol.* **2011**, *62*, 567–590. [CrossRef]
39. Enquist-newman, M.; Faust, A.M.E.; Bravo, D.D.; Santos, C.N.S.; Raisner, R.M.; Hanel, A.; Sarvabhowman, P.; Le, C.; Regitsky, D.D.; Cooper, S.R.; et al. Efficient ethanol production from brown macroalgae sugars by a synthetic yeast platform. *Nature* **2014**, *505*, 239–243. [CrossRef]
40. Fleurence, J. Seaweed proteins. In *Proteins in Food Processing*; Woodhead Publishing Limited.: Cambridge, UK, 1999; pp. 197–213.
41. Dumay, J.; Morançais, M. Proteins and Pigments. In *Seaweed in Health and Disease Prevention*; Elsevier Inc.: Amsterdam, The Netherlands, 2016; pp. 275–318.
42. Fleurence, J. The enzymatic degradation of algal cell walls: A useful approach for improving protein accessibility? *J. Appl. Phycol.* **1999**, *11*, 313–314. [CrossRef]
43. Fujiwara-arasakjl, T.; Mino, N.; Kuroda, M. The protein value in human nutrition of edible marine algae in Japan. *Hydrobiologia* **1984**, *116/117*, 513–516. [CrossRef]
44. Baghel, R.S.; Kumari, P.; Reddy, C.R.K.; Bhavanath, J. Growth, pigments, and biochemical composition of marine red alga Gracilaria crassa. *J. Appl. Phycol.* **2014**, *26*, 2143–2150. [CrossRef]
45. Denis, C.; Morançais, M.; Li, M.; Deniaud, E.; Gaudin, P.; Wielgosz-collin, G.; Barnathan, G.; Jaouen, P.; Fleurence, J. Study of the chemical composition of edible red macroalgae Grateloupia turuturu from Brittany (France). *Food Chem.* **2010**, *119*, 913–917. [CrossRef]
46. Dumay, J.; Morançais, M.; Munier, M.; Cecile, G.; Fleurence, J. Phycoerythrins: Valuable Proteinic Pigments in Red Seaweeds. In *Advances in Botanical Research*; Elsevier: Amsterdam, The Netherlands, 2014; Volume 71, pp. 321–343.
47. Baweja, P.; Kumar, S.; Sahoo, D.; Levine, I. Biology of Seaweeds. In *Seaweed in Health and Disease Prevention*; Elsevier: Amsterdam, The Netherlands, 2016; pp. 41–106.
48. South, G.R.; Whittick, A. *Introduction to Phycology*; John Wiley & Sons, Inc.: Hoboken, NJ, USA, 2009; pp. 1–352.

49. Macartain, P.; Gill, C.I.R.; Brooks, M.; Campbell, R.; Rowland, I.R. Nutritional Value of Edible Seaweeds. *Nutr. Rev.* **2007**, *1*, 535–543. [CrossRef]
50. Xiu-geng, F.; Ying, B.; Shan, L. Seaweed Cultivation: Traditional Way and its Reformation. *Chin. J. Oceanol. Limnol.* **1999**, *17*, 193–199. [CrossRef]
51. Monagail, M.M.; Cornish, L.; Morrison, L.; Araújo, R.; Critchley, A.T. Sustainable harvesting of wild seaweed resources. *Eur. J. Phycol.* **2017**, *52*, 371–390. [CrossRef]
52. ISO 2006a. *Environmental Management: Life Cycle Assessment: Principles and Framework*; ISO: Geneva, Switzerland, 2007; Volume 2, pp. 1–20.
53. ISO 2006b. *Environmental Management—Life Cycle Assessment—Requirements and Guidelines*; ISO: Geneva, Switzerland, 2007; Volume 2, pp. 1–20.
54. Holdt, S.L.; Edwards, M.D. Cost-effective IMTA: A comparison of the production efficiencies of mussels and seaweed. *J. Appl. Phycol.* **2014**, *26*, 933–945. [CrossRef]
55. Taelman, S.E.; Champenois, J.; Edwards, M.D.; De Meester, S.; Dewulf, J. Comparative environmental life cycle assessment of two seaweed cultivation systems in North West Europe with a focus on quantifying sea surface occupation. *Algal Res.* **2015**, *11*, 173–183. [CrossRef]
56. Seghetta, M.; Goglio, P. Life Cycle Assessment of Seaweed Cultivation Systems. *Methods Mol. Biol.* **2018**, *1980*, 103–119.
57. Seghetta, M.; Romeo, D.; Este, M.D.; Alvarado-morales, M.; Bastianoni, S.; Thomsen, M. Seaweed as innovative feedstock for energy and feed—Evaluating the impacts through a Life Cycle Assessment. *J. Clean. Prod.* **2017**, *150*, 1–15. [CrossRef]
58. Van Oirschot, R.; Thomas, J.E.; Gröndahl, F.; Fortuin, K.P.J.; Brandenburg, W.; Potting, J. Explorative environmental life cycle assessment for system design of seaweed cultivation and drying. *Algal Res.* **2017**, *27*, 43–54. [CrossRef]
59. Aitken, D.; Bulboa, C.; Godoy-faundez, A.; Turrion-gomez, J.L.; Antizar-ladislao, B. Life cycle assessment of macroalgae cultivation and processing for biofuel production. *J. Clean. Prod.* **2014**, *75*, 45–56. [CrossRef]
60. Langlois, J.; Sassi, J.-F.; Jard, G.; Steyer, J.-P.; Delgenes, J.-P.; Helias, A. Life cycle assessment of biomethane from offshore-Cultivated Seaweed. *Biofuels Bioprod. Biorefining* **2012**, *6*, 387–404. [CrossRef]
61. Ghosh, A.; Anand, K.G.V.; Seth, A. Life cycle impact assessment of seaweed based biostimulant production from onshore cultivated Kappaphycus alvarezii (Doty) Doty ex Silva—Is it environmentally sustainable ? *Algal Res.* **2015**, *12*, 513–521. [CrossRef]
62. Anand, K.G.V.; Eswaran, K.; Ghosh, A. Life cycle impact assessment of a seaweed product obtained from Gracilaria edulis—A potent plant biostimulant. *J. Clean. Prod.* **2018**, *170*, 1621–1627. [CrossRef]
63. Ku, H.; Wang, H.; Pattarachaiyakoop, N.; Trada, M. A review on the tensile properties of natural fiber reinforced polymer composites. *Compos. Part B* **2011**, *42*, 856–873. [CrossRef]
64. Bouafif, H.; Koubaa, A.; Perré, P.; Cloutier, A. Effects of fiber characteristics on the physical and mechanical properties of wood plastic composites. *Compos. Part A* **2009**, *40*, 1975–1981. [CrossRef]
65. Burg, S.W.; Van Den, K.; Van Duijn, A.P.; Bartelings, H.; Van Krimpen, M.M.; Poelman, M. The economic feasibility of seaweed production in the North Sea. *Aquac. Econ. Manag.* **2016**, *20*, 235–252. [CrossRef]
66. Dave, A.; Huang, Y.; Rezvani, S.; Mcilveen-wright, D.; Novaes, M.; Hewitt, N. Techno-economic assessment of biofuel development by anaerobic digestion of European marine cold-water seaweeds. *Bioresour. Technol.* **2013**, *135*, 120–127. [CrossRef]
67. Zimmermann, A.W.; Wunderlich, J.; Müller, L.; Buchner, G.A.; Marxen, A.; Michailos, S.; Armstrong, K.; Naims, H.; Mccord, S.; Styring, P.; et al. Techno-Economic Assessment Guidelines for CO_2 Utilization. *Front. Energy Res.* **2020**, *8*, 1–23. [CrossRef]
68. FAO. Declaration of the world Summit on Food Security. In *World Summit on Food Security*; FAO: Quebec City, QC, Canada, 2009; pp. 1–7.
69. Van Hal, J.W.; Huijgen, W.J.J.; Lopez-Contreras, A.M. Opportunities and challenges for seaweed in the biobased economy. *Trends Biotechnol.* **2014**, *32*, 231–233. [CrossRef]
70. Geldermann, J.; Kolbe, L.M.; Krause, A.; Mai, C.; Militz, H.; Osburg, V.-S.; Schöbel, A.; Schumann, M.; Toporowski, W.; Westpha, S. Improved Resource Efficiency and Cascading Utilisation of Renewable Materials. *J. Clean. Prod.* **2015**, *110*, 1–8. [CrossRef]
71. Mendis, E.; Kim, S. Present and Future Prospects of Seaweeds in Developing Functional Foods. In *Marine Medicinal Foods*; Elsevier: Amsterdam, The Netherlands, 2011; Volume 64, pp. 1–15.
72. Kumar, S.; Sahoo, D.; Levine, I. Assessment of nutritional value in a brown seaweed Sargassum wightii and their seasonal variations. *Algal Res.* **2015**, *9*, 117–125. [CrossRef]
73. Nisizawa, K.; Noda, H.; Kikuchi, R.; Watanabe, T. The main seaweed foods in Japan. *Hydrobiologia* **1987**, *29*, 5–29. [CrossRef]
74. Glicksman, M. Utilization of seaweed hydrocolloids in the food industry. *Hydrobiologia* **1987**, *151/152*, 31–47. [CrossRef]
75. Fleurence, J. Seaweeds as Food. In *Seaweed in Health and Disease Prevention*; Elsevier Inc.: Amsterdam, The Netherlands, 2016; pp. 149–167.
76. Li, H.; Ren, J.; Zhong, L.; Sun, R.; Liang, L. Production of furfural from xylose, water-insoluble hemicelluloses and water-soluble fraction of corncob via a tin-loaded montmorillonite solid acid catalyst. *Bioresour. Technol.* **2015**, *176*, 242–248. [CrossRef]
77. Bhaumik, P.; Dhepe, P.L. Effects of careful designing of SAPO-44 catalysts on the efficient synthesis of furfural. *Catal. Today* **2014**, *251*, 66–72. [CrossRef]
78. Climent, M.J.; Corma, A.; Iborra, S. Cponversion of biomass platform mpolecules into fuel additives and liquid hydrocarbon fuels. *Green Chem.* **2014**, *16*, 516–547. [CrossRef]

79. Mamman, A.S.; Lee, J.; Kim, Y.; Hwang, I.T.; Park, N.; Hwang, Y.K.; Chang, J.; Hwang, J.-S. Furfural: Hemicellulose/xylose-derived biochemical. *Biofuels Bioprod. Biorefining* **2008**, *2*, 438–454. [CrossRef]
80. Park, G.; Jeon, W.; Ban, C.; WOO, H.C.; Kim, D.H. Direct catalytic conversion of brown seaweed-derived alginic acid to furfural using 12-tungstophosphoric acid catalyst in tetrahydrofuran/water co-solvent. *Energy Convers. Manag.* **2016**, *118*, 135–141. [CrossRef]
81. Xia, A.; Cheng, J.; Ding, L.; Lin, R.; Song, W.; Su, H.; Zhou, J.; Cen, K. Substrate consumption and hydrogen production via co-fermentation of monomers derived from carbohydrates and proteins in biomass wastes. *Appl. Energy* **2015**, *139*, 9–16. [CrossRef]
82. Sambusiti, C.; Bellucci, M.; Zabaniotou, A.; Beneduce, L.; Monlau, F. Algae as promising feedstocks for fermentative biohydrogen production according to a biorefinery approach: A comprehensive review. *Renew. Sustain. Energy Rev.* **2015**, *44*, 20–36. [CrossRef]
83. Pierra, M.; Trably, E.; Godon, J.; Bernet, N. Fermentative hydrogen production under moderate halophilic conditions. *Hydrog. Energy* **2013**, *39*, 1–10. [CrossRef]
84. De Vrije, T.; Mars, A.E.; Budde, M.A.W.; Lai, M.H.; Dijkema, C.; De, W.P.; Claassen, P.A.M. Glycolytic pathway and hydrogen yield studies of the extreme thermophile Caldicellulosiruptor saccharolyticus. *Appl. Microbiol. Biotechnol.* **2007**, *74*, 1358–1367. [CrossRef]
85. Fang, H.H.P.; Liu, H. Effect of pH on hydrogen production from glucose by a mixed culture. *Bioresour. Technol.* **2002**, *82*, 87–93. [CrossRef]
86. Matsumura, Y.; Sato, K.; Al-saari, N.; Nakagawa, S.; Sawabe, T. Enhanced hydrogen production by a newly described heterotrophic marine bacterium, Vibrio tritonius strain AM2, using seaweed as the feedstock. *Int. J. Hydrog. Energy* **2014**, *39*, 7270–7277. [CrossRef]
87. Xia, A.; Jacob, A.; Herrmann, C.; Tabassum, M.R.; Murphy, J.D. Production of hydrogen, ethanol and volatile fatty acids from the seaweed carbohydrate mannitol. *Bioresour. Technol.* **2015**, *193*, 488–497. [CrossRef]
88. Wang, J.J.; Bensmail, H.; Yao, N.; Gao, X. Discriminative sparse coding on multi-manifolds. *Knowl. Based Syst.* **2013**, *54*, 199–206. [CrossRef]
89. Chung, I.K.; Beardall, J.; Mehta, S.; Sahoo, D.; Stojkovic, S. Using marine macroalgae for carbon sequestration: A critical appraisal. *J. Appl. Phycol.* **2011**, *23*, 877–886. [CrossRef]
90. Kumar, S.; Gupta, R.; Kumar, G.; Sahoo, D.; Kuhad, R.C. Bioethanol production from Gracilaria verrucosa, a red alga, in a biorefinery approach. *Bioresour. Technol.* **2013**, *135*, 150–156. [CrossRef]
91. McHugh, D.J. *A Guide to the Seaweed Industry*; ISO: Geneva, Switzerland, 2003; pp. 1–118.
92. Yeon, J.-H.; Lee, S.-E.; Choi, W.Y.; Kang, D.H.; Lee, H.-Y.; Jung, K.-H. Repeated-batch operation of surface-aerated fermentor for bioethanol production from the hydrolysate of seaweed Sargassum sagamianum. *J. Microb. Biotechnol.* **2016**, *21*, 323–331. [CrossRef]
93. Yanagisawa, M.; Nakamura, K.; Ariga, O.; Nakasaki, K. Production of high concentrations of bioethanol from seaweeds that contain easily hydrolyzable polysaccharides. *Process Biochem.* **2011**, *46*, 2111–2116. [CrossRef]
94. Kim, H.M.; Wi, S.G.; Jung, S.; Song, Y.; Bae, H. Efficient approach for bioethanol production from red seaweed Gelidium amansii. *Bioresour. Technol.* **2015**, *175*, 128–134. [CrossRef] [PubMed]
95. Council for Agricultural Science and Technology (CAST). *Animal Feed vs. Human Food: Challenges and Opportunities in Sustaining Animal Agriculture Toward 2050*; Iowa State University, Department of Economics: Ames, IA, USA, 2013; pp. 1–16.
96. Van Den Burg, S.; Stuiver, M.; Veenstra, F.; Bikker, P.; Contreras, A.L.; Palstra, A.; Broeze, J.; Jansen, H.; Jak, R.; Gerritsen, A.; et al. *A Triple P Review of the Feasibility of Sustainable Offshore Seaweed Production in the North Sea*; Wageningen UR: Wageningen, The Netherlands, 2013; pp. 1–108.
97. Reilly, P.; Doherty, J.V.O.; Pierce, K.M.; Callan, J.J.; Sullivan, J.T.O.; Sweeney, T. The effects of seaweed extract inclusion on gut morphology, selected intestinal microbiota, nutrient digestibility, volatile fatty acid concentrations and the immune status of the weaned pig. *Anim. Int. J. Anim. Biosci.* **2008**, *2*, 1465–1473. [CrossRef]
98. Gardiner, G.E.; Campbell, A.J.; Doherty, J.V.O.; Pierce, E.; Lynch, P.B.; Leonard, F.C.; Stanton, C.; Ross, R.P.; Lawlor, P.G. Effect of Ascophyllum nodosum extract on growth performance, digestibility, carcass characteristics and selected intestinal microflora populations of grower—Finisher pigs. *Anim. Feed Sci. Technol.* **2008**, *141*, 259–273. [CrossRef]
99. Kolb, N.; Vallorani, L.; Milanovi, N.; Stocchi, V. Evaluation of Marine Algae Wakame (Undaria pinnatifida) and Kombu (Laminaria digitata japonica) as Food Supplements. *Food Technol. Biotehnol.* **2004**, *42*, 57–61.
100. Gouveia, L.; Batista, A.P.; Sousa, I.; Raymundo, A.; Bandarra, N.M. Microalgae in Novel Food Products. In *Food Chemistry Research Development*; Nova Science Publishers: New York, NY, USA, 2008; pp. 1–37.
101. Funahashi, H.; Imai, T.; Tanaka, Y.; Tsukamura, K.; Hayakawa, Y.; Kikumori, T.; Mase, T.; Itoh, T.; Nishikawa, M.; Hayashi, H.; et al. Wakame Seaweed Suppresses the Proliferation of 7, 12-Dimethylbenz (a)- anthracene-induced Mammary Tumors in Rats. *Jpn. J. Cancer Res.* **1999**, *90*, 922–927. [CrossRef] [PubMed]
102. Lee, H.; Kim, J.; Kim, E. Fucoidan from Seaweed Fucus vesiculosus Inhibits Migration and Invasion of Human Lung Cancer Cell via PI3K-Akt-mTOR Pathways. *PLoS ONE* **2012**, *7*, e50624. [CrossRef]
103. Nelson, T.E.; Lewis, B.A. Separation and characterization of the soluble insoluble components of insoluble Laminaran. *Carbohydr. Res.* **1974**, *33*, 63–74. [CrossRef]

104. Park, E.; Pezzuto, J.M. Antioxidant Marine Products in Cancer Chemoprevention. *Antioxid. Redox Signal.* **2013**, *19*, 115–140. [CrossRef]

105. Déléris, P.; Nazih, H.; Bard, J. Seaweeds in Human Health. In *Seaweed in Health and Disease Prevention*; Elsevier Inc.: Amsterdam, The Netherlands, 2016; pp. 319–367.

106. Paradis, M.; Couture, P.; Lamarche, B. A randomised crossover placebo-controlled trial investigating the effect of brown seaweed (Ascophyllum nodosum and Fucus vesiculosus) on postchallenge plasma glucose and insulin levels in men and women. *Appl. Physiol. Nutr. Metab.* **2011**, *36*, 913–919. [CrossRef] [PubMed]

107. Ching, S.H.; Bansal, N.; Bhandari, B. Alginate gel particles—A review of production techniques and physical properties. *Crit. Rev. Food Sci. Nutr.* **2017**, *57*, 1133–1152. [CrossRef] [PubMed]

108. Couteau, C.; Coiffard, L. Seaweed Application in Cosmetics. In *Seaweed in Health and Disease Prevention*; Elsevier Inc.: Amsterdam, The Netherlands, 2016; pp. 423–441.

109. Wang, H.D.; Chen, C.; Huynh, P.; Chang, J. Exploring the potential of using algae in cosmetics. *Bioresour. Technol.* **2015**, *184*, 355–362. [CrossRef] [PubMed]

110. Price, R.D.; Berry, M.G.; Navsaria, H.A. Hyaluronic acid: The scientific and clinical evidence. *J. Plast. Recontstruction Asthetic Surg.* **2007**, *60*, 1110–1119. [CrossRef]

111. Pangestuti, R.; Kim, S. Biological activities and health benefit effects of natural pigments derived from marine algae. *J. Funct. Foods* **2011**, *3*, 255–266. [CrossRef]

112. Pallela, R.; Na-young, Y.; Kim, S. Anti-photoaging and Photoprotective Compounds Derived from Marine Organism. *Mar. Drugs* **2010**, *8*, 1189–1202. [CrossRef] [PubMed]

113. Food and Agriculture Organization of the United Nations. *Fishery and Aquaculture Statistics*; FAO: Quebec City, QC, Canada, 2016; pp. 1–108.

114. Adams, J.M.M.; Toop, T.A.; Donnison, I.S.; Gallagher, J.A. Seasonal variation in Laminaria digitata and its impact on biochemical conversion routes to biofuels. *Bioresour. Technol.* **2011**, *102*, 9976–9984. [CrossRef] [PubMed]

115. International Energy Agency (IEA) Bioenergy. *Biorefineries: Adding Value to the Sustainable Utilization of Biomass*; International Energy Agency (IEA) Bioenergy: Paris, France, 2009; pp. 1–16.

116. Laurens, L.M.; McMillan, J.D.; Baxter, D.; Cowie, A.L.; Saddler, J.; Barbosa, M.; Murphy, J.; Drosg, B.; Elliot, D.C.; Sandquist, J.; et al. *State of Technology Review—Algae Bioenergy: An IEA Bioenergy Inter-Task Strategic Project*; International Energy Agency (IEA) Bioenergy: Paris, France, 2017; pp. 1–158.

117. Roesijadi, G.; Jones, S.B.; Snowden-Swan, L.J.; Zhu, Y. *Macroalgae as a Biomass Feedstock: A Preliminary Analysi*; Pacific Northwest National Lab.: Richland, WA, USA, 2010; pp. 1–50.

118. Vera, J.; Castro, J.; Gonzalez, A.; Moenne, A. Seaweed Polysaccharides and Derived Oligosaccharides Stimulate Defense Responses and Protection Against Pathogens in Plants. *Mar. Drugs* **2011**, *9*, 2514–2525. [CrossRef]

119. Bikker, P.; Van Krimpen, M.M.; Van Wikselaar, P.; Houweling-tan, B.; Scaccia, N.; Van Hal, J.W.; Huijgen, W.J.J.; Cone, J.W.; Lopez-Contreras, A.M. Biorefinery of the green seaweed Ulva lactuca to produce animal feed, chemicals and biofuels. *J. Appl. Phycol.* **2016**, *28*, 3511–3525. [CrossRef] [PubMed]

120. Morales, M.; Collet, P.; Lardon, L.; Hélias, A.; Steyer, J.; Bernard, O. Life-cycle assessment of microalgal-based biofuel. In *Biomass, Biofuels and Biochemicals*; Elsevier B.V.: Amsterdam, The Netherlands, 2019; pp. 507–550.

121. Urban, R.A.; Bakshi, B.R. 1, 3-Propanediol from Fossils versus Biomass: A Life Cycle Evaluation of Emissions and Ecological Resources. *Ind. Eng. Chem. Resour.* **2009**, *48*, 8068–8082. [CrossRef]

122. Gnansounou, E.; Raman, J.K. Life cycle assessment of algae biodiesel and its co-products. *Appl. Energy.* **2016**, *161*, 300–308. [CrossRef]

123. Corona, A.; Biddy, M.J.; Vardon, D.R.; Birkved, M.; Houschild, M.; Beckkam, T. Life Cycle Assessment of Adipic Acid Production from Lignin. *Green Chem.* **2018**, *20*, 3857–3866. [CrossRef]

124. Montazeri, M.; Eckelman, M.J. Life cycle assessment of UV-Curable bio-based wood flooring coatings. *J. Clean. Prod.* **2018**, *192*, 932–939. [CrossRef]

125. Smidt, M.; Hollander, J.D.; Bosch, H.; Xiang, Y.; Van der Graaf, M.; Lambin, A.; Duda, J.-P. Life Cycle Assessment of Biobased and Fossil Based Succinic Acid. In *Sustainability Assessment of Renewables-Based Products: Methods and Case Studies*; John Wiley & Sons, Inc.: Hoboken, NJ, USA, 2016; pp. 307–321.

126. Brentner, L.B.; Eckelman, M.J.; Zimmerman, J.B. Combinatorial Life Cycle Assessment to Inform Process Design of Industrial Production of Algal Biodiesel. *Environ. Sci. Technol.* **2011**, *45*, 7060–7067. [CrossRef] [PubMed]

127. Raman, J.K.; Gnansounou, E. Life Cycle Assessment of Vetiver- Based Biorefinery With Production of Bioethanol and Furfural. In *Life-Cycle Assessment of Biorefineries*; Elsevier B.V.: Amsterdam, The Netherlands, 2017; pp. 147–165.

128. Pilicka, I.; Blumberga, D.; Romagnoli, F. Life Cycle Assessment of Biogas Production from Marine Macroalgae: A Latvian Scenario. *Sci. J. Riga Tech. Univ.* **2011**, *6*, 69–78. [CrossRef]

129. Alvarado-morales, M.; Boldrin, A.; Karakashev, D.B.; Holdt, S.L.; Angelidaki, I.; Astrup, T. Life cycle assessment of biofuel production from brown seaweed in Nordic conditions. *Bioresour. Technol.* **2013**, *129*, 92–99. [CrossRef]

130. Ertem, F.C.; Neubauer, P.; Junne, S. Environmental life cycle assessment of biogas production from marine macroalgal feedstock for the substitution of energy crops. *J. Clean. Prod.* **2017**, *140*, 977–985. [CrossRef]

131. Seghetta, M.; Hou, X.; Bastianoni, S.; Bjerre, A.; Thomsen, M. Life cycle assessment of macroalgal biore fi nery for the production of ethanol, proteins and fertilizers e A step towards a regenerative bioeconomy. *J. Clean. Prod.* **2016**, *137*, 1158–1169. [CrossRef]
132. Cappelli, A.; Gigli, E.; Romagnoli, F.; Simoni, S.; Palerno, M.; Guerriero, E. Co-digestion of macroalgae for biogas production: An LCA-based environmental evaluation. *Energy Procedia* **2015**, *72*, 3–10. [CrossRef]
133. Czyrnek-delêtre, M.M.; Rocca, S.; Agostini, A.; Giuntoli, J.; Murphy, J.D. Life cycle assessment of seaweed biomethane, generated from seaweed sourced from integrated multi-trophic aquaculture in temperate oceanic climates. *Appl. Energy* **2017**, *196*, 34–50. [CrossRef]
134. Giwa, A. Comparative cradle-to-grave life cycle assessment of biogas production from marine algae and cattle manure biorefineries. *Bioresour. Technol.* **2017**, *244*, 1470–1479. [CrossRef]
135. Parsons, S.; Allen, M.J.; Abeln, F.; Mcmanus, M.; Chuck, C.J. Sustainability and life cycle assessment (LCA) of macroalgae-derived single cell oils. *J. Clean. Prod.* **2019**, *232*, 1272–1281. [CrossRef]
136. Pérez-lópez, P.; Balboa, E.M.; González-garcía, S.; Domínguez, H.; Feijoo, G.; Moreira, M.T. Comparative environmental assessment of valorization strategies of the invasive macroalga Sargassum muticum. *Bioresour. Technol.* **2014**, *161*, 137–148. [CrossRef]
137. Plouviez, M.; Shilton, A.; Packer, M.A.; Guieysse, B. N$_2$O emissions during microalgae outdoor cultivation in 50 L column photobioreactors. *Algal Res.* **2017**, *26*, 348–353. [CrossRef]
138. Fagerstone, K.D.; Quinn, J.C.; Bradley, T.H.; De Long, S.K.; Marchese, A.J. Quantitative Measurement of Direct Nitrous Oxide Emissions from Microalgae Cultivation. *Environ. Sci. Technol.* **2011**, *45*, 9449–9456. [CrossRef]
139. Soleymani, M.; Rosentrater, K.A. Techno-Economic Analysis of Biofuel Production from Macroalgae (Seaweed). *Bioengineering* **2017**, *4*, 1–10. [CrossRef]
140. Clarens, A.F.; Resurreccion, E.P.; White, M.A.; Colosi, L.M. Environmental Life Cycle Comparison of Algae to Other Bioenergy Feedstocks. *Environ. Sci. Technol.* **2010**, *44*, 1813–1819. [CrossRef]
141. Jorquera, O.; Kiperstok, A.; Sales, E.A.; Embiruçu, M.; Ghirardi, M.L. Comparative energy life-cycle analyses of microalgal biomass production in open ponds and photobioreactors. *Bioresour. Technol.* **2010**, *101*, 1406–1413. [CrossRef]

 *sustainability*

Article

Improvement of the Crude Glycerol Purification Process Derived from Biodiesel Production Waste Sources through Computational Modeling

Matheus Oliveira [1], Ana Ramos [2], Eliseu Monteiro [2,3] and Abel Rouboa [2,3,4,*]

[1] Department of Mechanical Engineering, University of Trás-os-Montes and Alto Douro, Quinta de Prados, 5000-801 Vila Real, Portugal; projetos.matheusoliveir@gmail.com
[2] LAETA-INEGI, Associated Laboratory for Energy, Transports and Aeronautics, Institute of Science and Innovation in Mechanical and Industrial Engineering, R. Dr. Roberto Frias, 4200-465 Porto, Portugal; aramos@inegi.up.pt (A.R.); emonteiro@fe.up.pt (E.M.)
[3] Faculty of Engineering, University of Porto, R. Dr. Roberto Frias, 4200-465 Porto, Portugal
[4] Department of Mechanical Engineering and Applied Mechanics, School of Engineering, University of Pennsylvania, Philadelphia, PA 19104, USA
* Correspondence: rouboa@fe.up.pt

Abstract: Considering waste as a possible new resource for useful purposes is one of the strategies included in the circular economy principles. In fact, industrial processes are seen as great contributors to the formation of waste streams. With the aim to attain more sustainable and resilient systems, in this study, a process flow chart was elaborated in an Aspen Plus computer simulator, to obtain the production of pure glycerol from crude glycerol (a by-product of biodiesel production). This process occurs through fractional vacuum distillation, the methanol recovery route in the deacidification process and the removal of methanol from the reaction medium. The separation stages of the crude glycerol implemented enabled a degree of purification of 99.77%, meeting the specifications of the pharmaceutical use. The developed model allowed for the optimization of the purification process, raising by 40% the mass flow rate of pure glycerol. A conclusion could be drawn that the use of crude glycerol is an excellent option for the development of new products with greater added-value, contributing to the zero waste principles and to the circular economy.

Keywords: glycerol; biodiesel wastes; purification process modeling; Aspen Plus

Citation: Oliveira, M.; Ramos, A.; Monteiro, E.; Rouboa, A. Improvement of the Crude Glycerol Purification Process Derived from Biodiesel Production Waste Sources through Computational Modeling. *Sustainability* **2022**, *14*, 1747. https://doi.org/10.3390/su14031747

Academic Editor: Nicola Saccani

Received: 21 December 2021
Accepted: 31 January 2022
Published: 2 February 2022

Publisher's Note: MDPI stays neutral with regard to jurisdictional claims in published maps and institutional affiliations.

Copyright: © 2022 by the authors. Licensee MDPI, Basel, Switzerland. This article is an open access article distributed under the terms and conditions of the Creative Commons Attribution (CC BY) license (https://creativecommons.org/licenses/by/4.0/).

1. Introduction

The main factor in decreasing the availability of natural resources is disorderly population growth, associated with economic development, which in turn encourages overconsumption, exceeds the limits of the availability of natural resources and affects the balance of the ecosystem [1]. In the last few decades, however, reducing environmental pollution has become a global objective. The social-political-environmental scenario has been stimulating the replacement of the fossil fuel matrix with fuels from renewable sources, in order to reduce the emission of various greenhouse gases (GHG) that cause the greenhouse effect [2–4]. Improving environmental conditions, especially in large metropolitan centers, also means reducing government health costs for citizens. Environmental concerns about the use of fossils began to mark history with important events, such as the following: Toronto Conference on the Changing Atmosphere in 1988, IPCC's First Assessment Report in 1990 and ECO-92, culminating in the Kyoto Protocol in 1997 [5]. In addition, the adoption of environmental agreements makes it possible to plan international financing under favorable conditions, in the carbon credit market, under the mechanism of clean development.

In this context, biodiesel is considered as a potential and promising alternative to replace fossil fuels and its production has been growing worldwide [6–8]. Bearing in mind that fossil fuel reserves are not renewable, there are uncertainties regarding the market in the future, in addition to the availability of the resource [5]. Therefore, the paradigm between development and sustainability becomes a fertile environment to promote technological development, namely for the production of chemicals from renewable resources, such as biomass [6]. Bio-renewables in chemical commodities, including environmentally friendly biofuels, are mobilizing scientists from industries and universities around the world [9]. Biodiesel has emerged as a sustainable alternative to diesel, and its use has been encouraged by many countries, representing an economic, social, and environmental strategy [7,10].

The increase in biodiesel production has generated a surplus of crude glycerol in the market, which represents a major bottleneck in the biodiesel production chain and creates new challenges for its sustainable use; therefore, new technologies are required for the treatment of abundant residual glycerol that has great potential to be an important raw material, for the production of products with high added-value [7,8,11,12]. This residue cannot be deposited in landfills and the industries that use it as raw material will not be able to absorb this excess produced [8,13]. Therefore, finding new uses for glycerol is very important to ensure the sustainability of the global production of biodiesel.

Biodiesel refers to fuel formed by esters of fatty acids, methyl, ethyl or propyl esters of long chain carboxylic acids. It is a renewable and biodegradable fuel, commonly obtained from the chemical reaction of lipids, oils or fats, of animal or vegetable origin, with an alcohol in the presence of a catalyst (reaction known as transesterification) [1,12]. There are also several biodiesel production processes from renewable sources, such as agricultural products and microalgae, in the presence of a suitable catalyst. Numerous studies are available on the production of microalgae-based biodiesel [14,15], as well as through the thermal cracking reactions and esterification processes. This work addresses the purification of glycerol from production by transesterification, one of the main production routes for biodiesel in the world. In fact, it has lower energetic requirements, needs less time, and less quantities of alcohol for the reaction to occur [7,8,12,16]. Even with a higher water content, the proposed process presents satisfactory results, milder conditions being considered the most usual.

The products of oil and fats transesterification are esters of fatty acids (80–90%) and glycerol (10–20%, by-product of biodiesel production) [3,7,8,17]. The biodiesel production process is composed of the following steps: preparation of the raw material, transesterification reaction, phase separation, recovery and dehydration, alcohol distillation, glycerol distillation and purification of this renewable fuel, as well as the purification of water as a residue [12]. For every 100 liters of biodiesel, 10 kilograms of crude glycerol are produced, generating an average of 60 liters of waste water [18]. There are several options for purification procedures, such as treatment with ethanol and activated carbon, pH adjustment, solvent extraction, and precipitation of the fatty acids with calcium, ion exchange resins, membrane separation and distillation [19,20]. The process using ion exchange resins becomes unfeasible when the crude glycerin has a high content of dissolved salts. Distillation and membrane separation are used to obtain glycerol with a high degree of purity, the first process being the most effective. A highly employed process is vacuum distillation in an inert atmosphere, providing 99% content in glycerol [19]. Other processes can also be applied, such as neutralization, drying, saponification, polar solvent extraction and adsorption [21]. Purification with adsorbent materials has become an interesting alternative, as it eliminates the need to use water in the process. Another advantage is the avoidance of liquid effluents and the ability to reuse some adsorbents [22].

In many countries, including Brazil, one of the largest biodiesel producers in the world, most large-scale industrial biodiesel plants still do not effectively value glycerol. According to Freitas [23], the crude glycerol produced in the country is sold to refineries, and around 50% is exported to China. To comply with the pharmaceutical and food industry requirements, glycerol needs to undergo purification processes to obtain more competitive purity grades or valuable by-products [11,24]. Glycerol is a compound of extreme technical versatility. Due to its unique combination of properties, glycerol is used in many areas of the industry. Glycerol is an alcohol and viscous liquid, soluble in water, practically colorless, odorless, hygroscopic, virtually non-toxic to humans and nature, with a high boiling point (ebullition temperature of 290 °C) [5]. Due to this unusual combination of physical and chemical properties, glycerol has more than 2000 known end applications, including several large-scale applications [5,7,16]. A great diversity of research is being developed to reduce the impact of this waste on the environment, adding value to the production of biodiesel. Therefore, means of purification and transformation of glycerol are required in order to avoid future problems due to its accumulation, as well as to advance biodiesel production techniques, enabling higher competitiveness and viability [6]. The biodiesel industry was responsible for about 68% of glycerol produced worldwide in 2015 and glycerol production by transesterification is expected to grow about 6.8% by 2022 [25]. Thus, a technology that contributes to the storage or use of glycerol, giving value to this by-product, will also contribute to the production of biodiesel and renewable energies.

In this context, the present work aimed to evaluate glycerol's purification process, using computational modeling and simulation in Aspen Plus. The use of computational tools makes the analysis of quality, demand and cost easier and faster, with good accuracy [26,27]. Regarding simulators, the product portfolio of Aspen Technology Inc. has the optimal solution and process optimization tools on the market. These are based on mathematical and thermodynamic models, based on the basic principles of transport phenomena [28]. Scientific computational modeling applies computing to areas of knowledge in which it is impossible, or very expensive, to carry out experimental tests to analyze possible solutions for some processes, starting from experimental models or analytical solutions. Therefore, the development of existing processes is relevant and of interest, as it can help to identify previous problems and estimate whether what is being proposed is economically viable [29].

In this way, using Aspen Plus, it was possible to perform the chemical and thermodynamic modeling of the purification of glycerol from the biodiesel production co-product. The study of this process, through the Aspen Plus platform, allowed for analysis regarding the variables; the tempering and concentration of reagents in the process parameters that need to be optimized.

2. Materials and Methods

2.1. Glycerol Transesterification

Transesterification by basic catalysis is one of the main production routes for biodiesel in the world, among the countless ways of production [7,8,12,16,30]. Other production routes are, for example, acid catalysis [31], heterogeneous or enzymatic catalysis [32], ultrasonic radiation [33], or even thermal decomposition of the catalyzed oil [34,35]. Of these, transesterification by basic catalysis is the most used in commercial production, probably due to its high conversion rate of oil (triglycerides) into biodiesel (methyl esters), in a simple, short-term chemical reaction, presenting fewer problems related to equipment corrosion [36].

The methodology followed in this research addresses the purification of glycerol produced by transesterification by basic catalysis [7,8,12,16,30]. Commercially available purified common glycerol is manufactured to meet the requirements of the United States Pharmacopeia (USP) and the Food Chemicals Codex (FCC). Figure 1 shows the transesterification reaction.

Glycerol is normally classified in three categories, according to its purity, as seen in Table 1. Crude glycerol presents purity ranging between 40–88%. Glycerol with purity levels above this may be used in the transformation of products or chemical intermediates [7]. Technical glycerol, with purity greater than 96%, is used in industry to produce chemical compounds. Pharmaceutical glycerol shows purity levels higher than 99.7% and is used in the food and pharmaceutical industry, research and other high standard applications [37,38]. In practice, the glycerol obtained in the transesterification process also contains various impurities, such as methyl ester (ME), triglycerides, free fatty acids (FFA), methanol, water, inorganic salts, and other contaminating organic matter. The composition and, consequently, the properties of the crude glycerol obtained depend strongly on the type of process used and the quality of the raw material [8]. Thus, according to the combination of the process and the raw material used, crude glycerol, as a by-product from the transesterification of biodiesel production, can be considered to have a content between 30% and 60%. In the case studied, the glycerol had a low initial purity of 50% [37,38].

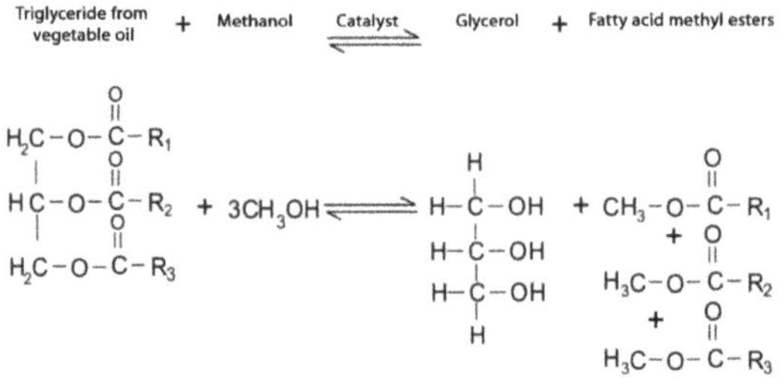

Figure 1. Transesterification Reaction by Catalysis.

Table 1. Specifications of Glycerol as to its Purity [37].

Degree	Degree I	Degree II	Degree III
Purity	~99.5% (Technical degree)	96-99.5% (USP * degree)	99.5-99.7% (Kosher or USP/FCC **)
Manufacturing and Use	Prepared by synthetic process and used in chemical industry, but not applicable to food or drug formulation.	Prepared from sources of animal fat or vegetable oil, suitable for food, pharmaceutical and cosmetic products.	Prepared from vegetable oil sources, suitable for use in kosher food and beverages.

* USP—United States Pharmacopeia, ** FCC—Food Chemicals Codex.

Table 2 shows a typical composition of crude glycerol derived from the biodiesel production used in this study [38]. The purity of the glycerol produced from the synthesis of biodiesel is of crucial importance as it increases the added-value of the product. Ashes are composed of dissolved inorganic species, formed mostly by sodium ions (due to the excess of catalyst in the production of biodiesel), chlorides and other species present in used oils. Non-glycerol organic matter (NGOM) in this case is represented by a mixture of many organic compounds such as free fatty acids, unconverted glycerides and other residual organic compounds present in the raw material [5].

Table 2. Composition of Crude Glycerol [38].

Component	Wt. (%)
Glycerol	50
Methanol	35
Potassium hydroxide	10
Methyl oleate	5
Sulfuric acid	0
Water	0

The use of modeling and simulation stands out as process improvement, since the use of computational tools makes quality analysis simpler, faster and with good accuracy. Therefore, the study of improvements through operational models is original and of relevant interest, as it can help to optimize the process, identifying problems and estimating economic viability. The use of modeling from the experimental data obtained for the esterification stage, allows us to predict the trend of the reaction behavior. Using the Aspen

Plus software, it was possible to model the reaction kinetics, in addition to the simulation of the fractional distillation process, being important for the analysis of the variables involved such as concentration, temperature and pressure, establishing the necessary conditions, demands, equipment and results.

For the purification of crude glycerol, a vacuum distillation method was used, which can be divided into several stages, as seen in Figure 2.

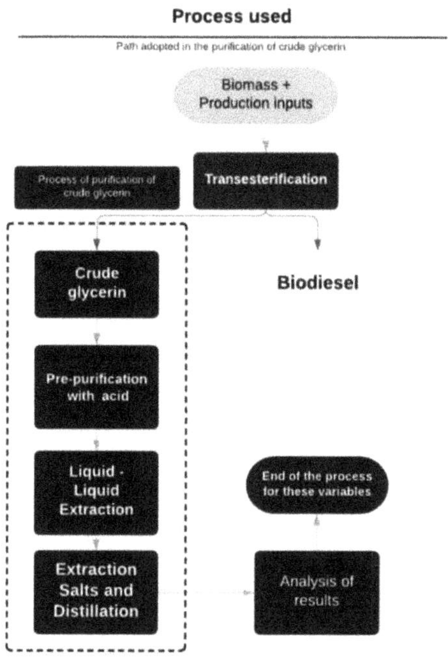

Figure 2. Vacuum Glycerol Purification Process.

Neutralization of the concentrated solution (potassium hydroxide) occurs through the addition of a strong acid, sulfuric acid (H_2SO_4). The mass flow was designed so that all the base present in the crude glycerol oil was consumed, forming a mixture of salt and water. The commercial software Aspen Plus allows changes in different variables; temperature and concentration of reagents, improving the efficiency of the process. The main parameters were the temperature and the reagent concentration. In this way, the minimum and maximum values were defined to obtain a sensitive variation, in order to avoid the degradation or polymerization of glycerol into polyglycerol, occurring at high temperatures [8].

2.2. Aspen Plus Model

In the present study, the modeling process was performed using Aspen Plus, and can be seen in Figure 3.

Maintaining a fixed feed flow, simulations were carried out involving variations in the flow of the distiller. At each interaction the feed flow rate remained constant, for each variation presented, the simulator automatically recalculated the flow rate of the bottom product.

The properties of all components were taken from the Aspen Plus library. As with the base model, the simulation involves ionic species (potassium hydroxide and sulfuric acid) and polar components (glycerol and methanol). As such, the thermodynamic model chosen for the purification processing was electrolyte non–random two liquid (ENRTL).

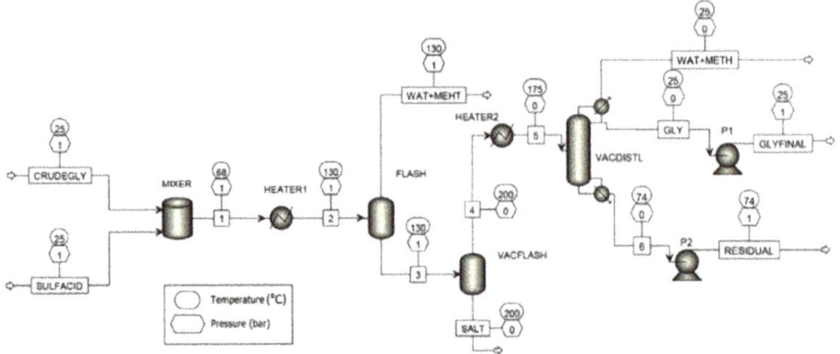

Figure 3. Process Flow Diagram of The Glycerol Treatment, in Aspen Plus Model.

This is a vacuum distillation process, where the potassium hydroxide is neutralized using sulfuric acid and the methanol is removed in a vacuum (flash) separator. The description of this initial stage of the process is in Figure 4.

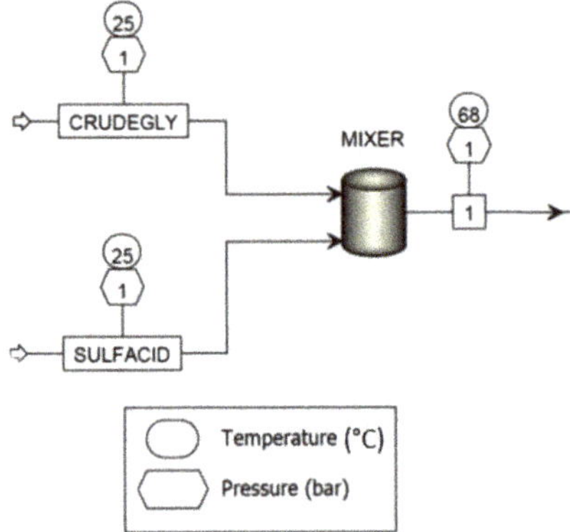

Figure 4. Mixing Tank.

Crude glycerol (CRUDEGLY) and sulfuric acid (SULFACID) enter the process at atmospheric pressure and at room temperature and are neutralized in the mixing tank (MIXER). Their compositions can be seen in Table 3 while the resulting flow is shown in Table 4.

Table 3. Properties and Composition of the Mixing Process Inlet Streams.

Property	Crude Glycerol	Sulfuric Acid
Temperature (°C)	25	25
Pressure (kPa)	101.325	101.325
Molar flow (kmol/h)	24.10	10.99
Mass flow (kg/h)	1200	283.78
Component Mass Fraction (%)		
Glycerol	50.0	0.0
Methanol	35.0	0.0
Potassium hydroxide	10.0	0.0
Methyl oleate	5.0	0.0
Sulfuric acid	0.0	37.0
Water	0.0	63.0

Table 4. Composition of the Mixing Process Outlet Streams.

Component	Mass Fraction (%)
Glycerol	2.22
Methanol	41.39
Potassium hydroxide	0.0
Methyl oleate	4.34×10^{-6}
Sulfuric acid	9.88×10^{-4}
Water	56.38

The mixture resulting from the reaction first stage of the process (1), is then heated to 130 °C by a heat exchanger (HEATER1) and goes to the separator (FLASH), seen in Figure 5.

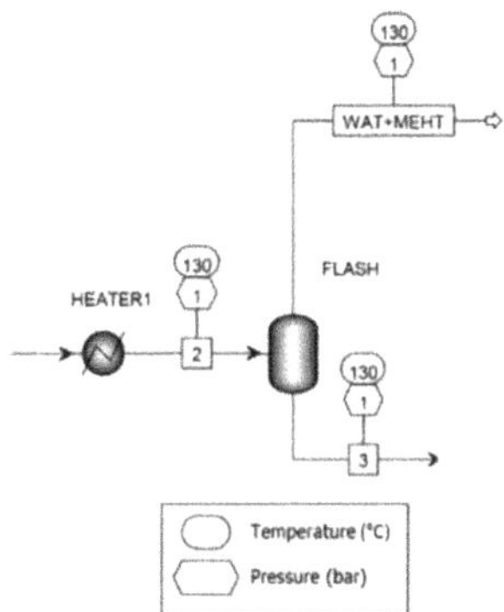

Figure 5. Heating and Separation Stage Stream (METH + WAT).

Methanol and water are separated in the stream (METH + WAT) and the main mixture flows through flow (3) to the vacuum separator (VACFLASH). In this component there is the separation of potassium hydroxide from the mixture. Some operational obstacles arose during the dimensioning of the temperature in this separator, which did not allow a resulting flow for temperatures below 310 °C. The temperature established for the separation was 200 °C. Figure 6 shows the resulting streams from the process of salt separation.

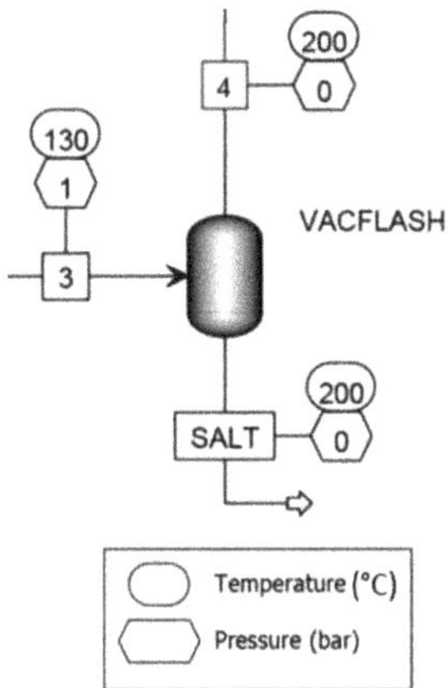

Figure 6. Flash for Salt Separation.

The resulting mixture was then sent to the final stage of the purification process in the vacuum distillation column (VACDISTL), after passing through a heat exchanger (HEATER2) that decreased its temperature to 200 °C. In the distillation column we had three products, two in liquid state and one in steam state. Figure 7 depicts a product (WAT + METH) consisting practically of water and methanol, another one (RESIDUAL) that presents residual material, but that is of interest for reuse to generate more purified glycerol, and finally the desired product, glycerol with 99.77% purity (GLYFINAL) as expressed in Table 5.

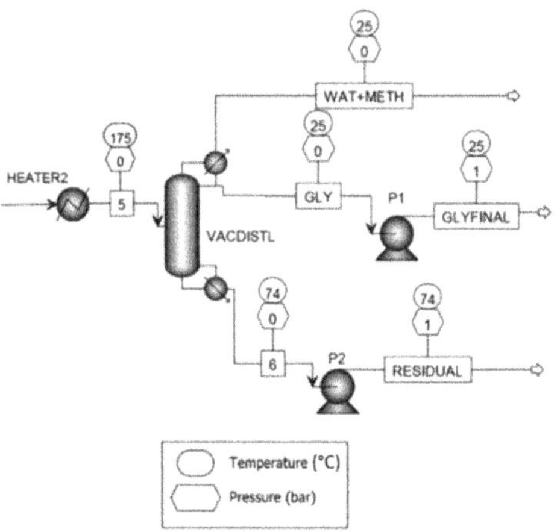

Figure 7. Vacuum Distillation Stage.

Table 5. Properties and Composition of the Outlet Streams.

Property	Glycerol	Residual	Wat + Meth
Molar flow (kmol/h)	6.69	2.70	3.30
Mass flow (kg/h)	616.53	867.25	84.20
Component Mass Fraction (%)			
Glycerol	99.768	20.64	2.22
Methanol	2.47×10^{-3}	1.56×10^{-3}	41.39
Potassium hydroxide	0.0	0.0	0.0
Methyl oleate	5.17×10^{-3}	23.82	23.82×10^{-10}
Sulfuric acid	0.206	55.53	4.33×10^{-10}
Water	0.018	8.23×10^{-5}	56.38

3. Results and Discussion

The purification process in the distillation column was accomplished in several stages. The simulation results have shown that the process is sensitive to only five levels, since in additional levels, the degree of purification was inexpressive, as shown in Figure 8.

Through these procedures, a purification level of 99.77% was reached, as shown in Table 6. In this study, an improved process is suggested, with the addition of a heat exchanger after the distillation column, following the main line, to avoid the presence of vapors in the pump. Results may be seen in Table 6.

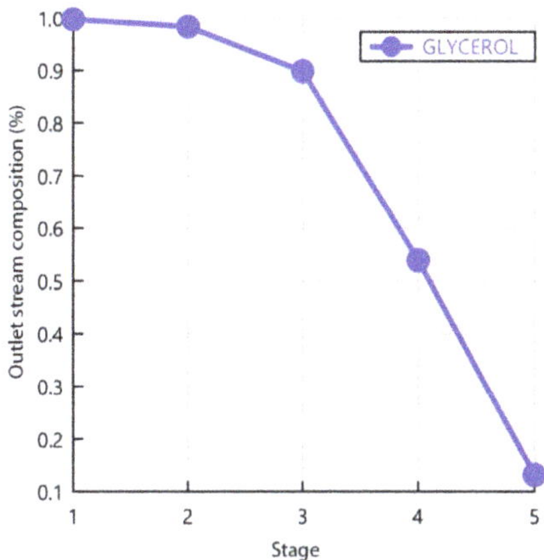

Figure 8. Composition and Stage of the Vacuum Distillation Process.

Table 6. Composition of the Outlet Streams in Each Stage of the Vacuum Distillation Process.

Stage	Glycerol (%)	Sulfuric Acid (%)	Methyl Oleate (%)	Methanol (%)	Potassium Hydroxide (%)	Water (%)
1	13.19	37.79	49.01	3.46×10^{-6}	4.83×10^{-6}	1.03×10^{-5}
2	53.95	26.19	19.86	6.43×10^{-6}	3.65×10^{-18}	2.01×10^{-5}
3	89.80	7.44	2.75	8.43×10^{-6}	0.0	2.69×10^{-5}
4	98.37	1.39	0.24	9.44×10^{-6}	0.0	3.04×10^{-5}
5	99.77	0.21	0.01	8.59×10^{-4}	0.0	3.46×10^{-3}

The present model was based on the model of Arora et al. [38], in which the authors performed simulations to obtain a flow of glycerol with a purity content of 99%. Seeking to maintain the separation conditions, such as number of stages and column sizing, a sensitivity analysis in the operational variables and the addition of a heat exchanger that provides a substantial additional income is proposed. Table 7 shows a comparison of the results of the product glycerol obtained in this study and Arora's [38].

Table 7. Comparison Between the Obtained Results and Literature Results.

Property	Arora et al. [38]	This Work
Glycerol (%)	99	99.77
Methanol (%)	0.047	2.47×10^{-3}
Potassium hydroxide (%)	0.0	0.0
Methyl oleate (%)	3.61×10^{-6}	5.17×10^{-3}
Sulfuric acid (%)	3.60×10^{-6}	0.206
Water (%)	0.072	0.018
Mass flow (kg/h)	436.90	616.53
Molar flow (kmol/h)	4.76	6.69

The mass fraction of the six components in the product glycerol is very similar. The purity of the glycerol is 99% in the Arora [38] study and 99.77% in the current study. The most relevant result of this comparison is the 40% increase in mass and molar flow, which can be justified by the improvement of the process promoted in this study, by the inclusion of an additional heat exchanger in the vacuum separator component or in the distillation column. These values depend on the temperature in the separator, as well as the required flow in the distillation column and its reflux degree. The higher the applied temperature, the greater the separation and, consequently, the greater the flow. In this way, the separator is the equipment most sensitive to temperature variations. Future works may provide further knowledge in reducing the energy requirements, using pinch analysis of the heat exchangers network, and optimization of the operating conditions.

4. Conclusions

In this study, a numerical methodology for the improvement of the purification process of the biodiesel by-product glycerol was developed in the Aspen Plus process simulator. It was shown that glycerol can be produced from the crude glycerol by-products of biodiesel, with a theoretical purity level of 99.77%. The developed model allows the improvement of the purification process of glycerol by the vacuum distillation route, by the inclusion of an additional heat exchanger, leading to around 40% higher mass of glycerol, in relation to the reference case study. This study demonstrates that the use of crude glycerol, as a by-product of biodiesel production, is an excellent option for the development of new products, with greater added-value and a consequent decrease in the production cost of the main product, contributing to the zero waste principles and circular economy. This result is even more relevant in the actual and future scenario of the biodiesel and glycerol market. These markets are expected to grow significantly in the years to come, which presents good business opportunities for the related companies. A cost analysis is envisioned for the next steps, as a follow-up study, to assess the viability of the proposed improvements.

Author Contributions: Conceptualization, M.O. and A.R. (Abel Rouboa); methodology, M.O. and A.R. (Abel Rouboa); software, M.O.; validation, A.R. (Abel Rouboa), E.M. and A.R. (Ana Ramos); formal analysis, M.O. and E.M.; investigation, M.O.; resources, M.O. and A.R. (Abel Rouboa); data curation, E.M.; writing—original draft preparation, M.O.; writing—review and editing, E.M. and A.R (Ana Ramos); visualization, A.R. (Ana Ramos); supervision, A.R (Abel Rouboa).; project administration, A.R. (Abel Rouboa); funding acquisition, A.R. (Abel Rouboa) and E.M. All authors have read and agreed to the published version of the manuscript.

Funding: This research was funded by FCT (Portuguese Foundation for Science and Technology), grant number PCIF/GBV/0169/2019 and by The European Union's Horizon 2020 research and innovation program, grant number 818012, and co-funded by FITEC, Interface Programme.

Institutional Review Board Statement: Not applicable.

Informed Consent Statement: Not applicable.

Data Availability Statement: Not applicable.

Conflicts of Interest: The authors declare no conflict of interest.

References

1. Pereira Gabriel, C.K. Produção de Biodiesel a Partir de Óleo de Palma. Master's Thesis, Instituto Superior Técnico, Lisbon, Portugal, 2015; pp. 1–117.
2. Malek, L. Simulation of Propionaldehyde Production from Glycerol. Master's Thesis, Department of Chemical Engineering, Lund University, Lund, Sweden, 2012.
3. Goh, B.H.H.; Chong, C.T.; Ge, Y.; Ong, H.C.; Ng, J.-H.; Tian, B.; Ashokkumar, V.; Lim, S.; Seljak, T.; Józsa, V. Progress in utilisation of waste cooking oil for sustainable biodiesel and biojet fuel production. *Energy Convers. Manag.* **2020**, *223*, 113296. [CrossRef]
4. Ramos, A. and A. Rouboa. Renewable energy from solid waste: Life cycle analysis and social welfare. *Environ. Impact Assess. Rev.* **2020**, *85*, 12. [CrossRef]
5. Miscenco, D. *Recuperação e Valorização de Glicerol Bruto da Produção de Biodiesel a Partir de Óleos Alimentares Usados*; Faculdade de Ciências e Tecnologia da Universidade de Coimbra: Coimbra, Portugal, 2016; pp. 1–140.
6. Alves, A.D.P.; Rodrigues Filho, G.M.; Mendes, M.F. Avaliação técnica de diferentes processos de separação para purificação do glicerol como subproduto—Revisão. *Rev. Bras. Energ. Renov.* **2017**, *6*, 1–29. [CrossRef]
7. Monteiro, M.R.; Kugelmeier, C.L.; Pinheiro, R.S.; Batalha, M.O.; da Silva César, A. Glycerol from biodiesel production: Technological paths for sustainability. *Renew. Sustain. Energy Rev.* **2018**, *88*, 109–122. [CrossRef]

8. Ardi, M.; Aroua, M.; Hashim, N.A. Progress, prospect and challenges in glycerol purification process: A review. *Renew. Sustain. Energy Rev.* **2015**, *42*, 1164–1173. [CrossRef]
9. Beatriz, A.; Araújo, Y.J.K.; De Lima, D.P. Glicerol: Um breve histórico e aplicação em sínteses estereosseletivas. *Quim. Nova* **2011**, *34*, 306–319. [CrossRef]
10. Peiter, G.C.; Alves, H.J.; Sequinel, R.; Bautitz, I.R. Alternativas para o uso do glicerol produzido a partir do biodiesel. *Rev. Bras. Energ. Renov.* **2016**, *5*, 1–19. [CrossRef]
11. Alves, I.R.; Mahler, C.F.; Oliveira, L.B.; Reis, M.M.; Bassin, J.P. Assessing the use of crude glycerol from biodiesel production as an alternative to boost methane generation by anaerobic co-digestion of sewage sludge. *Biomass Bioenergy* **2020**, *143*, 105831. [CrossRef]
12. Mehrpooya, M.; Ghorbani, B.; Abedi, H. Biodiesel production integrated with glycerol steam reforming process, solid oxide fuel cell (SOFC) power plant. *Energy Convers. Manag.* **2020**, *206*, 112467. [CrossRef]
13. Milessi, T.S.S.; Tabuchi, S.C.T.; Antunes, F.A.F.; da Silva, S. Variação da temperatura no tratamento do glicerol residual de biodiesel visando sua aplicação como fonte de carbono em processos fermentativos. In Proceedings of the 5° BIOCOM, Rio De Janeiro, Brazil, 21–23 March 2012; pp. 4–5.
14. Jacob, A.; Ashok, B.; Alagumalai, A.; Chyuan, O.H.; Le, P.T.K. Critical review on third generation micro algae biodiesel production and its feasibility as future bioenergy for IC engine applications. *Energy Convers. Manag.* **2020**, *228*, 113655. [CrossRef]
15. Ong, H.C.; Tiong, Y.W.; Goh, B.H.H.; Gan, Y.Y.; Mofijur, M.; Fattah, I.R.; Chong, C.T.; Alam, M.A.; Lee, H.V.; Silitonga, A. Recent advances in biodiesel production from agricultural products and microalgae using ionic liquids: Opportunities and challenges. *Energy Convers. Manag.* **2020**, *228*, 113647. [CrossRef]
16. Domingos, A.M.; Pitt, F.D.; Chivanga Barros, A. Purification of residual glycerol recovered from biodiesel production. *S. Afr. J. Chem. Eng.* **2019**, *29*, 42–51.
17. do Corgo e Silva, S. *Breve Enciclopédia do Biodiesel*; Vida Económica: Porto, Portugal, 2009.
18. Palomino-Romero, J.A.; Leite, O.M.; Barrios Eguiluz, K.I.; Salazar-Banda, G.R.; Silva, D.P.; Cavalcanti, E.B. Tratamentos dos efluentes gerados na produção de biodiesel. *Quimica Nova* **2012**, *35*, 367–378. [CrossRef]
19. Pott, R.W.; Howe, C.J.; Dennis, J.S. The purification of crude glycerol derived from biodiesel manufacture and its use as a substrate by Rhodopseudomonas palustris to produce hydrogen. *Bioresour. Technol.* **2014**, *152*, 464–470. [CrossRef]
20. Rios, E.A.M. *Avaliação de Metodologia de Purificação da Glicerina Gerada Como Coproduto na Produção de Biodiesel*; Universidade Tecnológica Federal do Paraná: Curitiba, Brazil, 2016.
21. Stojković, I.J.; Stamenković, O.S.; Povrenović, D.S.; Veljković, V.B. Purification technologies for crude biodiesel obtained by alkali-catalyzed transesterification. *Renew. Sustain. Energy Rev.* **2014**, *32*, 1–15. [CrossRef]
22. Contreras-Andrade, I.; Avella-Moreno, E.; Sierra-Cantor, J.F.; Guerrero-Fajardo, C.A.; Sodre, J.R. Purification of glycerol from biodiesel production by sequential extraction monitored by 1H NMR. *Fuel Process. Technol.* **2015**, *132*, 99–104. [CrossRef]
23. de Carvalho Freitas, E.S. *Produção de Biodiesel a Partir do Sebo Bovino: Proposta de um Sistema de Logística Reversa*; Universidade Federal da Bahia: Salvador, Brazil, 2016.
24. Rivaldi, J.D.; Sarrouh, B.F.; Fiorilo, R.; Silva, S. Glicerol de biodiesel: Estratégias biotecnológicas para o aproveitamento do glicerol gerado da produção de biodiesel. *Biotecnol. Ciênc. Desenvolv.* **2016**, *37*, 44.
25. Candeias, D. Valorization of crude glycerin from biodiesel production by acetalization. *Téc. Lisboa* **2017**, 1–10.
26. Ramos, A.; Monteiro, E.; Rouboa, A. Numerical approaches and comprehensive models for gasification process: A review. *Renew. Sustain. Energy Rev.* **2019**, *110*, 188–206. [CrossRef]
27. Silva, V.B.; Rouboa, A. Optimizing the DMFC operating conditions using a response surface method. *Appl. Math. Comput.* **2012**, *218*, 6733–6743. [CrossRef]
28. Birchal, M.A.S.; Birchal, V.S. Automação De Uma Planta De Produção De Biodiesel. *E-Xacta* **2013**, *6*, 139–145. [CrossRef]
29. Dos Santos, J.S.; de Araújo, A.C.B. Modelagem matemática do processo reativo e de. In Proceedings of the XXI Congresso Brasileiro de Engenharia Química, Fortaleza, Brazil, 25–29 December 2016.
30. Bastos, R.R.C.; da Luz Corrêa, A.P.; da Luz, P.T.S.; da Rocha Filho, G.N.; Zamian, J.R.; da Conceição, L.R.V. Optimization of biodiesel production using sulfonated carbon-based catalyst from an amazon agro-industrial waste. *Energy Convers. Manag.* **2020**, *205*, 112457. [CrossRef]
31. Marchetti, J.; Errazu, A. Esterification of free fatty acids using sulfuric acid as catalyst in the presence of triglycerides. *Biomass Bioenergy* **2008**, *32*, 892–895. [CrossRef]
32. Lam, M.K.; Lee, K.T.; Mohamed, A.R. Homogeneous, heterogeneous and enzymatic catalysis for transesterification of high free fatty acid oil (waste cooking oil) to biodiesel: A review. *Biotechnol. Adv.* **2010**, *28*, 500–518. [CrossRef]
33. Nakpong, P.; Wootthikanokkhan, S. High free fatty acid coconut oil as a potential feedstock for biodiesel production in Thailand. *Renew. Energy* **2010**, *35*, 1682–1687. [CrossRef]
34. Hameed, B.H.; Lai, L.; Chin, L. Production of biodiesel from palm oil (*Elaeis guineensis*) using heterogeneous catalyst: An optimized process. *Fuel Process. Technol.* **2009**, *90*, 606–610. [CrossRef]
35. Quesada-Medina, J.; Olivares-Carrillo, P. Evidence of thermal decomposition of fatty acid methyl esters during the synthesis of biodiesel with supercritical methanol. *J. Supercrit. Fluids* **2011**, *56*, 56–63. [CrossRef]
36. Ma, F.; Hanna, M.A. Biodiesel production: A review. *Bioresour. Technol.* **1999**, *70*, 1–15. [CrossRef]

37. Hudha, M.I.; Laksmana, D.I. Glycerin purification of biodiesel production side products by distillation method. In Proceedings of the Seminar Nasional Kimia-National Seminar on Chemistry (SNK 2018), State University of Surabaya, Jawa Timur, Indonesia, 7–12 August 2018; pp. 27–29.
38. Arora, P.; Baroi, C.; Dalai, A.K. Feasibility study of crude glycerol purification processes. *J. Basic Appl. Eng. Res.* **2015**, *2*, 60–63.

Article

Superheated Steam Torrefaction of Biomass Residues with Valorisation of Platform Chemicals—Part 1: Ecological Assessment

Baharam Roy [1], Peter Kleine-Möllhoff [2],* and Antoine Dalibard [3]

[1]   Reutlingen Research Institute (RRI), Reutlingen University, 72762 Reutlingen, Germany; Md_Baharam.Sk@Reutlingen-University.DE or roybaharam@gmail.com
[2]   ESB Business School, Reutlingen University, 72762 Reutlingen, Germany
[3]   Fraunhofer Institute for Interfacial Engineering and Biotechnology IGB, 70569 Stuttgart, Germany; antoine.dalibard@igb.fraunhofer.de
*   Correspondence: peter.kleine-moellhoff@reutlingen-university.de; Tel.: +49-712-1271-5009

Abstract: Within the last decade, research on torrefaction has gained increasing attention due to its ability to improve the physical properties and chemical composition of biomass residues for further energetic utilisation. While most of the research works focused on improving the energy density of the solid fraction to offer an ecological alternative to coal for energy applications, little attention was paid to the valorisation of the condensable gases as platform chemicals and its ecological relevance when compared to conventional production processes. Therefore, the present study focuses on the ecological evaluation of an innovative biorefinery concept that includes superheated steam drying and the torrefaction of biomass residues at ambient pressure, the recovery of volatiles and the valorisation/separation of several valuable platform chemicals. For a reference case and an alternative system design scenario, the ecological footprint was assessed, considering the use of different biomass residues. The results show that the newly developed process can compete with established bio-based and conventional production processes for furfural, 5-HMF and acetic acid in terms of the assessed environmental performance indicators. The requirements for further research on the synthesis of other promising platform chemicals and the necessary economic evaluation of the process were elaborated.

Keywords: biorefinery; superheated steam torrefaction; environmental assessment; volatile recovery; platform chemicals

Citation: Roy, B.; Kleine-Möllhoff, P.; Dalibard, A. Superheated Steam Torrefaction of Biomass Residues with Valorisation of Platform Chemicals—Part 1: Ecological Assessment. *Sustainability* **2022**, *14*, 1212. https://doi.org/10.3390/su14031212

Academic Editor: Ana Ramos

Received: 23 December 2021
Accepted: 18 January 2022
Published: 21 January 2022

Publisher's Note: MDPI stays neutral with regard to jurisdictional claims in published maps and institutional affiliations.

Copyright: © 2022 by the authors. Licensee MDPI, Basel, Switzerland. This article is an open access article distributed under the terms and conditions of the Creative Commons Attribution (CC BY) license (https://creativecommons.org/licenses/by/4.0/).

1. Introduction

What would the world look like if we could realise the production of chemical products in a bio-based circular economy? What impact would bio-based chemistry have on the climate and the environment? How can the torrefaction of biomass be used for environmentally friendly chemicals production?

1.1. General Context

Sustainable biomass will play a significant role in meeting the 2030 target to reduce greenhouse gas emissions, as well as the objective of climate neutrality by 2050 in the European Green Deal [1,2]. Biomass can be used as a renewable energy source, a material substitute and as a carbon sink, thus contributing towards negative emissions. A recent study shows that the current trend in EU biomass use has to be corrected in order to achieve a net zero economy. Traditional bioenergy applications will become less and less competitive due to new future options based on increasing electrification and hydrogen share. Instead, material uses of biomass must increase significantly with a special focus on high-value applications (chemicals, textiles, etc.). The use of biomass for energy application must be reserved for special niches (e.g., industrial heat, fuels for aviation, etc.) [3]. Therefore, innovative technologies that allow for this strategic change are strongly required.

Bio-based products can be used in small, specialised but also large-volume markets. In the area of fine and speciality chemicals as well as active pharmaceutical ingredients, bio-based products are already competitive to some extent due to their functionality and thus offer worthwhile investment targets [4]. Biddy et al. (2016) presented a study in which 12 promising chemicals were identified that can already be produced in the near future from renewable sources, such as sugar, lignocellulose or algae [5]. For example, 5-hydroxymethylfurfural (HMF) and furfural are two interesting platform chemicals for a bio-based chemical production economy, and their current bio-based production has been studied in terms of their environmental footprint [5,6]. Some industrial companies in the chemical industry have already adapted their business model to a bio-based circular economy [7–10].

1.2. Biomass Torrefaction

In this context, biomass torrefaction is a promising technology since it allows one to upgrade low-quality biomass to higher quality products to be used either directly or further processed into high-value products [11,12]. Up to now, torrefied biomass is essentially used for energy applications, either directly for electricity generation (co-firing with coal in power plant) or as feedstock for further conversion into high quality biofuels (pyrolysis, gasification, catalytic synthesis, etc.). The increasing number of scientific investigations within the last decade in this field focused mainly on improving the energy density of biomass, the production of synthetic fuels and the integration into existing production and industrial structures for cascaded use [13–15]. In particular, the research works focused on application areas such as agriculture and food, the paper industry, energy suppliers, the steel industry, but also on the production of pyrolysis products, as well as liquefaction and gasification [4,11,15–18].

Torrefaction is a mild form of pyrolysis in which biomass is usually heated to about 200–300 °C. During the process, the three main constituents of woody biomass (cellulose, hemicellulose and lignin) are thermally decomposed at different degrees, leading to the formation of non-condensable gases (CO, CO_2) and condensable volatiles [11,18,19]. Beside water, the condensable volatile fraction contains valuable chemical substances (e.g., furfural, acetic acid, methanol, formic acid) that can be used as platform chemicals [19]. In this regard, the use of superheated steam as torrefaction agent is very interesting because it allows for a fast and uniform process and an easy recovery of volatiles [20–22]. In addition, the condensation heat can be recovered in order to make the whole process more energy efficient.

However, the production of value-added chemicals from torrefaction condensate is a challenging task due to the low concentrations and the resulted complex mixture of water, aldehydes, carboxylic acids, furans, ketones and alcohols. The organic fraction comprises innumerable substances with extensive distribution of molecular weight and polarity, which affects the effective separation of chemicals [23]. Therefore, new concepts are required in order to achieve an economical valorisation of these chemicals at the market-required purity.

For the reasons mentioned above, the authors of the present paper have dealt with the recovery and separation of valuable platform chemicals from torrefied biomass. In the context of this work, the ecological footprint of the newly developed process is investigated, which can be used to obtain valuable platform chemicals from biomass residues by torrefaction using superheated steam under atmospheric conditions. This analysis is intended to provide a statement on the ecological footprint of the newly developed process and how it compares to competing processes.

2. Superheated Steam Torrefaction

2.1. History

The core technology of the developed biorefinery concept is the superheated steam (SHS) processing under atmospheric pressure. This technology has been originally developed for drying purposes in the early 1990s [24], then further developed by Fraunhofer IGB for torrefaction within the EU project SteamBio (https://www.steambio.eu/ (accessed on 6 December 2021)). The use of SHS and thus the absence of oxygen permits an inert processing, prevents oxidation of the product and significantly reduces the risk of explosion.

2.2. Process Principle

The process principle is based on a system, which is hermetically closed at the top and atmospherically open at the bottom. The material to be processed is introduced to an SHS atmosphere, which is maintained in a superheated state through the supply of heat (see Figure 1 below). Enhanced heat transfer is achieved convectively by recirculating the SHS in a closed loop. The vapours (moisture, volatiles) released from the material during the thermal processing are extracted/condensed in order to maintain the system atmospheric pressure. The energy contained in the excess vapours (at temperature above 100 °C) can be recovered in other processes of the plant, which results in a high overall energy efficiency. Energy recovery can be conducted, e.g., by means of condensation, which allows volatile organic compounds (VOCs) to be condensed out with the excess steam. These condensable organic substances can be further separated and valorised as value-added products.

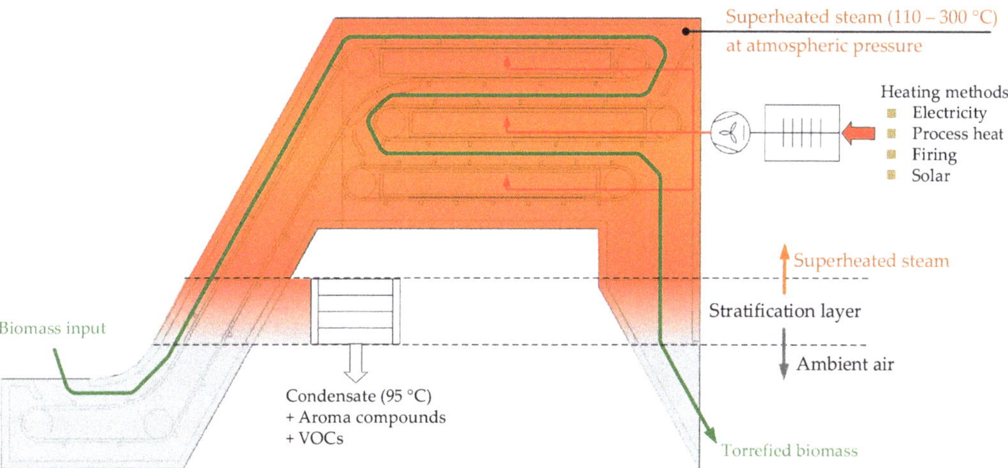

Figure 1. Principle design concept of the SHS drying and torrefaction process.

3. Materials and Methods

Within this paper, a new process for extracting valuable products from biomass that have been steam dried and torrefied under ambient pressure is analysed. This study is based on experimental results obtained within a German public-funded research project on an innovative biorefinery concept (see funding section). The experiments were conducted in a pilot-scale SHS-drying/torrefaction unit and laboratory-scale batch rectification and extraction set-ups.

3.1. Developed Refinery Concept

Figure 2 shows the different steps of the developed biorefinery concept, which was investigated in terms of its environmental impacts, as presented in this paper.

The process units with a grey background represent the reference case. In addition to this reference case, another variant was investigated in which the torrefied biomass is fed to an incineration plant (hatched background and dashed lines) that serves to supply heat internally to the biorefinery process and externally to potential consumers.

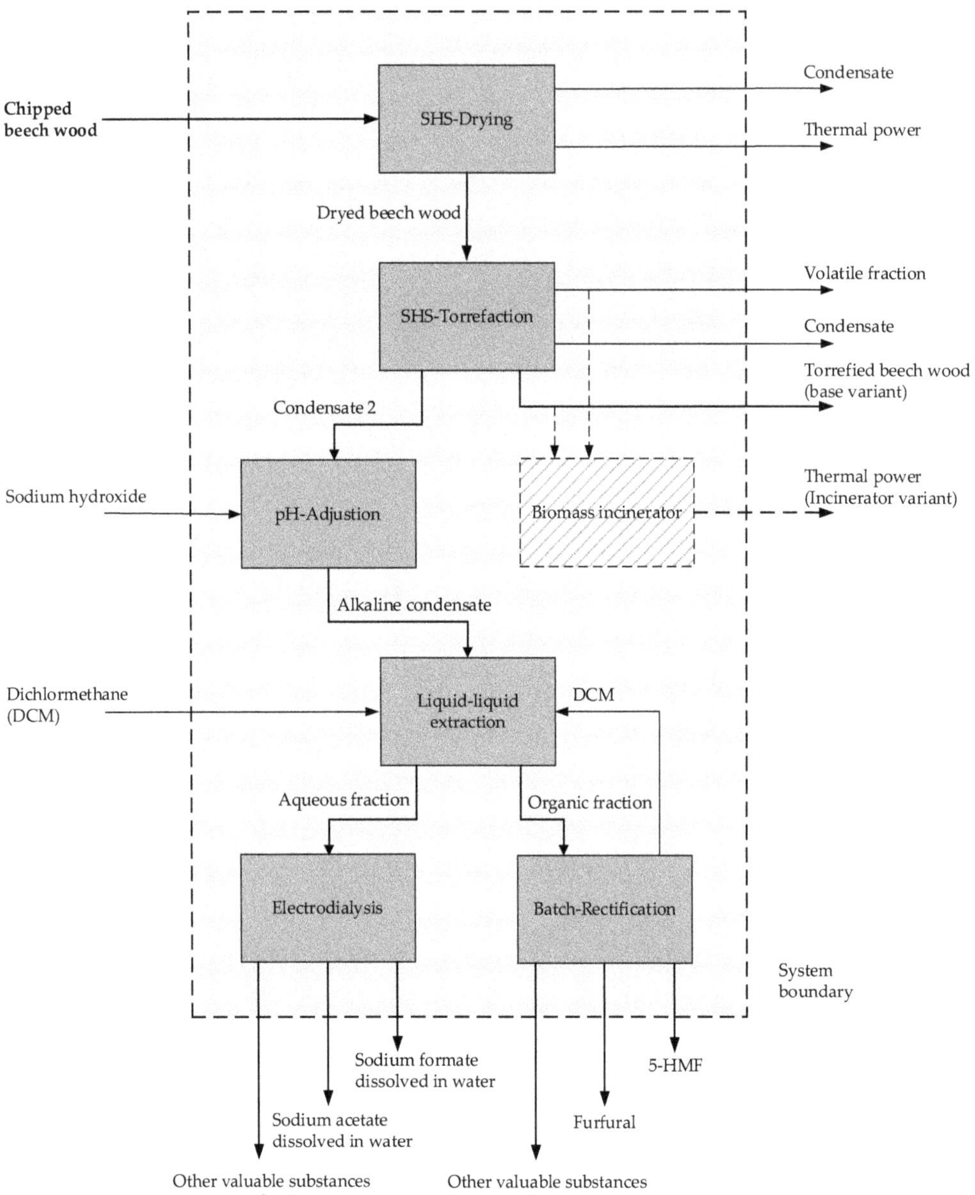

Figure 2. Biorefinery process set up.

3.2. Methods, Framework and Assumptions

Figure 3 shows the basic methodology used in this work. For each process step, the energy and material balances were drawn up on the basis of thermodynamic and fluid mechanics principles as well as experimental results. These balances formed the basis for the assessment of the environmental impact. The environmental impacts were carried out in accordance with ISO 14040 and 14044.

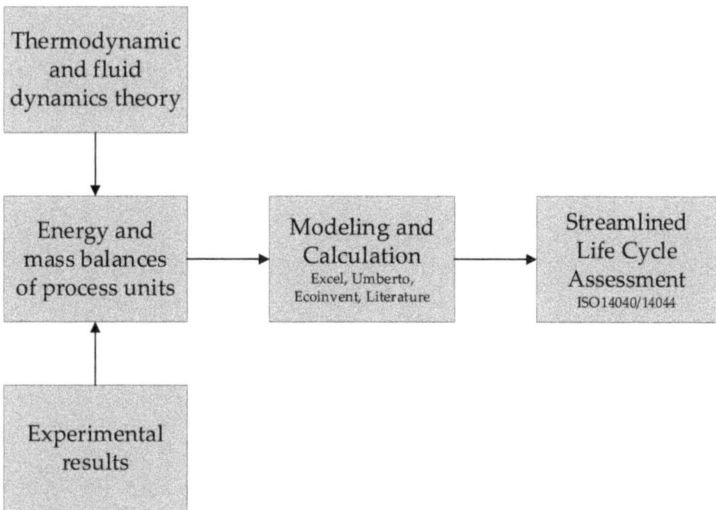

Figure 3. Assessment methodology.

The system boundary is shown in Figure 2. The upstream supply chains for biomass, energy and auxiliary materials were considered for the analysis. Table 1 lists the assumptions and parameters for the ecological analysis that allow for comparison with other relevant studies, such as [6,11,25].

Table 1. Assumptions and parameter for the ecological analysis.

Designation	Value	Unit	Description
Functional unit	1	kg	Biomass input
Design input	1000	kg/h	Biomass input
Allocation			Mass related
Plant location			Germany
Base year	2019		Before COVID-19
External energy supply			Electricity and natural gas
Reference biomass type			Shredded beechwood
Sensitivity analysis	20–45	% (dry basis)	Moisture content biomass input
	150–600	$kgCO_2$/kWh	Specific CO_2-emission electricity
	22–28	MJ/kg	Higher heating value torrefied biomass
	0 and 5	%	Steam loss (referred to evaporated water)
	100–300	km	Biomass transportation distance
	Straw	/	Biomass input alternative

In total, 1000 kg/h of biomass input material was chosen for the size of the plant and one kilogram of processed biomass as the functional unit to maintain the comparability with other studies. The allocation of the primary energy demand and CO_2 emissions to the different products was nevertheless also investigated.

The process structure with the energy and material flows was first calculated, then used as input for the modelling in the life cycle assessment software Umberto and supplemented with life cycle assessment inventory data obtained, e.g., from the Ecoinvent 3.7.1 database. The environmental impact assessment was carried out using the ReCiPe midpoint (H) w/o LT method [26]. For the reference case and the process variant, the impact categories climate change and primary energy consumption were considered. In addition, for two different biomass inputs, the impact categories human toxicity, freshwater ecotoxicity and freshwater eutrophication were identified as significant and therefore considered. Germany was assumed as the location for the production plant. For the manual calculations and verification of the results from the Umberto software, primary energy factors and specific CO_2 emissions values for Germany from 2019 were used because the years 2020 and 2021 are not representative due to the COVID-19 pandemic. Nevertheless, the influence of different specific CO_2 emissions of electricity generation was investigated and presented to allow for a transfer of the results to other locations or other energy supply structures.

The incineration plant of the analysed variant was considered as a system extension in order to be able to present environmental credits in terms of primary energy demand and CO_2 emissions. A sensitivity analysis was carried out with regard to the factors listed in Table 1. The results obtained were compared and evaluated with available information from competing processes.

The considerations end at the gate of the process under investigation. This means that product transports or the onward transmission of the generated thermal energy were not considered. The research in this paper does not include any economic considerations, which will be evaluated in a future paper.

4. Results

4.1. Energy and Material Flows

Figure 4 shows the energy and material flows of the reference case for an assumed steam loss of 5% (related to the evaporated mass) during SHS drying and torrefaction.

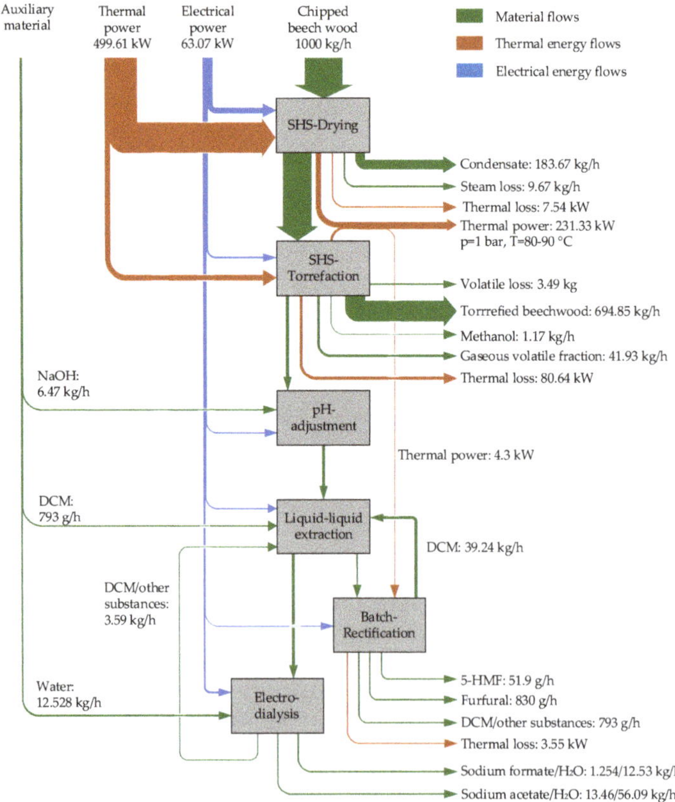

Figure 4. Energy and material input and output flows (reference case, 5% steam loss).

The width of the arrows in the Sankey diagram is not scaled exactly. They are only intended to illustrate which flows are significant. Therefore, the absolute amounts of the flows are indicated. The thermal and electrical energy flows are shown separately. The heat losses through the walls of the individual apparatus are not shown. Heat losses of 5% were assumed for the calculations. The energy flows due to the chemical internal energy of the materials fed into and removed from the system, i.e., in particular the woodchips and the torrefied biomass, are not shown in the Sankey diagram. In the Sankey diagram, the energy flow rates are given in kW, that is, kJ/s. All data refer to a thermal energy supply with natural gas, a chipped beech wood input flow of 1000 kg/h and 35 weight % (dry) water content of the biomass input for a typical production environment. In the following, the term "energy demand" is used for ease of reading when referring to the percentage shares of the individual energy flow rates.

The thermal energy demand of the dryer, which is operated at a temperature range between 150 and 180 °C, represents the largest energy demand, with 92.65% of the total energy demand, followed by the torrefaction unit, with 6.27%. The thermal energy demand of the rectification column is very small in comparison and can be covered internally from the torrefaction. The discharge of the non-condensable gaseous volatile fraction causes the most thermal losses in the torrefaction unit. The thermal output stream from the SHS-drying plant due to the condensation of the steam expelled from the biomass accumulates at a temperature level of approx. 95 °C (see Figure 1) and can be utilised for external purposes.

The fan that circulates the steam in the SHS-drying unit has the greatest demand for electrical energy, with 9.23% of the total energy consumption, followed by the fan of the torrefaction unit, with 0.69%. The electrodialysis unit consumes 2.78% of the total electrical energy. All other electrical energy consumers together require 10.90% of the total energy consumption.

In terms of mass flow quantity, with an input of 1000 kg/h of beech wood chips, the torrefied biomass represents the largest recyclable material output, with almost 70%, followed by the condensate from the dryer (approximately 18%), which can be thermally utilised.

The torrefied biomass has a higher heating value (HHV), approx. 22 MJ/kg, and can be utilised either as a very clean fuel with coal-like characteristics, as a raw material for further use in activated carbon production or for the production of synthesis gas. The non-condensable volatile fraction (CO, CO_2, CH_4, H_2) from torrefaction cannot be further utilised and is released to the environment in the reference case. The valuable substances 5-HMF, furfural, sodium formate and sodium acetate are produced in small quantities but high purities (>95%). The pilot tests have shown that about 20 different chemical substances can be extracted. However, only the above-mentioned platform chemicals are shown here, as they represent the largest share in terms of quantity. Dichloromethane is used as extraction agent and can be recovered up to 98% in the rectification unit.

The data presented up to this point are based on a steam loss of 5%. However, in a well-designed real plant, it can be expected that the steam loss will be close to zero. Reducing the steam loss has a direct effect on the product yield and on the thermal losses, especially of the dryer. The product yields, the total power demand and thermal losses at 5 and 0% steam loss are shown in Table 2.

In the reference case, the entire energy supply of the biorefinery is provided externally via electricity and natural gas. The torrefied biomass produced is materially utilised as output in the reference case. The resulting non-condensable volatile fraction of the torrefaction process is not thermally utilised in the reference case and is released to the atmosphere. Therefore, a second variant was investigated within the scope of this work, which operates self-sufficiently in terms of thermal energy. For this purpose, an incineration plant is added to the biorefinery, in which the entire torrefied biomass produced and the volatile fraction are thermally utilised. The Sankey diagram of this second variant is shown in Figure 5.

Table 2. Product yields, thermal and electrical power demand and losses for 5 and 0% steam loss (reference case).

Output	5% Steam Loss	0% Steam Loss
Total thermal power demand	499.61 kW	499.79 kW
Total electrical power demand	63.07 kW	64.04 kW
Thermal loss drying	7.54 kW	0 kW
Thermal loss torrefaction	80.64 kW	77.70 kW
Torrefied biomass	694.85 kg/h	694.85 kg/h
5-HMF	51.9 g/h	54.7 g/h
Furfural	830 g/h	873 g/h
Sodium formate	1.254 kg/h	1.32 kg/h
Sodium acetate	13.46 kg/h	14.17 kg/h
Methanol	1.169 kg/h	1.231 kg/h

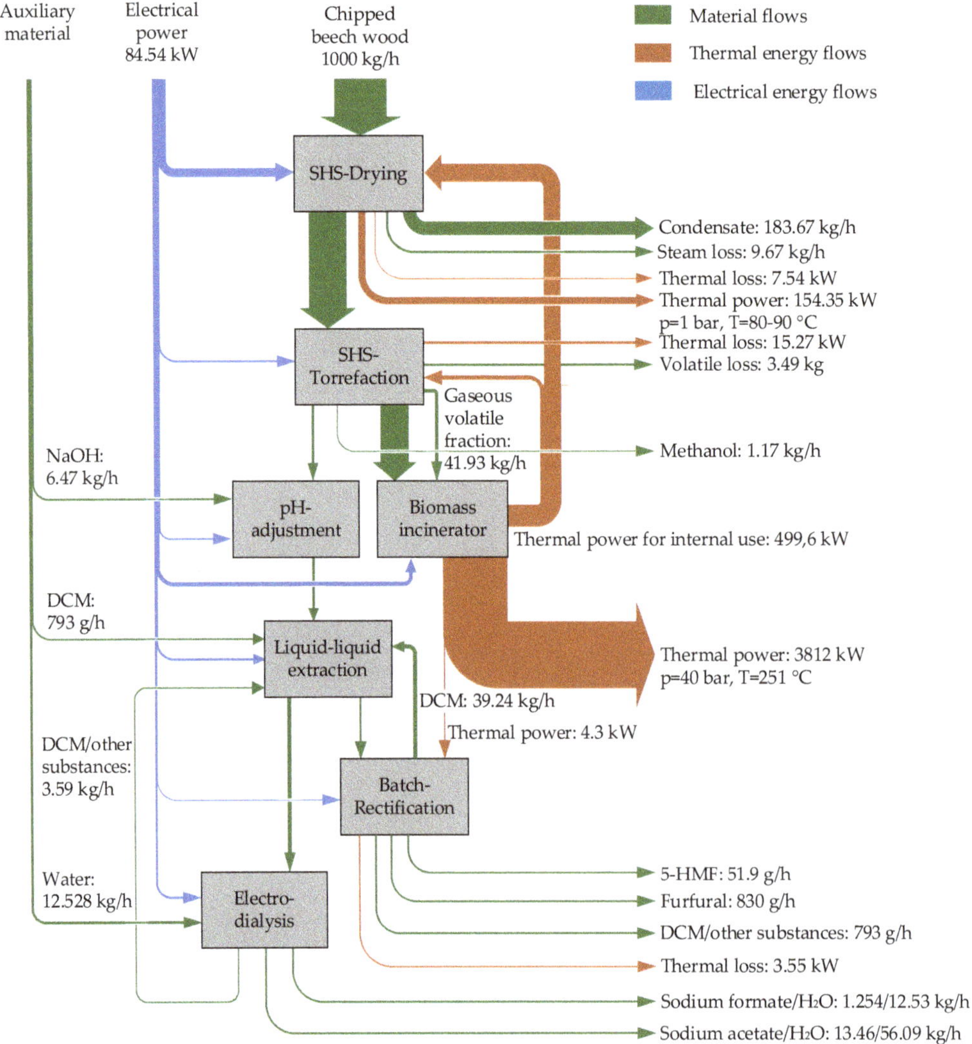

Figure 5. Energy and material input and output flows (incineration variant, 5% steam loss).

By combustion of the torrefied biomass and the volatile fraction, a thermal energy flow rate of approx. 4.3 MW can be generated assuming a calorific value of 22 MJ/kg. Since the plant's own thermal demand at 5% steam loss amounts to slightly less than 500 kW, approx. 3.8 MW of thermal power can be provided externally at 250 °C. In addition, 0.15 MW heat at 90 °C can be recovered from the SHS dryer. The reference case has a thermal loss of 92 kW, whereas the process variant has a thermal loss of only 26.36 kW. The process variant is particularly suitable for integration into existing, locally close production networks that have a heat demand at the temperature levels mentioned above. In terms of production volumes, the ratios at 5 and 0% steam loss in the process variant are the same as in the reference case. The process variant, however, causes higher electrical power consumption than the reference case, primarily due to the additional operation of a combustion fan and an exhaust gas fan for the combustion system. This additional consumption of electrical power totals approx. 21 kW.

4.2. Primary Energy Demand and CO_2 Emissions from a Process View

The energy embodied in natural resources prior to any human-made conversion is referred to as primary energy. The primary energy demand is the non-renewable portion of the primary energy required by the process to produce the end products; this is also referred to as cumulative non-renewable energy consumption.

The results presented here are based on a biomass input of 1000 kg per hour. The units 'kWh' and 'kgCO$_{2\text{-eq}}$' denote the total primary energy demand and CO_2 emission for processing 1000 kg of biomass per hour. The specific values of CO_2 emission and primary energy demand for the operation with 5% steam loss for the reference case and process variants are shown in this section. The results are categorized into three groups: upstream, thermal, and electrical. Upstream processes included the supply of auxiliary materials such as NaOH and DCM, as well as the supply chain of wood chips. Table 3 displays the values for the primary energy factor (PEF) and CO_2 footprint for the materials and energy carriers used in the process under investigation.

Table 3. The primary energy factor and CO_2 footprint of various materials and energy inputs.

	Years	Primary Energy Factor (PEF)		CO_2 Footprint	
NaOH [1] [27]	-	6.11 [27]	kWh$_{PE}$/kg	1.2 [1]	kgCO$_{2\text{-eq}}$/kg
DCM [1]	-	11.26	kWh$_{PE}$/kg	3.42	kgCO$_{2\text{-eq}}$/kg
Wood Supply chain [1]	-	0.217	kWh$_{PE}$/kg	0.0547	kgCO$_{2\text{-eq}}$/kg
Electricity [28,29]	2018	1.71	kWh$_{PE}$/kWh$_{el}$	0.471	kgCO$_{2\text{-eq}}$/kWh
	2019	1.55	kWh$_{PE}$/kWh$_{el}$	0.408	kgCO$_{2\text{-eq}}$/kWh
	2020	1.37	kWh$_{PE}$/kWh$_{el}$	0.366	kgCO$_{2\text{-eq}}$/kWh
Natural Gas [28,30]	2019	1.1 [30]	kWh$_{PE}$/kWh$_{el}$	0.202 [28]	kgCO$_{2\text{-eq}}$/kWh

[1] Umberto calculation, ecoinvent 3.7.1.

Umberto LCA models were developed to calculate the total non-renewable energy consumption and CO_2 footprint of the upstream processes. Printouts of the models can be found as Supplementary Materials to this publication. Suitable activities were chosen from the Ecoinvent 3.7.1 database. These activities included all upstream activities, starting from the cradle and ending with the product's reception at the consuming entity, as well as the average transportation requirements. The region Europe (RER) was used for DCM, and the DE region with 'transport, freight, and lorry 16–32 metric ton EURO 4' was chosen for wood chip supply. The global region had to be chosen for NaOH, and the calculated value was 19.33 MJ/kg. According to a report on sodium hydroxide eco-profiles, the gross primary energy required to produce 1 kg of NaOH is 22.04 MJ/kg (6.11 kWh/kg) [27]. This value is used in the calculations. The mass of the supplied product and the transportation requirements determine the upstream primary energy demand and CO_2 footprint. Because the average transportation distance of 100 km is already factored into the activities and the mass of biomass input is constant across all variants, the values of primary energy demand

(0.217 kWh/kg) and CO_2 emissions (0.0547 $kgCO_{2\text{-eq}}$/kg) are also constant. The required supply of auxiliary materials is determined by the percentage of steam/volatile loss during the process; the lower the volatile loss, the greater the mass supply, and thus the greater the primary energy demand and CO_2 footprint from these auxiliary materials.

Figures 6 and 7 depict the distribution of primary energy demand and CO_2 emissions to the individual processes, where 'others' refers to the sum of pH adjustment, liquid–liquid extraction, rectification, and electrodialysis processes. DCM and NaOH are separately indicated.

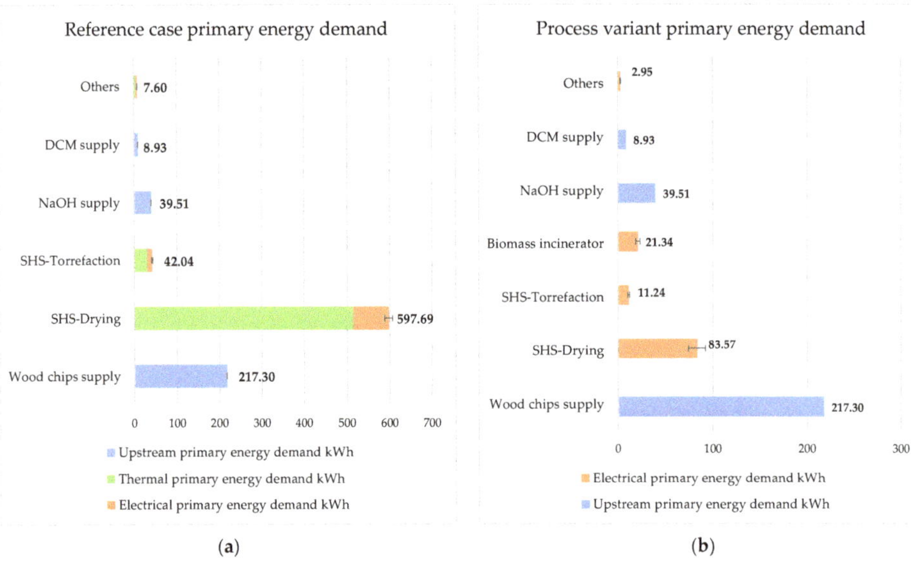

Figure 6. VALORKON primary energy demand with 5% steam loss: (**a**) reference case; (**b**) process variant.

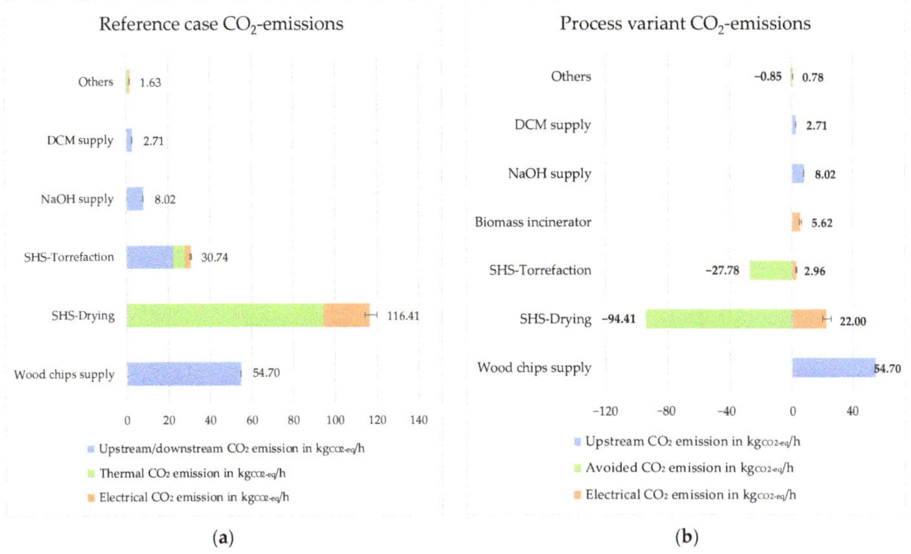

Figure 7. VALORKON CO_2 emission with 5% steam loss: (**a**) reference case; (**b**) process variant.

The thermal primary energy demand and CO_2 emissions were calculated by multiplying the PEF and CO_2 footprint of natural gas by the total amount of thermal energy required. Because the PEF of natural gas is constant (PEF = 1.1, where it includes the energy required for processing and distribution, i.e., 10%), as is the CO_2 footprint, there is no deviation in the bar charts for thermal primary energy demand and CO_2 emission. The electrical primary energy demand, on the other hand, is determined by the country's electricity mix. The German electricity mix for 2019 (prior to COVID-19) is considered here. The deviation shown in the black bar is derived from different PEF and CO_2 footprint values for the years 2018 and 2020, as shown in Table 3.

If the primary energy consumptions for processing 1000 kg of biomass per hour for both variants are added up counting 5% steam loss, the result is 913.07 kWh for the reference case and 384.84 kWh for the process variant. The significant difference in primary energy demand is due to the fact that in the process variant the required thermal energy is provided internally by a biomass incinerator. Figure 6 also shows the distribution of primary energy demand for both variants when varying the primary energy demand for the individual process steps according to Table 3. The SHS-drying process consumes the most primary energy in the reference case, followed by the biomass supply chain, whereas the biomass supply chain consumes the most primary energy in the process variant, followed by the SHS drying. The process variant includes an additional electrical energy consumer, a biomass incinerator with a primary energy demand of 21.34 kWh per hour.

Figure 7 illustrates the CO_2 emissions and their distribution due to the variation of the primary energy factors for electric power according to Table 3 for the different process stages and the supply chain. During the torrefaction process, the non-condensable gaseous volatile fraction, which also contains methane, is released into the environment in the reference case and fed to a biomass incinerator in the process variant case. Since methane is released into the environment, the corresponding downstream CO_2 equivalents in the torrefaction process must be added for the reference case. For this purpose, a CO_2 equivalence factor of 25 was applied according to [31]. Total CO_2 emissions are 214.21 $kgCO_{2\text{-eq}}$ per ton of biomass input for the reference case and 96.78 $kgCO_{2\text{-eq}}$ per ton of biomass input for the process variant. The calculation with Umberto resulted in CO_2 emissions of 218.55 $kgCO_{2\text{-eq}}$ per ton of biomass input for the reference case, thus verifying the calculations. Since the biomass incinerator provides the necessary thermal energy for the process, an emission of 123.04 $kgCO_{2\text{-eq}}$ per hour is avoided, which also includes the avoided CO_2 emission from the combustion of non-condensable gaseous volatile fraction. The negative side of the bar chart represents avoided CO_2 emissions, while the positive side represents caused CO_2 emissions in the process variant. The avoided CO_2 emissions are calculated under the assumption that the combustion of natural gas is replaced in the biorefinery. For the reference case, the SHS-drying process emits the most CO_2, followed by the wood supply chain, and for the process variant, drying has the second highest caused CO_2 emission but avoids most CO_2 emission through the internal heat supply.

4.3. Primary Energy Demand and CO_2 Emissions from a Product View (Allocation)

To know how much primary energy is required or how much CO_2 is emitted to produce one kilogram of the respective end product, the total primary energy demand and CO_2 emission must be allocated to the end product. To begin, the end products are set mass based in relation to the total output of a respective sub-process. The primary energy demand/CO_2 emissions of the individual processes can then be allocated to the corresponding end products. The total primary energy requirement/CO_2 emissions to produce the respective end product can then be calculated by adding the individual values. This can be divided by the respective product output to obtain the primary energy demand/CO_2 emission per kg of end product.

Figure 8a represents a comparative representation of the primary energy demand required to produce one kilogram of the respective end product for both variants. It clearly shows for the reference case that the products 5-HMF and furfural have the highest primary energy requirement per kilogram, with 9.6 kWh each, followed by sodium formate and sodium acetate, with 4.5 kWh each. It can be seen that the primary energy demand for the process variant is significantly lower because the required thermal energy is covered by the incinerator; the same is true for CO_2 emission. Figure 8b represents for both variants the CO_2 emissions that occur when producing one kilogram of end product. The highest emission per kilogram occurs when producing 5-HMF and furfural for the reference case, and when producing sodium acetate and sodium formate for the process variant. The effect of the process variant is more significant on 5-HMF and furfural, because the thermal energy requirement during rectification has a significant weight on 5-HMF and furfural, and because this can be supplied internally in the process variant, the value of primary energy demand and CO_2-emission is further reduced. The higher CO_2 emission and primary energy demand of furfural, 5-HMF, sodium acetate, and sodium formate can be explained by the small amount of product produced and the number of steps required to obtain these products. Torrefied beech wood, on the other hand, emits the least amount of CO_2 per kilogram. This is due to the small amount of beech wood input required to produce one kilogram of torrefied beech wood.

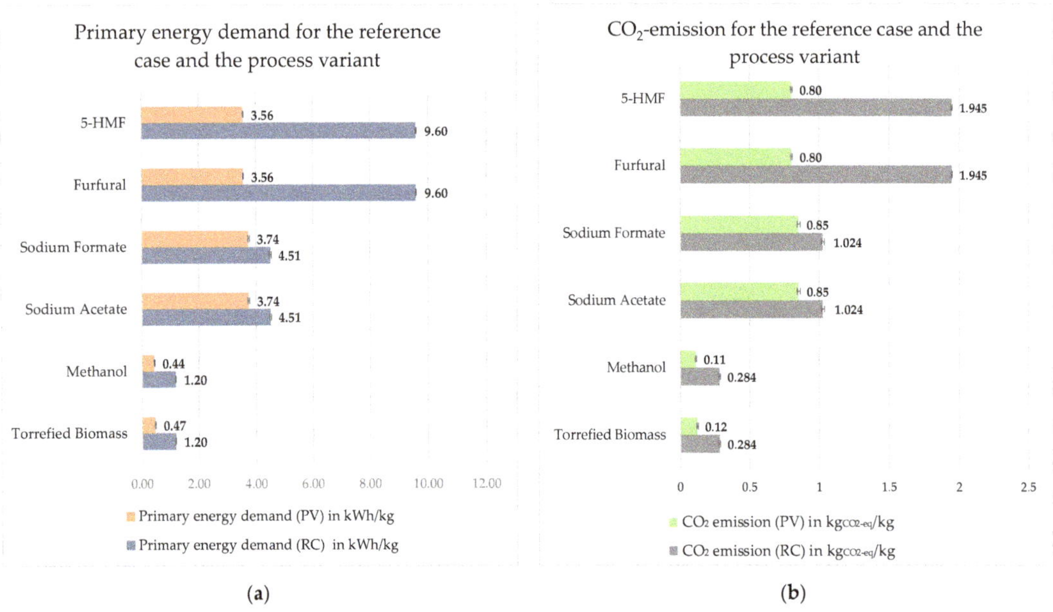

Figure 8. Comparison from product perspective with 5% steam loss: (**a**) primary energy demand for the reference case and the process variant; (**b**) CO_2 emission for the reference case and the process variant.

When the steam loss is reduced to 0%, it has a minor effect on the reference case and almost no effect on the process variant's primary energy demand and CO_2 emissions. The primary energy demand and CO_2 emissions for the reference case with 0% steam loss are shown in the Table 4 below, where the value of primary energy demand for furfural and 5-HMF has been reduced to 9.55 kWh/kg, while the values for other products have remained the same as with 5% steam loss. The CO_2 emissions, on the other hand, have been slightly reduced for all products.

Table 4. Primary energy demand and CO_2 emission for the reference case with 0% steam loss.

Products	Primary Energy Demand	CO_2 Emission
	kWh/kg	$kgCO_{2\text{-eq}}/kg$
Torrefied biomass	1.20	0.25
Methanol	1.20	0.25
Sodium acetate	4.51	0.99
Sodium formate	4.51	0.99
Furfural	9.55	1.90
5-HMF	9.55	1.90

4.4. Further Environmental Impacts due to Changes in the Biomass Feedstock

As outlined in the methodology chapter, it was investigated for the reference case how the conversion of the biomass input from beech wood chips to straw residues affects the environmental impacts. For this purpose, the Umberto model on the reference case was used to investigate which other environmental indicators change significantly when switching to a biomass residue originating from intensive agriculture. Only the supply chain for the biomass was altered, but not the product streams obtained, as their concentrations had so far only been investigated experimentally using beech wood. This analysis showed that the following categories in particular are changing strongly:

- Climate change, GWP 100 a in $kgCO_{2\text{-eq}}$
- Human toxicity w/o LT, HTPinf w/o LT in kg $1,4\text{-DCB}_{\text{eq}}$
- Freshwater eutrophication w/o LT, FEP w/o LT in kg $P_{\text{-eq}}$
- Freshwater ecotoxicity w/o LT, FETPinf w/o LT in kg $1,4\text{-DCB}_{\text{eq}}$

The results represented here are based on 1000 kg of biomass input of the two biomass residues considered. Figure 9a shows the global warming potential. The methodology used for the impact assessment was IPCC 2013, which takes the impacts of emissions over a period of 100 years into account. The total emission for the beechwood biomass is 218.55 $kgCO_{2\text{-eq}}$. When using straw residues, the total emission is 254.73 $kgCO_{2\text{-eq}}$. The additional 36.18 $kgCO_{2\text{-eq}}$ is caused by the straw supply chain, which has a CO_2 footprint of 0.0908 $kgCO_{2\text{-eq}}/kg$ of straw, whereas the beechwood has a CO_2 footprint of 0.0547 $kgCO_{2\text{-eq}}/kg$ of beechwood.

Figure 9b depicts human toxicity, freshwater ecotoxicity and freshwater eutrophication potential for the two investigated biomass types. For the impact estimation method, ReCiPe Midpoint (H) w/o LT [26] has been used. All substances which are considered to be toxic are standardized at 1,4-dichlorobenzene (DCB). Total emissions from substances toxic to humans from the reference case are 7.729 kg $1,4\text{-DCB}_{\text{eq}}$ for beechwood and 23.599 kg $1,4\text{-DCB}_{\text{eq}}$ for straw, while total emissions having an ecotoxic impact on freshwater ecosystems are 0.079 kg $1,4\text{-DCB}_{\text{eq}}$ for beechwood and 3.103 kg $1,4\text{-DCB}_{\text{eq}}$ for straw. All eutrophication-potential substances are converted to the same amount of phosphorous (P) with the same eutrophication impact. The total accumulation of excess nutrients in a body of water is 0.010 kg $P_{\text{-eq}}$ for beechwood and 0.014 kg $P_{\text{-eq}}$ for straw residues.

Agricultural cultivation is associated with considerable environmental burdens due to the use of pesticides, fertilisers, machinery and increased water consumption, among other things. Since agricultural cultivation is responsible for most of the impact, beech wood as a biomass residue input has a lower environmental impact than straw residues.

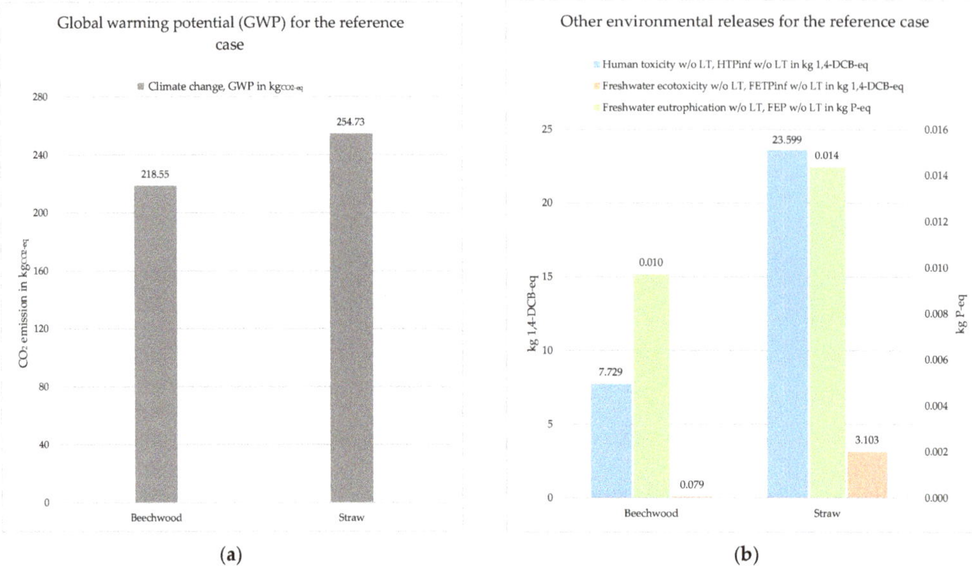

Figure 9. Life cycle impact assessment of the reference case for different biomass inputs: (**a**) global warming potential (GWP); (**b**) other environmental releases.

4.5. Sensitivity Analysis

According to reference [32], the HHV of torrefied biomass ranges from 22 to 28 MJ/kg due to an increase in carbon content. The heat generated during biomass incineration and the thermal power available to external customers vary depending on the HHV of torrefied biomass. Figure 10a shows a sensitivity analysis concerning the HHV of torrefied biomass. The blue line represents the excess thermal power that could be sold to an external customer, and the green line represents the avoided CO_2 emissions from supplying excess thermal energy in the process variant, assuming that it replaces natural gas consumption. Within the scope of this study, an HHV of 22 MJ/kg was considered. This value can be used as a reference value in the diagram, and all process variant calculations are based on it. At this value, the excess thermal power and avoided CO_2 emissions are 3.81 MW and 769 $kgCO_2$. From this reference point, an elevation of 2 MJ/kg of HHV increases thermal power and CO_2 emissions by 10.12%.

Another important factor that must be considered is the moisture content of the input biomass. The moisture content of biomass varies depending on the type of biomass and amount of time it has been stored. The SHS dryer contributes to the majority of the overall biorefinery's primary energy demand. The latter is highly sensitive to the moisture content of biomass, as shown in Figure 10b, where the moisture content, based on dry matter ranges from 20 to 45%. These changes would have an impact on the total energy demand as well as the CO_2 emissions. As shown in the diagram, the effect of this parameter on both variants is different. In the reference case, the SHS dryer's total primary energy demand includes both electrical and thermal primary energy demand. The SHS heater is the main energy consumer, and as the moisture content decreases, it requires less superheated steam supply to dry the biomass input, resulting in a decrease in energy consumed by the SHS heater. In the case of a process variant, however, the total primary energy demand consists solely of the electrical primary energy demand. The steam fan is the largest electrical energy consumer, accounting for 96% of the total electrical primary energy demand. As the amount of steam decreases with decreasing moisture content, so does the energy required by the fan to recirculate the steam. The basic calculation was carried out for a moisture content of 35%.

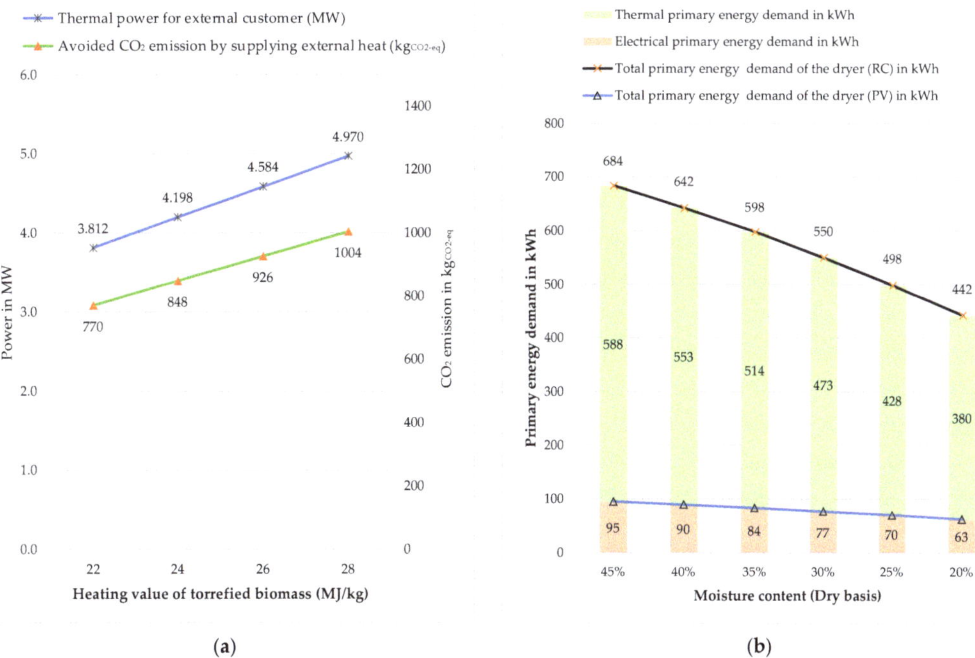

Figure 10. Sensitivity analysis: (**a**) heat value of torrefied biomass; (**b**) moisture content of biomass (dry basis).

The composition of electricity generation, the so-called electricity mix, was related to Germany in the basic calculations (see also Table 3). The electricity mix is not a constant factor. The electricity mix depends on the deployment of renewable energy sources and thus is country and time dependent. Figure 11a shows the CO_2 emissions for both variants with different electricity mixes. This graph can be used to estimate total CO_2 emissions for various countries, as well as in the future when the electricity mix changes.

The biomass supply chain is divided into two parts: wood harvesting/sawmill and transport. An LCA model was created to calculate the cumulative non-renewable energy demand or primary energy demand and CO_2 emissions of the biomass supply chain, which included the transportation activity "transport, freight, lorry 16–32 metric ton EURO 4 [RER]". The primary energy demand and CO_2 emission due to harvesting and sawmilling are constant, as the study is based on 1000 kg of biomass processing, and the values are 143.61 kWh and 38.17 kgCO_2-eq, respectively. Hence, the total primary energy demand and CO_2 emission of the biomass supply chain are primarily determined by the supply distance. Figure 11b shows the variation in primary energy demand and CO_2 emissions based on the supply distances ranging from 100 to 300 km. The reference distance is 100 km, resulting in a primary energy demand of 217.13 kWh and CO_2 emission of 54.7 kgCO_2-eq for the reference case. From the reference point, an increase of 50 km increases the primary energy demand by 17.16% and CO_2 emission by 15.07%.

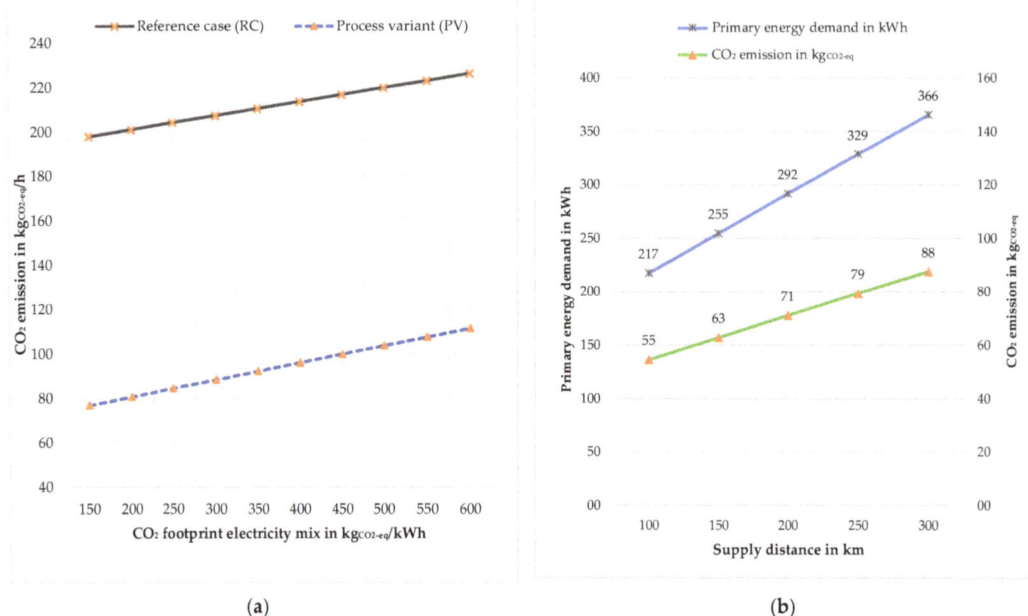

Figure 11. Sensitivity analysis: (**a**) CO_2-emission for the reference case and process variant with different electricity mix; (**b**) biomass supply distance.

5. Discussion

As part of the research project to develop the biorefinery presented here, competing processes for the production of 5-HMF and furfural were also investigated. In reference [6], the results of a first investigation of the process of reference [33] for the production of 5-HMF were presented. The process under investigation was at the laboratory stage. The total CO_2 emissions determined ranged from 326 to 1160 $kgCO_2/kg_{HMF}$, with the main contributor being the use of dichloromethane [6]. In the meantime, AvaBiochem, in close cooperation with the Karlsruhe Institute of Technology, has brought another process for the production of 5-HMF to market maturity [34,35]. For this process, CO_2 emissions between approx. 9 and 49 $kgCO_2/kg_{HMF}$ (depending on the degree of energy recovery and the energy source used) were determined in this project. The newly developed biorefinery presented in this publication is ecologically advantageous compared to the competing processes with 1.945 $kgCO_2/kg_{HMF}$ for the reference case and 0.80 $kgCO_2/kg_{HMF}$ for the variant with biomass combustion. The supply chains are considered in all studies.

For furfural, investigations by reference [6] identified a range of approx. 13 to 14 $kgCO_2/kg_{furfural}$ for the Huaxia-Westpro process. Here, all input chains were included except for the biomass raw material. The main CO_2 driver here is the provision of heat for the production process, accounting for approx. 90%. Further investigations in this research project concerning that process showed that the CO_2 emissions of the Huaxia-Westpro process can be reduced to 1.6 to 3.7 $kgCO_2/kg_{furfural}$ by heat recovery and switching to energy sources with lower CO_2 emissions. The upper value refers to production in China. These values include all input supply chains and transports, including those of the biomass raw material maize. The allocated CO_2 emission values for the new biorefinery presented here are at the same level as the values mentioned above for 5-HMF, because the same process steps have to be passed through for the production of both platform chemicals. In the reference case, the CO_2 emissions of the investigated biorefinery (1.945 $kgCO_2/kg_{furfural}$) are slightly above the best value of the Huaxia-Westpro process. In the process variant with its own thermal energy supply, the new biorefinery causes less than 50% of the CO_2 emissions of the competing process. Here, too, all preceding supply chains are considered.

Sodium acetate and sodium formate are follow-on products from the reaction of acetic acid with sodium hydroxide. For conventional fossil-based acetic acid production, the authors of reference [36] give a value of 1.846 $kgCO_2/kg_{acetic\ acid}$. The authors' own examination of the upstream chains of methanol and carbon monoxide with natural gas as raw material results in 0.89 $kgCO_2/kg_{acetic\ acid}$ for methanol ([36]: 0.36 $kgCO_2/kg_{acetic\ acid}$) and 0.994 $kgCO_2/kg_{acetic\ acid}$ for carbon monoxide ([36]: 1.065 $kgCO_2/kg_{acetic\ acid}$). For classical acetic acid production, e.g., by the Celanese company in Texas, CO_2 emissions of approx. 2.3 $kgCO_2/kg_{acetic\ acid}$ can therefore be assumed. For sodium acetate and sodium formate combined, the biorefinery investigated in this study produces a total of 2.048 $kgCO_2/kg$ for the reference case and 1.7 $kgCO_2/kg$ for the process variant with thermal auto-supply. The deduction of the CO_2 proportion of the sodium hydroxide solution used for comparison with acetic acid is omitted here because the new biorefinery does not cause higher specific CO_2 emissions than the classic competing process based on natural gas.

This comparison applies to operation with beech wood chips with a humidity content of 35% (dry). If the moisture content can be further reduced through natural solar drying and storage without using non-renewable energy carriers, the CO_2 emissions of the new biorefinery will decrease further. The above comparisons do not include CO_2 credits from the sale of excess thermal energy and thus the external substitution of fossil energy sources, such as natural gas.

The CO_2 emissions determined are mainly based on the use of fossil non-renewable raw materials (except limestone) and thus also give a reflection of the energy demand of the processes considered. The newly developed biorefinery is thus competitive from an ecological and energetic point of view compared to the bio-based competing processes for the production of 5-HMF and furfural and the classic natural gas-based process for the production of acetic acid.

The laboratory tests with the batch-rectification column have shown a loss of dichloromethane of approx. 1% or less. However, a loss of 2% was conservatively assumed for the calculations. Accounting for 2.71 $kgCO_{2\text{-eq}}/h$, dichloromethane causes 1.27% of the total CO_2 emissions in the reference case and 2.8% in the process variant with thermal self-supply. If the dichloromethane losses in a commercial plant were similar to those in the laboratory tests, the CO_2 emissions caused by the consumption of dichloromethane would be cut in half.

The biorefinery's steam loss has almost no impact on total energy consumption and total CO_2 emissions. However, the steam loss affects the product yield and thus also the consumption of sodium hydroxide solution and dichloromethane. The steam loss was assumed to be 5%, but given the design conditions and the associated phase separation, it should be between 0 and 1% for a well-designed system. The increased product yield and the additional consumption of auxiliary materials can be extrapolated straightforwardly from the values shown in Figures 4 and 5. The product-related, allocated CO_2 emission values are reduced accordingly. Reducing the steam losses only marginally changes the overall picture presented so far.

Thus, the moisture content of the biomass represents the greatest lever for reducing energy demand and CO_2 emissions if the reduction of the humidity level is achieved in a natural way, e.g., by dry storage for several months or as described above. A reduced moisture content reduces the demand for thermal energy but also for electrical energy to drive the fan to circulate the superheated steam.

However, the transport distance for the biomass raw material also has a significant influence on the primary energy demand and CO_2 emissions. The baseline calculations were carried out for a distance of 100 km. Distances up to 300 km were considered in the sensitivity analysis. Since the plant concept under consideration is based on relatively small input quantities, the objective should be, from an environmental point of view, to obtain the biomass residue as locally as possible, with a distance of well under 100 km.

The nature and the quantity of the expected compounds in the condensable volatile fraction depend mainly on the composition (cellulose, hemicellulose and lignin) of the biomass [11]. Examples of promising feedstock are: hay, straw, wood chips, digestate, manure or other bio-based residues. Orive et al. (2020) showed that olive residues are a promising feedstock for the extraction of high-value chemicals [37].

The process parameters (temperature, residence time and flow velocity) and the particle size of the used feedstock have a great influence on the process liquid and solid fraction outputs [19,38]. The process parameters (temperature and residence time) have been selected based on experimental data in order to guarantee a good compromise between obtained liquid and solid fractions.

A mass-based allocation of the primary energy demand and the CO_2 emissions to the generated products was shown as an example. A value-based allocation would be possible in principle. However, since the monetary values of the products can change dynamically depending on demand, availability and product quality, and the biorefinery can extract further valuable materials, a value-based allocation was not carried out in the context of this work.

Apart from the possibility of supplying thermal energy via the combustion of torrefied biomass, no other process integration scenarios were considered within the scope of this work. The biorefinery could, for example, be integrated into an industrial symbiosis in which electrical power is generated in an environmentally friendly manner in addition to thermal energy. The pulp and paper industry, the chemical industry, the pharmaceutical industry and the steel producing industry, for example, would be suitable for such a symbiosis. It would also be conceivable to further process the torrefied biomass into activated carbon or to generate syngas.

Commercially, torrefaction is at an early stage of development. Several technology companies are aiming for a commercial launch, but are struggling with technical problems in demonstration plants. Non-oxidative torrefaction has higher commercialization potential compared to other developed methods because the yield is higher and the process operation is safer. The information currently available on the practical applications of biomass torrefaction in industry is still insufficient [11].

The transition to a bioeconomy is increasingly being demanded and supported politically. In 2018, the EU Commission presented an action plan for the development of a sustainable and cycle-oriented bioeconomy which, among other factors, provides considerable financial resources for the establishment of biorefineries in Europe [39]. In their strategy development, companies should analyse potential conflicts of objectives with regard to ecological, economic and social sustainability at an early stage. Ecological conflicts of interest should be discussed here first, because ecological sustainability is the primary claim with which bio-based products and the bio-economy as an economic system are promoted [4]. The paper presented here clearly demonstrates that the newly developed biorefinery can make a positive contribution on the way to climate neutrality from an ecological point of view.

6. Conclusions and Outlook

In the context of this work, the energy demand, CO_2 footprint and other environmental parameters were analysed for a newly developed biorefinery, which works on the principle of torrefaction with superheated steam. The whole supply chain of biomass and auxiliary materials were included in the assessment. The results show that the biorefinery can compete with established bio-based and conventional production processes for furfural, 5-HMF and acetic acid in terms of its environmental performance indicators. The biorefinery is versatile enough to produce various high-quality platform chemicals and torrefied biomass from biomass residues.

The main drivers for further optimisation were identified. In this context, the plant is particularly interesting for an industrial symbiosis with actors from several different sectors. The biorefinery particularly supports the EU's 2050 goal of climate neutrality by using biomass residues in an environmentally friendly way to produce high-quality platform chemicals.

The research has shown that it is very time consuming to collect reliable data for the environmental analysis, especially for the competing processes. Wherever possible, the results were verified by means of the authors' own experiments, calculations and estimates.

In the second part of this publication series, the economic performance and possible commercialisation strategies are presented. Further works will focus on the continuous separation of chemicals, the flexible extraction of different valorisation products and the upscaling of the SHS-based drying/torrefaction with different biomass residues.

Supplementary Materials: The following are available online at https://www.mdpi.com/article/10.3390/su14031212/s1: Umberto-models: Roy-etal-MDPI-2022-Umberto-model-Reference-Case-Beechwood.jpg, Roy-etal-MDPI-2022-Umberto-model-Process-Variant-Beechwood.jpg.

Author Contributions: Conceptualization, B.R., P.K.-M. and A.D.; methodology, P.K.-M. and A.D.; software, B.R.; validation, B.R.; formal analysis, B.R., P.K.-M. and A.D.; investigation, B.R.; resources, P.K.-M.; data curation, B.R.; writing—original draft preparation, B.R. and P.K.-M.; writing—review and editing, P.K.-M. and A.D.; visualization, B.R. and P.K.-M.; supervision, P.K.-M. and A.D.; project administration, P.K.-M. and A.D.; funding acquisition, P.K.-M. and A.D. All authors have read and agreed to the published version of the manuscript.

Funding: This research was funded by the FEDERAL MINISTERY OF EDUCATION AND RESEARCH (BUNDESMINISTERIUM FÜR BILDUNG UND FORSCHUNG), grant number 031B0664.

Institutional Review Board Statement: Not applicable.

Informed Consent Statement: Not applicable.

Data Availability Statement: Data are contained within the article or Supplementary Materials.

Conflicts of Interest: The authors declare no conflict of interest.

References

1. Cătuți, M.; Elkerbout, M.; Alessi, M.; Egenhofer, C. *Biomass and Climate Neutrality*; CEPS: Brussels, Belgium, 2020.
2. Tsiropoulos, I.; Nijs, W.; Tarvydas, D.; Ruiz, P. *Towards Net-Zero Emissions in The Eu Energy System by 2050: Insights from Scenarios in Line With the 2030 and 2050 Ambitions of the European Green Deal*; JRC Technical Reports JRC118592; JRC: Luxembourg, 2020.
3. Material Economics. EU Biomass Use In A Net-Zero Economy: A Course Correction for EU Biomass. 2021. Available online: https://materialeconomics.com/latest-updates/eu-biomass-use (accessed on 25 November 2021).
4. Kircher, M. *Bioökonomie im Selbststudium: Unternehmensstrategie und Wirtschaftlichkeit*; Springer: Berlin/Heidelberg, Germany, 2020; ISBN 978-3-662-61004-6.
5. Biddy, M.J.; Scarlata, C.; Kinchin, C.A. Market Assessment of Bioproducts with Near-Term Potential. Available online: http://www.osti.gov/scitech (accessed on 8 October 2021).
6. Schöppe, H.; Kleine-Möllhoff, P.; Epple, R. Energy and Material Flows and Carbon Footprint Assessment Concerning the Production of HMF and Furfural from a Cellulosic Biomass. *Processes* **2020**, *8*, 119. [CrossRef]
7. proFagus. pro Qualität-pro Natur: Sustainability. Our Purity Law. Available online: https://profagus.de/en/sustainability/ (accessed on 8 October 2021).
8. Kaserer, W. Fasern für fast alles im Leben. In *CSR und Klimawandel*; Sihn-Weber, A., Fischler, F., Eds.; Springer: Berlin/Heidelberg, Germany, 2020; pp. 381–393. ISBN 978-3-662-59747-7.
9. Chandel, A.K.; Garlapati, V.K.; Singh, A.K.; Antunes, F.A.F.; da Silva, S.S. The path forward for lignocellulose biorefineries: Bottlenecks, solutions, and perspective on commercialization. *Bioresour. Technol.* **2018**, *264*, 370–381. [CrossRef]
10. Gabriella, N.A. Delphi Study of Factors Affecting Forest Biorefinery Development in the Pulp and Paper Industry: The Case of Bio-Based Products. Master's Thesis, University of Graz, Graz, Austria, 2016.
11. Chen, W.-H.; Lin, B.-J.; Lin, Y.-Y.; Chu, Y.-S.; Ubando, A.T.; Show, P.L.; Ong, H.C.; Chang, J.-S.; Ho, S.-H.; Culaba, A.B.; et al. Progress in biomass torrefaction: Principles, applications and challenges. *Prog. Energy Combust. Sci.* **2021**, *82*, 100887. [CrossRef]
12. Mamvura, T.A.; Danha, G. Biomass torrefaction as an emerging technology to aid in energy production. *Heliyon* **2020**, *6*, e03531. [CrossRef] [PubMed]

13. Ribeiro, J.; Godina, R.; Matias, J.; Nunes, L. Future Perspectives of Biomass Torrefaction: Review of the Current State-Of-The-Art and Research Development. *Sustainability* **2018**, *10*, 2323. [CrossRef]
14. Tumuluru, J.S.; Ghiasi, B.; Soelberg, N.R.; Sokhansanj, S. Biomass Torrefaction Process, Product Properties, Reactor Types, and Moving Bed Reactor Design Concepts. *Front. Energy Res.* **2021**, *9*, 728140. [CrossRef]
15. Tumuluru, J.S.; Sokhansnj, S.; Wright, C.T.; Boardman, R.D. Biomass Torrefaction: Process Review and Moving Bed Torrefaction System Model Development. Available online: https://www.osti.gov/servlets/purl/991885 (accessed on 16 November 2021).
16. Donner, M.; Gohier, R.; De Vries, H. A new circular business model typology for creating value from agro-waste. *Sci. Total. Environ.* **2020**, *716*, 137065. [CrossRef] [PubMed]
17. Habte, Y.; Hector, D. Business Model for Black Pellets Production in Sweden. Bachelor's Thesis, KTH Royal Institute of Technology, Stockholm, Sweden, 2017.
18. Nhuchhen, D.; Basu, P.; Acharya, B. A Comprehensive Review on Biomass Torrefaction. *IJREB* **2014**, *2014*, 1–56. [CrossRef]
19. Tumuluru, J.S.; Sokhansnj, S.; Hess, R.S.; Wright, C.T.; Boardman, R.D. A Review on Biomass Torrefaction Process and Product Properties for Energy Applications. Available online: https://www.liebertpub.com/doi/pdfplus/10.1089/ind.2011.7.384 (accessed on 16 November 2021).
20. Zhang, D.; Han, P.; Yang, R.; Wang, H.; Lin, W.; Zhou, W.; Yan, Z.; Qi, Z. Fuel properties and combustion behaviors of fast torrefied pinewood in a heavily loaded fixed-bed reactor by superheated steam. *Bioresour. Technol.* **2021**, *342*, 125929. [CrossRef] [PubMed]
21. Mustaza, M.N.F.; Mizan, M.N.; Yoshida, H.; Izhar, S. Torréfaction of Mangrove Wood by Introducing Superheated Steam for Biochar Production. *IOP Conf. Ser. Earth Environ. Sci.* **2021**, *765*, 12027. [CrossRef]
22. Isemin, R.; Muratova, N.; Kuzmin, S.; Klimov, D.; Kokh-Tatarenko, V.; Mikhalev, A.; Milovanov, O.; Dalibard, A.; Ibitowa, O.A.; Nowotny, M.; et al. Characteristics of Hydrochar and Liquid Products Obtained by Hydrothermal Carbonization and Wet Torrefaction of Poultry Litter in Mixture with Wood Sawdust. *Processes* **2021**, *9*, 2082. [CrossRef]
23. Zhang, X.-S.; Yang, G.-X.; Jiang, H.; Liu, W.-J.; Ding, H.-S. Mass production of chemicals from biomass-derived oil by directly atmospheric distillation coupled with co-pyrolysis. *Sci. Rep.* **2013**, *3*, 1120. [CrossRef] [PubMed]
24. Stubbing, T.J. Airless Drying-Developments Since IDS'94. *Dry. Technol.* **1999**, *17*, 1639–1651. [CrossRef]
25. Nitzsche, R.; Budzinski, M.; Gröngröft, A.; Majer, S. Bewertungsansätze bei der Optimierung von Bioraffinerie-Konzepten. In Michael Nelles (Ed.): Bioenergie. Vielseitig, Sicher, Wirtschaftlich, Sauber?! With Assistance of Andrea Thrän, Jan Liebetrau, Elena Angelova, Franziska Müller-Langer, Volker Lenz. DBFZ-Jahrestagung am 01./02. Oktober 2014. Leipzig, 01./02. Oktober 2014. DBFZ Deutsches Biomasseforschungszentrum gemeinnützige GmbH. 2014, pp. 57–68. Available online: https://www.researchgate.net/publication/290920013_Bewertungsansatze_bei_der_Optimierung_von_Bioraffinerie-Konzepten (accessed on 17 September 2021).
26. Huijbregts, M.A.J.; Steinmann, Z.J.N.; Elshout, P.M.F.; Stam, G.; Verones, F.; Vieira, M.; Zijp, M.; Hollander, A.; van Zelm, R. ReCiPe2016: A harmonised life cycle impact assessment method at midpoint and endpoint level. *Int. J. Life Cycle Assess* **2017**, *22*, 138–147. [CrossRef]
27. Boustead, I. Eco-Profiles of the European Plastics Industry: Sodium Hydroxide. 2005. Available online: http://www.inference.org.uk/sustainable/LCA/elcd/external_docs/naoh_31116f0a-fabd-11da-974d-0800200c9a66.pdf (accessed on 6 December 2021).
28. Petra Icha. Entwicklung der Spezifischen Kohlendioxid Emissionen des Deutschen Strommix in den Jahren 1990–2020. Climate Change 45/2021. Dessau-ROßlau. 2021. Available online: https://www.umweltbundesamt.de/sites/default/files/medien/5750/publikationen/2021-05-26_cc-45-2021_strommix_2021.pdf (accessed on 6 December 2021).
29. Fritsche, U.R.; Greß, H.W. Der Nichterneuerbare Kumulierte Energieverbrauch und THG-Emissionen des Deutschen Strom-Mix im Jahr 2020 Sowie Ausblicke auf 2030 und 2050. Bericht für die HEA-Fachgemeinschaft für, Darmstadt, Berlin. 2021. Available online: www.iinas.org (accessed on 6 December 2021).
30. BDEW Bundesverband der Energie- und Wasserwirtschaft e.V. Primärenergiefaktoren: Der Zusammenhang von Primärenergie und Endenergie in der Energetischen Bewertung. Available online: https://www.bdew.de/media/documents/20150422_Grundlagenpapier-Primaerenergiefaktoren.pdf (accessed on 6 December 2021).
31. Solomon, S.; Qin, D.; Manning, M.; Chen, Z.; Marquis, M.; Averyt, K.B.; Tignor, M.; Miller, H.L. Contribution of Working Group I to the Fourth Assessment Report of the Intergovernmental Panel on Climate Change; Climate Change 2007-The Physical Science Basis, Cambridge, United Kingdom and New York, NY, USA. 2007. Available online: https://www.ipcc.ch/site/assets/uploads/2018/05/ar4_wg1_full_report-1.pdf (accessed on 6 December 2021).
32. Yun, H.; Wang, Z.; Wang, R.; Bi, X.; Chen, W.-H. Identification of Suitable Biomass Torrefaction Operation Envelops for Auto-Thermal Operation. *Front. Energy Res.* **2021**, *9*, 636938. [CrossRef]
33. Kougioumtzis, M.A.; Marianou, A.; Atsonios, K.; Michailof, C.; Nikolopoulos, N.; Koukouzas, N.; Triantafyllidis, K.; Lappas, A.; Kakaras, E. Production of 5-HMF from Cellulosic Biomass: Experimental Results and Integrated Process Simulation. *Waste Biomass Valor* **2018**, *9*, 2433–2445. [CrossRef]
34. Kläusli, T. AVA Biochem: Commercialising renewable platform chemical 5-HMF. *Green Process. Synth.* **2014**, *3*, 235–236. [CrossRef]
35. Dümpelmann, R.; Nikulski, N. This is Custom Heading Element. Available online: https://baselarea.swiss/blog-post/muttenz-is-home-to-the-world-s-largest-plant-for-the-production-of-5-hmf-from-biomass/ (accessed on 30 November 2021).
36. Wernet, G.; Bauer, C.; Steubing, B.; Reinhard, J.; Moreno-Ruiz, E.; Weidema, B. The ecoinvent database version 3 (part I): Overview and methodology. *Int. J. Life Cycle Assess* **2016**, *21*, 1218–1230. [CrossRef]

37. Orive, M.; Cebrián, M.; Amayra, J.; Zufía, J.; Bald, C. Techneconomic assessment of a biorefinery plant for extracted olive pomace valorization. *Process. Saf. Environ. Prot.* **2021**, *147*, 924–931. [CrossRef]
38. Konsomboon, S.; Commandré, J.-M.; Fukuda, S. Torrefaction of Various Biomass Feedstocks and Its Impact on the Reduction of Tar Produced during Pyrolysis. *Energy Fuels* **2019**, *33*, 3257–3266. [CrossRef]
39. Europäische Kommission. Eine Neue Bioökonomie-Strategie für ein Nachhaltiges Europa. Available online: https://ec.europa.eu/commission/presscorner/detail/de/IP_18_6067 (accessed on 13 January 2022).

 *sustainability*

Article

Superheated Steam Torrefaction of Biomass Residues with Valorisation of Platform Chemicals Part—2: Economic Assessment and Commercialisation Opportunities

Baharam Roy [1], Peter Kleine-Möllhoff [2,*] and Antoine Dalibard [3]

[1] Reutlingen Research Institute (RRI), Reutlingen University, 72762 Reutlingen, Germany; roybaharam@gmail.com
[2] ESB Business School, Reutlingen University, 72762 Reutlingen, Germany
[3] Fraunhofer Institute for Interfacial Engineering and Biotechnology IGB, 70569 Stuttgart, Germany; antoine.dalibard@igb.fraunhofer.de
* Correspondence: peter.kleine-moellhoff@reutlingen-university.de; Tel.: +49-712-1271-5009

Abstract: Up to now biorefinery concepts can hardly compete with the conventional production of fossil-based chemicals. On one hand, conventional chemical production has been optimised over many decades in terms of energy, yield and costs. Biorefineries, on the other hand, do not have the benefit of long-term experience and therefore have a huge potential for optimisation. This study deals with the economic evaluation of a newly developed biorefinery concept based on superheated steam (SHS) torrefaction of biomass residues with recovery of valuable platform chemicals. Two variants of the biorefinery were economically investigated. One variant supplies various platform chemicals and torrefied biomass. The second variant supplies thermal energy for external consumers in addition to platform chemicals. The results show that both variants can be operated profitably if the focus of the platform chemicals produced is on high quality and thus on the higher-priced segment. The economic analysis gives clear indications of the most important financial influencing parameters. The economic impact of integration into existing industrial structures is positive. With the analysis, a viable business model can be developed. Based on the results of the present study, an open-innovation platform is recommended for the further development and commercialisation of the novel biorefinery.

Keywords: biorefinery; superheated steam torrefaction; economic assessment; volatile recovery; platform chemicals; commercialisation

Citation: Roy, B.; Kleine-Möllhoff, P.; Dalibard, A. Superheated Steam Torrefaction of Biomass Residues with Valorisation of Platform Chemicals Part—2: Economic Assessment and Commercialisation Opportunities. *Sustainability* **2022**, *14*, 2338. https://doi.org/10.3390/su14042338

Academic Editor: Ana Ramos

Received: 3 February 2022
Accepted: 16 February 2022
Published: 18 February 2022

Publisher's Note: MDPI stays neutral with regard to jurisdictional claims in published maps and institutional affiliations.

Copyright: © 2022 by the authors. Licensee MDPI, Basel, Switzerland. This article is an open access article distributed under the terms and conditions of the Creative Commons Attribution (CC BY) license (https://creativecommons.org/licenses/by/4.0/).

1. Introduction

What would our economies look like if we produced exclusively biobased chemical products with renewable resources? Can chemical production plants be operated economically independently of finite fossil raw materials in the sense of a circular economy? Is superheated steam (SHS) torrefaction suitable for the economic production of platform chemicals?

Revenues from circular economy products and services are not yet as profitable as their classic linear-oriented alternatives. Raising venture capital, promoting commercialisation and exporting to international markets are among the key future challenges that circular economy companies will have to face [1–3]. Biorefineries must increase their revenues in order to be competitive and attract future investments. However, the economic performance of biorefineries is influenced by many external factors, such as the development of energy prices, the cost of biomass feedstocks and the change in supplier markets due to changing competitive conditions.

1.1. Economic Aspects

Investigations by Ref. [4] show that biorefinery concepts are particularly hampered in their commercialisation by the following economic aspects: necessity of high investments, high operating costs, high overall costs, low return of investment, high capital costs, high logistical costs and volatile biomass costs [4]. Raising venture capital, promoting commercialisation and exporting to international markets are further challenges for the bioeconomy [2,3,5,6]. As long as biobased products are in direct competition with conventional alternatives, they can only be profitable in niche markets that reward sustainability in terms of price [7] or offer application potentials [8]. On the other side, the changed risk analysis of financial institutions with regard to sustainability supports the acquisition of funds for biorefineries. Business models that require non-renewable raw materials are increasingly excluded from financing [7]. Quality and product differentiation seem to be the preferred strategic response to compete successfully with biorefineries [9]. For instance, Ref. [8] reports on a profitable biorefinery concept for the valorisation of olive oil pomace that can produce high value products while improving the economic and environmental sustainability of the olive oil value chain [8].

From the market side, fluctuating prices of the end products and the raw materials used represent the most important economic challenges for a biorefinery [10]. The markets for bioproducts and biofuels are relatively new and volatile. This has an impact on demand and can significantly affect the selling prices of these products in the market [11].

Most publications assessing the competitiveness of biorefineries refer to the production of biomass fuels or chemicals sold in higher quantities, such as fibres or synthetic fuels. The few examples from the chemical industry show that the switch to circular value chains is preferable via commodity chemicals [12–16].

1.2. Pilot and Demonstration Scale, Commercialisation

The majority of biorefinery plants are still under development, at pilot or demonstration scale. Established research does not address the immediate challenges at this level of development. Even the literature on technology and innovation management is scarce in this regard. However, pilot and demonstration scale are important and necessary to raise industrial interests and encourage collaboration between different actors, i.e., industrial companies, academia, public authorities and equipment industry [17,18]. Robust biorefinery concepts can only be realised through a multi-layered approach that takes the relationship between biorefinery configurations, economic performance and future uncertainties into account [16]. Aspects of the innovation system in the biorefinery environment that have been weak so far are primarily resource mobilisation and market formation [9]. Therefore, new markets and businesses need to be developed and their potential needs to be assessed. This in turn requires that new market applications are established and new technologies are developed [19]. Ref. [19] conducted an exploratory case study specifically for a biorefinery in Germany. They focused on the company actors and found that the combination of leading actors, learning companies and gap fillers from the SME or start-up sector was an important success factor. Investments in new technologies with long payback periods need to be made by mature industries that have alternative investment opportunities available [9]. Established companies are seen as playing a crucial role in the commercialisation of new technologies because no market entry would take place without an adaptation of their strategies and business models [20]. The presence of "gap fillers" is an indication of partnership-based strategic system-building activities, where suppliers and manufacturers integrate complementary resources to jointly create products and markets [21]. The innovation system must be able to mobilise the necessary resources while creating supply and demand [20].

One of the biggest challenges in terms of commercialising a biorefinery is setting up a biomass supply system that can meet the biorefinery's long-term biomass demand in a cost-effective way. In addition to the fluctuations in the costs of the supply system, the volatilities of the market prices for the produced goods of the biorefinery should not be underestimated [10]. A diversification strategy developed in a network by fully valorising a biomass resource into a range of products with maximum total value, combined with a good business model, can improve the economic feasibility of biorefineries [22].

1.3. Industrial Symbiosis Aspects

Many of the higher value-added chemical compounds found in biomass residues are present in low quantities. The value proposition must therefore be to maximize the added value of biomass and resource efficiency. This strategy requires cost-effective, cascading biomass utilization that can fully utilize all components [1,23]. The potential for new process combinations has certainly not yet been exhausted and offers considerable market opportunities [7]. Thus, an optimised biorefinery realises maximum cascading value creation throughout the life cycle [24].

Recent studies show that flexibility on the feedstock and integration into existing production networks allows one to increase the economic performance of biorefineries [15,25–27]. Combining biorefinery concepts like torrefaction with other processes such as co-combustion, pyrolysis, gasification or industrial processes to further increase the value of the product outputs makes it more efficient and economical than using it as a stand-alone process [15,28,29]. Biorefinery technologies that are thermally integrated with cooperating industries offer opportunities to increase significantly the economic and energetic performance [30,31]. If heat integration is chosen, the optimal production capacity of the biorefinery can be matched to the demand of the consumers in the heat network [16]. However, the capacity of the biorefinery can also be designed in such a way that all biological residues produced by the hosting companies are used as feedstock [32], or the biorefinery production covers the fuel needs of the host firms [33].

For an environmentally friendly and economic use of biomass-residues, it is necessary to develop cross-sectoral valorisation visions [24]. It is therefore important to concentrate on high-quality products on the one hand and to increase the spectrum of products for the solid and liquid fractions obtained on the other [9]. It has been widely discussed in the literature that a diversification strategy and the extraction of a range of products with maximum total value from biomass, combined with a good business model, significantly improves the feasibility of biorefineries in economic terms [20,34,35]. Although research on biomass torrefaction is increasing recently, there is a lack of information on the integration of the technology within an overall biorefinery concept for specific applications [15].

1.4. Goal, Scope and Structure

This paper builds on Part 1 of the series, which examines a biorefinery concept under development for the simultaneous production of high-quality biochar and platform chemicals by SHS torrefaction of biobased residues. In the first part of the series, it was shown that the newly developed biorefinery is capable of producing several platform chemicals with a favourable environmental footprint. In this second part of the series, the process is evaluated economically, and the most important economic levers are identified. Based on the economic analysis and the technical possibilities of the investigated biorefinery as well as the state of the art in science, it is furthermore shown which path to commercialisation could be taken.

This publication is structured as follows. First, there is a detailed presentation of the methods and materials used. This is followed by a chapter on the economic analysis of the biorefinery studied. Based on this economic analysis and the state of the art presented in the introduction, a further section is devoted to the derivation of a possible commercialisation strategy. This is followed by a discussion of the economic results and the commercialisation aspects. The results of the work are then briefly summarised in the conclusion.

2. Materials and Methods

This study is based on experimental results obtained within a German public-funded research project on an innovative biorefinery concept (see funding section). The economic evaluation of the newly developed biorefinery builds on the process flow diagram as shown in Figure 1, and the material and energy balances developed in part 1 of this publication series. More information on the process steps is given there. The process units with grey background represent the reference case. In addition to this reference case, another variant was investigated in which the torrefied biomass is fed to an incineration plant (hatched background and dashed lines) that serves to supply thermal energy to the biorefinery process being investigated and to external consumers. A steam loss of 5% was considered for each of the two variants, which translates into a corresponding condensate accumulation and product output.

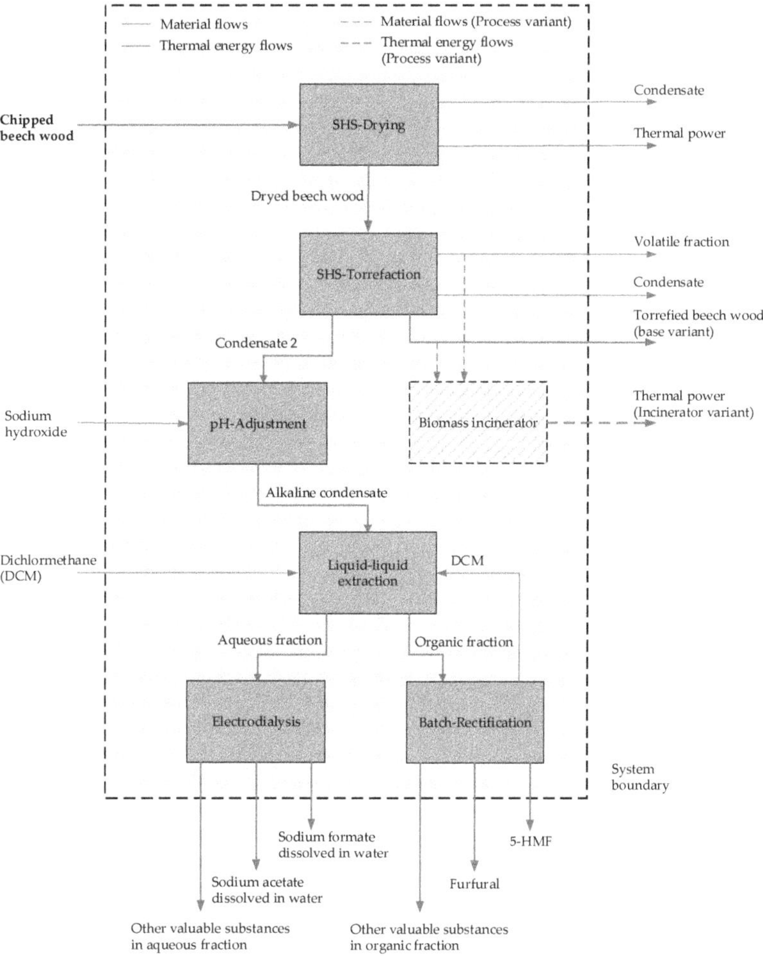

Figure 1. Biorefinery process set up.

Traditional discounted cash flow analysis is often used to provide an indication of how likely it is that an investment will be profitable. However, the shift to biorefinery concepts is a long-term investment and is subject to time uncertainties. The discounted cash flow analysis ignores the possibility for investors to adjust their investment strategy to changing market conditions and is therefore not necessarily suitable for evaluating biorefineries [36], especially when it comes to the flexible production of small quantities with high product quality. For this reason, no discounted valuation method was used for the present investigation.

The economic evaluation was carried out according to the method of [37], which estimates the costs of process plants with an accuracy of $+/-20\%$ and is based on financial calculation methods in chemical and process plant engineering [38–40]. Table 1 lists the assumptions and parameters for the economic analysis. The determination of the investment costs is based on a list of equipment that is developed from the process flow diagram [37] and Part 1 of our publication series in the Materials and Methods section [41], and thus on the mass and energy balance of the biorefinery. For this purpose, the process equipment was dimensioned for an input flow of 1000 kg/h of chipped beech wood residue within the scope of this project. Standard apparatuses such as motors, pumps, fans, heat exchangers, silos and conveying elements were queried from suppliers or determined via the internet. Special designs, such as the housings of the drying and torrefaction chambers, but also silos and tanks, were dimensioned, and the steel weights were estimated, by the authors themselves.

Table 1. Assumptions and parameters for the economic analysis.

Designation	Value	Unit	Description	Source
Evaluation model			according to references	[37–40]
Accuracy for the investment cost	+/−20	%		
Design input	1000	kg/h	Biomass input	
Operating time	5080	h/a		Authors' assumption
Plant location			Germany	
Indirect investment costs	40	%	of direct investment costs (reference case *)	[37–40]
Process investment costs	20	%	of direct and indirect investment costs (reference case *)	
Risk supplement	15	%		
Annual depreciation	5	%	of investment costs	Authors' assumption
Weighted average cost of capital (WACC)	6.6	%	according to reference	[42]
Maintenance costs	3	%	of investment costs	[37–40]
Property tax, municipal charges, insurance and license fees	4	%		
Development overheads and profit margin	10	%	of total annual operation costs	Authors' assumption
Sales and administrative overheads	7.5	%		
Sensitivity analysis			Material costs Energy costs Personnel costs Product revenues	Authors' decision

* for the process variant the absolute costs were estimated.

The material and processing costs were taken from specific sources for normal and stainless steel from [43,44]; the raw prices of these components were determined accordingly. All heat exchangers were designed using an online shortcut heat exchanger sizing tool from Ref. [45], and their weights were determined as well. From this, the raw prices of these components were determined for verification purposes using the method described above. This approach had to be taken because the offers were obtained in spring and summer 2021 during the COVID-19 pandemic, and offers with an unusually long delivery time in particular appeared to be overpriced.

The investment costs were determined from the apparatus costs using two different calculation methods. One method called Hand works with multiplication factors for different types of equipment. The other method, the Plant Component Ratio (PCR) method, uses cost distribution factors for equipment and other trades, such as the foundations, instrumentation or piping of a process plant [37–40]. The Hand method adds cost factors for indirect investments (e.g., buildings and infrastructure), process investments and risk supplement to the equipment costs to arrive at the total investment costs. The PCR method determines the total investment costs via cost distribution factors between the process engineering apparatus, process design costs, electrical and instrumentation and control technology, construction site costs, etc. [37–40]. Individual cost distribution factors, such as those for process design and construction site costs, were adjusted after the results were provided in order to arrive at realistic values. The absolute process design costs, for example, could be estimated by the authors themselves based on experience developing similar plants. The biorefinery should be constructed from as many standard components as possible, and the individual process units should be delivered to the site completely pre-assembled on frames (skid-mounted) in order to minimise the assembly effort and thus the costs at the construction site. For the process variant, only a completely supplied biomass incineration plant including flue gas cleaning is added in terms of investment. Due to the high additional costs for the incineration plant compared to the biorefinery, the surcharges for the Hand method were adjusted for the total investment costs of the variant so that realistic absolute values are given. The same applies to the variant for the cost allocation factors for the PCR method. Two offers were obtained for the biomass incineration plant and verified with the specific investment costs from Ref. [46]. The mean values for the investment costs were then calculated from the two investment cost methods according to Ref. [37] for the reference case and the process variant with biomass incinerator. These were depreciated over 20 years and assigned to the fixed operating costs. These annual depreciation costs represent the only CapEx factor.

Capital costs, maintenance, laboratory expenses, property taxes, municipal charges, insurance costs and license fees were determined according to Ref. [37] and allocated to fixed operating costs. For the capital costs, the weighted average cost of capital (WACC) for 2019/2020 in Germany according to Ref. [42] has been taken. The personnel costs were determined for the case that the biorefinery is an independent facility and thus must be fully staffed from the managing director to the cleaning staff and the security service. The variable operating costs were determined from the process-related material and energy balances. All costs have been calculated for the location Germany.

According to Ref. [37], research and development costs as well as sales and administration costs were added to the total operating costs to finally obtain the total annual operating costs of the biorefinery in the reference case and the variant with biomass incineration plant.

The total annual costs calculated in this manner were counterbalanced against the total annual revenues from the extracted platform chemicals, the torrefied biomass or, in the case of the process variant with an incineration plant, the sales revenue from the heat generated. Only the products shown in Figure 1 were considered for the revenues.

In addition, it was analysed how the investment costs are distributed over the entire process and which subprocesses represent the biggest cost drivers.

Within the course of a sensitivity analysis, the personnel, material and energy costs as well as the revenues from the platform chemicals, torrefied biomass and thermal energy generated were considered in particular.

A consideration of the cost allocation to the various saleable end products was carried out using the allocation methodology according to the Ref. [47].

A basic eco-efficiency analysis was conducted following references [48–50].

The economic considerations refer to a plant with 1000 kg/h input material and an assumed annual operating time of 5080 h and takes into consideration the development status of the plant. Only the two variants described above were considered. Further considerations, such as the combination with a combined heat and power plant, pyrolysis or synthesis gas production, were not carried out within the scope of this work. In addition, all investigations referred to conditions in Germany.

Based on the results of the economic analysis and the state of the art in science, a proposal for a possible development and commercialisation path was prepared.

3. Results

3.1. Investment Costs

Figure 2 shows the investment costs of the reference case and the process variant, which were determined using the Hand and PCR methods. These include all direct and indirect investment costs, risk surcharges, etc., as described in the methodology chapter. Details of the calculation are provided as Supplementary Material. The incineration plant for torrefied biomass represents a considerable additional investment in the process variant. However, it must be taken into account that the incineration plant with boiler and flue gas treatment is a proven technology with low additional risk, which causes significantly lower additional costs in percentage terms with regard to the indirect investments. These inter-relations were considered in the calculations. The two investment cost calculations according to Hand and PCR lead to results that differ by less than 5% for the two variants examined.

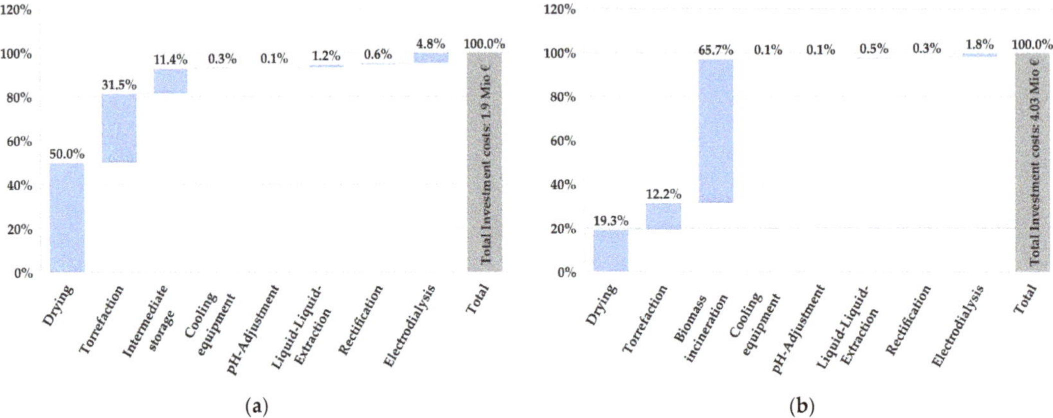

Figure 2. Investment costs: (**a**) reference case; (**b**) process variant.

In the reference case, it becomes clear that the processing units that handle the largest material flows also cause by far the largest investment costs. Dryer and torrefaction account for about 92% of the total investment. The separation of the different chemicals thus only causes about 8% of the total investment costs.

In the process variant, the incineration plant with boiler and flue gas treatment accounts for the largest share of costs (approx. 65.7%), followed by the dryer and torrefaction (approx. 31.5%). Here, the separation of chemicals accounts for an even smaller share (approx. 3% in total).

The investment costs are depreciated on a straight-line basis over a 20-year period and then included in the total operating costs.

3.2. Operating Costs

Figure 3 shows the total annual operating costs for both variants. These costs do not yet include the surcharges for research and development, profit, sales and administration.

In the reference case, 66% of the total operating costs have to be spent on fixed costs, and 34% have to be spent on variable operating costs. In the process variant, 77.5% are attributable to fixed costs and 22.5% are attributed to variable costs.

In both variants, personnel costs account for the largest share, followed by the costs for biomass woodchips residues. The energy costs burden the reference case with 10% and can be significantly decreased to less than 4% in the process variant with thermal self-supply. For electricity, a price of 18.25 Eurocents per kWh was used [51], and 3.96 Eurocents per kWh was used for natural gas [52]. In the process variant, the capital costs of more than 13% and depreciation of 10% of the total costs have a considerable impact due to the significantly higher investment costs. All other cost shares are well below 10%.

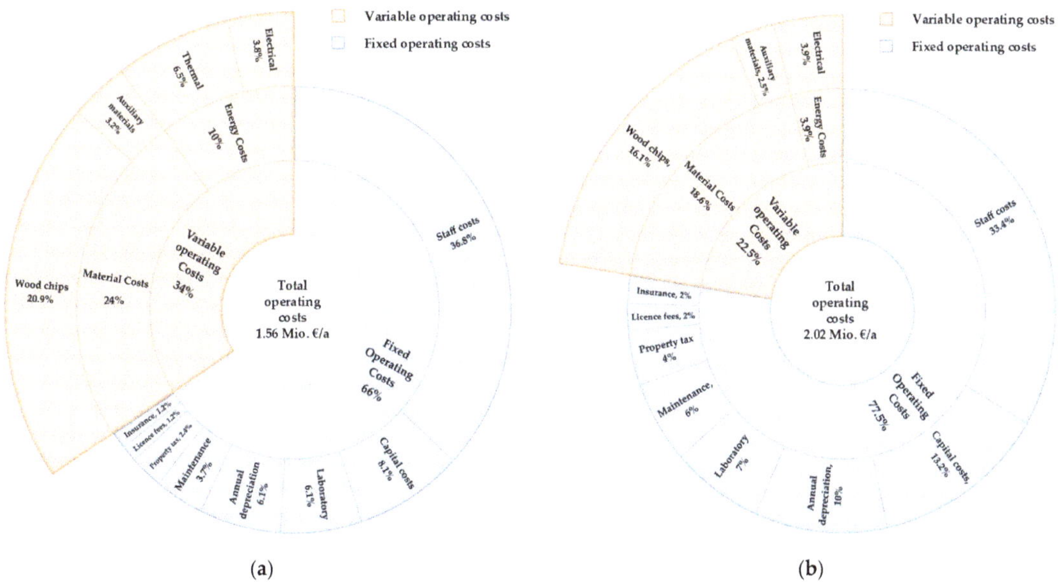

(**a**)   (**b**)

Figure 3. Annual Fixed and variable operating costs: (**a**) reference case; (**b**) process variant.

The total costs for both variants under investigation were then further charged with 10% for development overhead and profit margin, as well as 7.5% for sales and administration as indicated in Table 1.

3.3. Product Revenues and Profitability

Table 2 lists the products generated with an assumed steam loss of 5%, the market prices found and the revenues calculated from these. The steam loss should be below 1% in a well designed and constructed plant so that with this conservative assumption, the predicted chemical yields are realistic. The excess thermal energy found in the table represents the thermal surplus energy from the combustion of the torrefied biomass in the process variant after deduction of the plants' own thermal energy demand

Table 2. Product revenues.

Product	Amount		Market Price		Revenues		Source
					Reference Case	Process Variant	
Torrefied beechwood	694.85	kg/h	0.24	€/kg	847,161 €		[53]
Excess thermal energy	3812	kWh/h	0.075	€/kWh		1,452,372 €	[54]
Methanol	1.169	kg/h	0.40	€/kg	2376 €	2376 €	[55]
Sodium acetate	13.459	kg/h	6.93	€/kg	473,820 €	473,820 €	[56]
Sodium formate	1.254	kg/h	62.39	€/kg	397,444 €	397,444 €	[57]
Furfural	0.830	kg/h	21.97	€/kg	92,584 €	92,584 €	[58–60]
5-HMF	0.052	kg/h	1400	€/kg	369,113 €	369,113 €	[61]
Total revenues					**2,182,497 €**	**2,787,708 €**	

The market prices for the products extracted essentially depend on the quality and the sales quantities. Because the aim is to serve the market for the extracted chemicals with small production quantities and a high product quality, the corresponding prices were taken from the sources mentioned in the table and used as a basis for the calculations. For example, three types of furfural with purity greater than 98% are available on the mentioned sources. As a result, the market price for furfural is calculated as an average price per kilogramme of furfural. A higher heating value (HHV) of 22 MJ/kg was assumed for the torrefied biomass produced. In reality, however, this value can be as high as 28 MJ/kg depending on the feedstock and selected process parameters, resulting in higher revenues. For the thermal energy produced, it was assumed that there are several industrial consumers with different heat demands. A corresponding mixed price was set for the thermal energy delivered.

Market prices for sodium acetate and sodium formate have increased since 2021. The extent to which changed market prices affect the profitability of the plant is examined in the following chapter as part of the sensitivity analysis. For the calculation of profitability, it was assumed that the 10% cost surcharge for development overhead and profit are each divided 50% on average over the years of operation.

This results in a profitability of 19.8% for the reference case and 18.5% for the variant with biomass incineration. This means that the high investment costs for biomass incineration with the assumptions deteriorated the profitability of the plant by about one percentage point. However, it must also be considered that the calculation of the investment costs with the chosen method results in a target accuracy of plus-minus 20%. Thus, the result for profitability is also subject to a higher uncertainty than the above-mentioned difference regarding the profitability of the reference case and the process variant. For this reason, the cost parameters with the greatest influences were exposed to a sensitivity analysis.

3.4. Sensitivity Analysis

Sensitivity analyses were carried out for the cost blocks energy, material, personnel costs and revenues from the sale of the products.

Figure 4a shows the influence of the variation in energy costs on profitability. Given the recent energy price development in 2021, a stronger price increase rather than a price decrease is expected in the future. Therefore, a price escalation of up to plus 50% was considered. The sensitivity analysis shows that the reference variant in particular is affected by energy price increases in terms of profitability and that this does not drop by more than five percentage points.

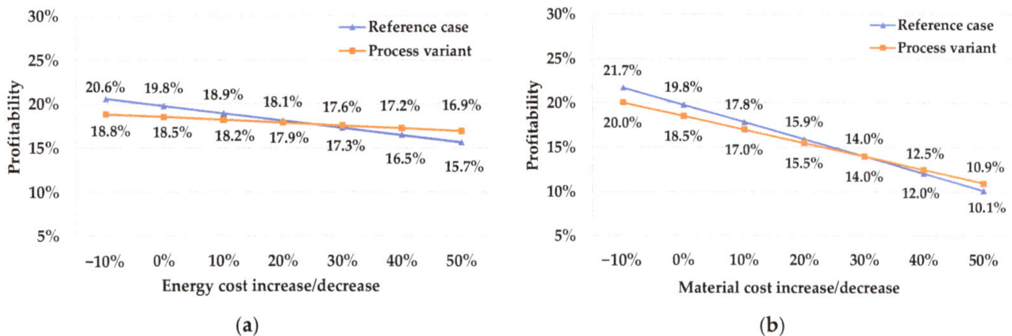

Figure 4. Sensitivity analysis: (**a**) energy costs; (**b**) material costs.

Because material costs are usually linked to energy costs, a variation between minus 10% and plus 50% was also considered here (see Figure 4b). Material costs have a much greater impact on profitability than energy costs. Here, as well, the reference variant is more affected by price increases. In the range considered, profitability decreases by about 11 percentage points for this variant and by almost 9 percentage points for the variant with an incineration plant. Even with a 50% increase in material prices, profitability still remains above 10%.

The sensitivity analysis also considered the case in which the biorefinery is operated in an industrial network. In this case, personnel capacities can be shared, and thus the proportional personnel costs can be reduced. Figure 5a shows the influence of personnel costs. The basic calculations assume a self-sufficient operation of the plant with complete staffing. A range of plus 20 to minus 50% was considered for the sensitivity analysis of personnel costs. An increase in personnel costs by 20% reduces profitability by about 5 to 6% in both variants. A 50% reduction in personnel costs increases profitability by 13 to 15%, whereby the basic variant benefits more from lower personnel costs.

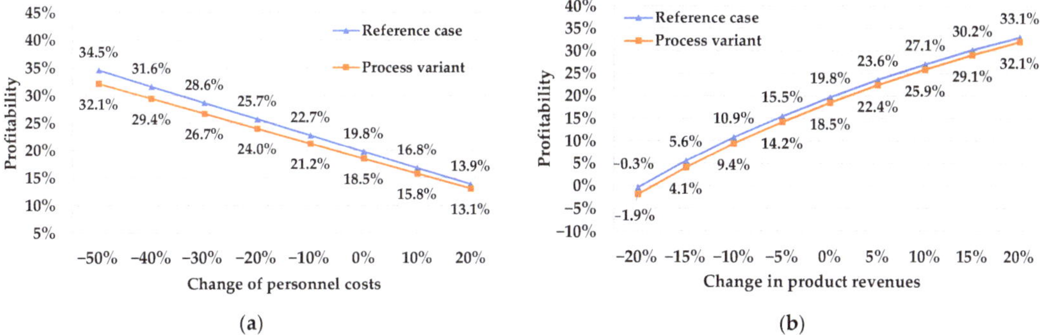

Figure 5. Sensitivity analysis: (**a**) personnel costs; (**b**) product revenues.

The market prices for the products produced have the greatest influence on the profitability of the plant (see Figure 5b). A variation of plus/minus 20% was considered for the cost block revenue. If total revenues fall by 20%, then profitability also falls by approx. 20% for both variants. The process variant falls below 5% profitability earlier than the reference case because its values are generally about 1 to 1.5% lower. At a revenue decline of approximately 15%, the plant reaches the 5% profitability threshold, and at a decline of 20%, the plant runs at a loss. On the other hand, an increase in the revenue situation by 20% causes an increase in the return by about 13%.

3.5. Allocation of Costs to Products

The first part of this publication series examined how the primary energy demand and CO_2 emissions of the new biorefinery are distributed among the individual products generated. In this economic part of this publication series, an analysis was therefore made of how the operating costs of the plant are distributed among the individual products according to Ref. [47]. Such an analysis reveals whether all products provide a contribution margin or whether particularly profitable products have to compensate for less or non-profitable ones. For this allocation, a distribution of product volumes and another distribution of relative market values were generated from Table 2. Torrefied biomass has the largest production mass share of all products at over 97%, but it has the smallest share of the total market price per ton at 0.016%. The torrefied biomass was set to an equivalent value of 1 due to its smallest market price share, and all other products were set relative to the market price share of this reference product. This results in an equivalence value of 5833 for the most expensive product 5-Hydroxymethylfurfural (5-HMF) which has a mass share of 0.0073% and a total market price share per ton of 93.84%. The sales units can be calculated by multiplying the individual equivalent values by the annual output of each product. The total cost per year is then divided by the total number of sales units to obtain the total cost per sales unit, which is 0.1925. The equivalence numbers can then be multiplied by the total cost per sales unit to determine the production costs for each product shown in blue in Figure 6. For instance, the equivalence value of 5-HMF is multiplied by 0.1925 to obtain a production price of 1123 €. The thermal energy produced had to be related to the torrefied biomass, which is energetically recovered by combustion, in order to keep the relation to one unit (kg/h). The thermal energy is then allocated using the aforementioned method, which has a market price share of 0.031%, nearly twice that of torrefied biomass. The allocation shows that all products provide a positive contribution to profit in the reference case as well as in the variant with incineration plant.

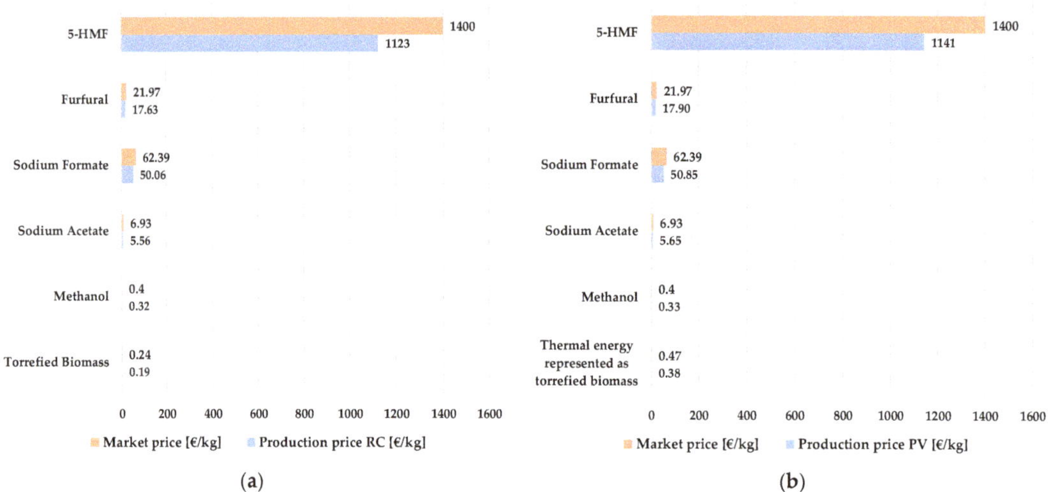

Figure 6. Cost allocation: (**a**) reference case; (**b**) process variant.

3.6. Eco-Efficiency Analysis

The results from the ecological analysis from the first part of this publication series were presented again for the operation of the biorefinery with beech woodchips in the form of a spider diagram for the reference case and the variant with biomass incineration in Figure 7a. The categories human toxicity (HTP), freshwater eutrophication (FEP) and freshwater ecotoxicity (FETP) are the same for both variants because only the biomass feedstock affects these three categories, and the operation with straw is not examined here.

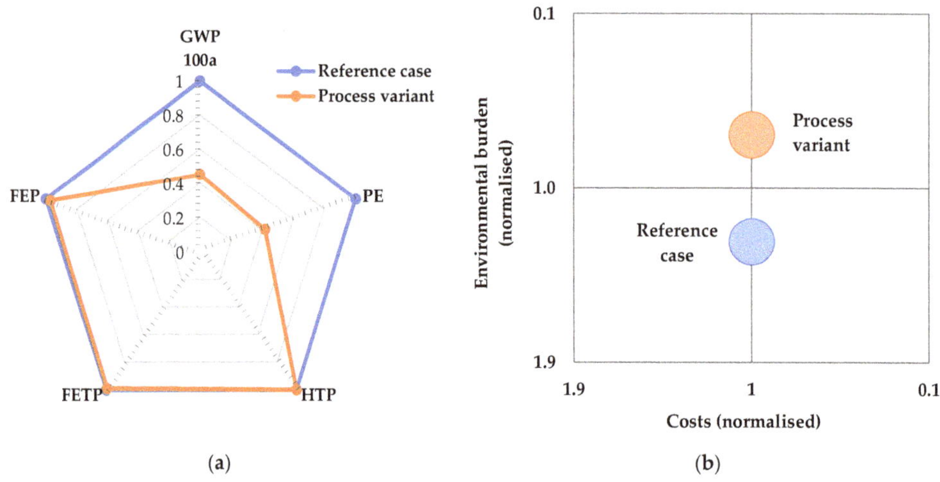

Figure 7. (**a**) normalised environmental effects; (**b**) eco-efficiency portfolio.

Figure 7b depicts the eco-efficiency portfolio of both operating variants considered. This profile was created according to references [48–50]. The values of the LCA and the sum of operating costs and revenues (indicated as "costs") are normalised in relation to each other, with the worst performance being assigned a value of 1 in each case. From this, the normalised mean values for the environmental impacts and the normalised mean values for the costs are calculated. Because there are five environmental impact categories, each must be weighted. Climate change (GWP) was weighted at 40%, primary energy demand (PE) at 30% and the remaining three at 10% each. The individual results are then related to the mean value and multiplied by the individual weighting factors in the case of environmental impacts.

The normalised costs are 1.018 in the reference case and 0.982 in the variant with incineration plant, i.e., the process variant performs slightly poorer under the selected boundary conditions. However, with a value of 0.724, the process variant performs significantly better in terms of environmental impacts than the reference case with 1.276.

4. Derivation of Commercialisation Opportunities

4.1. Context

In today's economy, fossil-based products set the economic standard. To be economically successful, biobased alternatives must offer at least the same quality and functionality as their conventional alternatives [7].

The bioeconomy is traditionally very strong in the food sector and in wood processing. However, it still has considerable growth potential in the economic sectors of construction materials, paper, textile fibres, chemicals, pharmaceuticals, heat, electricity and fuels [7]. In the processing sector, which currently still uses fossil carbon sources, biobased products have a potential for expansion (in some cases considerable) that goes far beyond the pure economic growth of these sectors [7].

Biobased products are in most cases more expensive than fossil carbon sources under current framework and market conditions. Due to the complex processing of biological raw materials, biobased alternatives are also less competitive than fossil-based products. Therefore, biobased products that are in direct competition with conventional alternatives can only be profitable in niche markets and only those that reward sustainability through price. For companies, this means adjusting their strategy, anticipating changing framework conditions at an early stage and adapting their product range accordingly [7]. Commercialisation opportunities for biorefineries may arise from the fact that new solutions and more sustainable alternatives are either mandated by regulations or demanded by customers [1]. Biobased alternatives must be assessed on a case-by-case basis for their suitability compared to their conventional alternatives, and if no suitable products are available today, appropriate research efforts can be successful [7]. However, if products manufactured with fossil raw materials are burdened by taxation or an expansion of trading in emission certificates, this can change the competitive situation in favour of biorefineries [7].

The literature research carried out within this work does not allow for a sound quantitative representation of the market sizes of different markets and the competitive positions of biobased products from biorefineries. However, the following markets seem to be of special interest for the investigated biorefinery:

- Plastics, base and fine/speciality chemicals
- Pharmaceuticals
- Active carbon, synthetic fuels and others
- Energy markets (electricity, heat)

The review showed that currently the market development for the bioeconomy cannot be assessed so easily, especially because neither the industrial companies nor the markets examine biobased products separately. Only in the chemical and pharmaceutical industries are the possible developments more foreseeable [62].

The novel biorefinery studied here extracts a relatively small amount of high-value platform chemicals compared to the feedstock and produces a relatively large amount of high-quality torrefied biomass (see Figure 8).

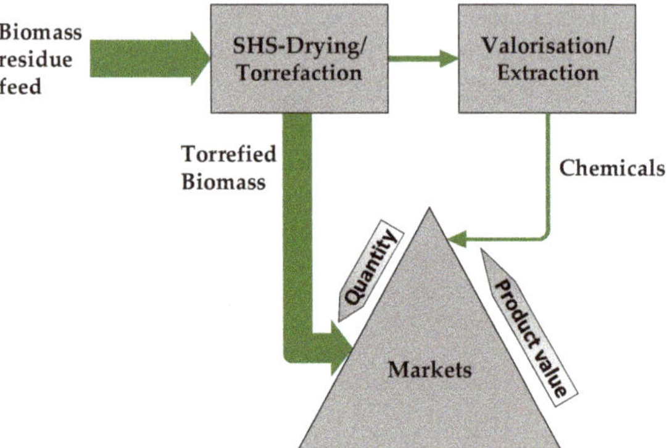

Figure 8. Biorefinery product-market-relationship.

Both product groups are interesting for different markets, which are discussed in the following.

4.2. Markets for the Platform Chemicals

Of particular importance in this development are biobased platform chemicals, bioplastics and biobased fine and speciality chemicals [62]. The chemical companies are most likely to see competitive advantages in biobased speciality chemicals and plan to focus more strongly here [62]. Chinese companies currently dominate the production and demonstration of marketable biobased products [62]. For biobased plastics, China, India and Southeast Asia are seen as the most important production locations of the future, and Brazil is also expanding its capacities [62]. The large chemical companies have little incentive to change the system towards biorefineries because they use proven, highly optimised technologies on a large scale, the infrastructure is built up and the process chains are established. The plants are often depreciated and therefore highly profitable [62]. Biobased chemical products are sold to the application-oriented industry, i.e., as intermediates. Market introduction was particularly successful when a "green premium" could be obtained for the intermediates [62,63].

Even though the use of renewable resources has only played a minor role in the pharmaceutical industry so far, innovative biotechnological approaches are of central importance for this branch of industry [62]. In Germany, a biopharmaceutical innovation system has been successfully established over the past decades and is networked with strong, internationally competitive players. A number of large companies and a significant and stable number of SMEs are active here and one finds a well-developed corporate landscape for biopharmaceutical activities [62].

The economic analysis shows that the investigated biorefinery can be operated economically under current market prices. However, the personnel costs and, in particular, the revenues must be viewed critically in terms of profitability.

The impact of personnel costs can be reduced by increasing the annual number of operating hours of the biorefinery and by integrating it into existing industrial networks.

The revenues from the sale of the products and demand must be considered volatile, especially for the extracted chemicals [11]. The quality and price of the chemicals produced must be kept at a high level to avoid jeopardising the economic viability of the plant. This means focusing on high quality products and even on application services for the speciality market rather than the mass market. Product and market diversification towards higher value products is seen as key for biorefineries to overcome commercialisation barriers [20].

These arguments are also supported by the fact that the plant concept of the biorefinery under investigation is not designed for large production volumes of chemicals and that biorefineries based on torrefaction for the production of mass products (biofuels) have not yet been successfully commercialised despite years of development with different approaches [15].

4.3. Markets for the Torrefied Biomass

The thermal integration of biorefineries in an internal or external production network promises a significant energy and economic improvement [31]. If heat integration is chosen, the optimal production capacity of the biorefinery can be matched to the demand of the consumers in the heat network [16]. These findings can also be transferred to the plant investigated here. Against the background of the results of the economic assessment, the optimal solution would be to integrate the investigated plant into an industrial network that already has its own heat supply or, in the best case, its own electricity generation. This would be particularly beneficial for the profitability of the investigated process variant and the ecological aspects.

The capacity of the biorefinery can be adapted so that it can best be integrated into an industrial symbiosis. It can be flexibly adapted in terms of the use of different biomass residues and energy recovery. In biorefineries that supply torrefied biomass as output, further processing by pyrolysis or gasification is possible in addition to thermal use through combustion. In this way, high quality products with technical advantages for iron production, adsorbents for environmental protection or feedstocks for the production of synthetic fuels can be generated [15].

4.4. Biomass Supply

An argument against mass production is the demand for biomass residues, which could possibly drive up prices. Long-term and the year-round availability of biomass at competitive prices is an important criterion for the operation of biorefineries. It is therefore advantageous if the plant can flexibly use several feedstocks per year depending on availability [13]. In the available studies on biomass torrefaction, different biomass materials have been used as feedstock. Lignocellulosic biomass and energy crops (e.g., bamboo, pine, spruce, willow, eucalyptus, banyan, larch, beech, birch, wood briquettes and wood pellets, etc.) are the most commonly used feedstocks. Torrefaction of agricultural residues and forestry wastes (e.g., sawdust, bagasse, coffee residues, rice husks and straw, wheat straw, empty fruit bunches from oil palms, etc.) has also been studied. However, special care is taken to ensure that there is no imbalance to the disadvantage of food supply and forest vegetation [15].

4.5. Recommendations for Commercialisation

A step-by-step approach for successful commercialisation in an open innovation system has proven to be viable to ensure resource mobilisation and market creation. Collaboration across the supply chain has proven to be beneficial because it can generate positive momentum. The innovation system must be able to mobilise the necessary resources while creating supply and demand [20].

As most business decisions are still made on the basis of the economic value of the main product in a value chain, awareness of value creation opportunities in alternative sectors is generally low from a marketing perspective. This is reinforced by the fact that costs and benefits of valorisation of biomass residues are assigned to different parties in the value chain. Therefore, new business models are needed that create an environment of equitable distribution of effort and benefit and that enable the development of innovative technologies across sectors [24]. It is important to involve public and private partners from the biomass supply sector and the end-product sector, and also to consider partners outside the value chain [64]. Typical customer–supplier relationships must be replaced by multi-actor relationships [24].

Therefore, a recommendation for the successful commercialisation of the investigated biorefinery concept is the operation of a flexibly usable demonstration plant in an open innovative network. The following sectors in particular should be considered as innovative partners for such a network, whereby the order does not represent a preference:

- Utilities with high innovation potential (e.g., from the municipal sector), which already have experience with the use of biomass
- Companies from the field of speciality chemicals, pharmaceuticals and food with a connection to the bioeconomy
- Innovative engineering companies with experience in biomass processing
- Consulting firms with roots in the bioeconomy and strong knowledge concerning supply chain aspects
- Start-ups and SMEs dedicated to application know-how and commercialisation aspects in the bioeconomy.

Companies from the paper industry or the large-scale chemical industry, for example, were deliberately not named because they are usually too entrenched in their existing structures and, after studying the state of the art, thus do not have the necessary innovation potential. On the other hand, they might be interested in buying green energy (heat or electricity) in the near future to reduce their carbon footprint and meet the new policy targets for CO_2 emissions.

5. Discussion

5.1. Economic Assessment

The biorefinery investigated in this publication series fulfills both the ecological [41] and the economic sustainability criteria, provided that product revenues do not fall by more than 15% compared to the current market price level.

Both cost estimation methods used in this analysis lead to comparable investment costs. The depreciation over 20 years leads to reasonable annual fixed costs in the reference case. In the process variant, the necessary investment in an incineration plant leads to more than a doubling of the annual depreciation costs. In the process variant, the premiums for the indirect investment costs (such as infrastructure), for the process investment costs and the risk supplement, were reduced compared to the reference case. It is assumed that the incineration plant represents a high technical maturity level with low risk and is delivered as a pre-assembled unit that requires only minor additional investments in infrastructure compared to the reference case. A capital recovery factor over a 20-year lifetime is a common practice [16], especially for mature components such as silos, conveyors, fans, ductwork and the biomass incineration plant, which are used in the biorefinery in this case and account for the largest share of the costs. A reduction of the depreciation period of the investment costs to 15 years would nevertheless not endanger the profitability of the plant.

In the economic evaluation, it was assumed that the facility under investigation is operated as an independent enterprise and must therefore have complete staffing. Full staffing is a particular financial burden as it makes up in the investigated case for more than one third of the total costs (see Figure 3). The economic integration of the biorefinery into an industrial network would help to reduce the proportion of personnel costs and thus improve significant profitability (see Figure 5a).

The presented analyses of the newly developed biorefinery assumes an hourly processing capacity of 1000 kg of biomass and an annual operating time of 5080 h. Because the fixed costs in the reference case and in the process variant account for 66 and 77.5% of the total operating costs, respectively, the annual operating time has a significant influence on profitability. An increased annual operating time raises the variable operating costs, but higher revenues are generated and the proportion of fixed costs is reduced.

The costs for the biomass residue raw material represent the second largest cost block in both variants after the personnel costs. These costs can be decreased significantly if low value feedstock, such as residues from forestry activities, is used. However, an increase in procurement costs has a significantly lower impact on the profitability of the plant than personnel costs (see Figure 4b).

The biorefinery has been significantly optimised in terms of energy as part of the process development to date, and the process variant allows fossil fuels to be abandoned for heat generation. The variation in energy costs has little effect on profitability in the reference case (<5%) and very little effect in the process variant (<2%) within the considered price range of −10 to +50% (see Figure 4a).

From the output side, two different categories of products must be delineated. Almost 70% of the input material is converted into torrefied biomass with a lower monetary specific value and just under 2% into platform chemicals with a high monetary specific value. In the scenarios considered, the torrefied biomass is made available as a high-grade fuel in the reference case or as feedstock for upgrading to activated carbon, synthesis gas or synthetic fuel. In the variant with incineration plant, the torrefied biomass covers the biorefinery's own thermal energy demand and provides a significant amount of thermal energy for external customers.

Given today's market prices for torrefied biomass and for thermal energy, between 38 and 52% of total revenues are generated directly from torrefied biomass or indirectly through thermal energy sales. Due to the small plant size, the sales volume of torrefied biomass or thermal energy is considered small, which can be seen as an advantage for the integration into existing industrial networks. The design of the dryers and the torrefaction systems is modular (250 kg/h input per unit). This means that the plant can be designed precisely to meet the needs of industrial consumers. Because heat with a temperature of up to 270 °C is generated, it could possibly be sold at a higher price than is currently assumed (7.5 Eurocents/kWh), which would increase the profitability of the biorefinery.

Due to their small quantities, the extracted platform chemicals can be viewed in isolation from the torrefied biomass produced concerning the plant design, the location of the extraction process steps and the market for chemicals. More than 20 extractable chemicals were identified in the experiments with beech wood chips. However, only five currently attractive chemicals were explicitly included in the present economic study. Further studies need to show what other chemicals can be extracted if other biomass residues are used and what prices may be obtained for them. With the values taken in this study, the platform chemicals contribute more than 60% to the value generation in the reference case and 48% in the variant with an incineration plant.

In the ideal case, customers from the chemical (speciality and fine chemicals), pharmaceutical, cosmetic or other specialty industries would be in direct local connection with the biorefinery. However, due to the small production volumes and the high value of the extracted chemicals, the transport costs for these play a subordinate role. In other words, the customers for the platform chemicals can be further away.

By varying the feedstock, temperature and residence time in the torrefaction chamber, the output of potential chemicals can be altered. This means that with the choice of biomass feedstocks used and the variation of plant parameters, one can flexibly extract different value-added chemicals and thus respond flexibly to market changes. The literature review has shown that the flexibility of a biorefinery in respect to constantly changing markets and the products that are in particular demand is a very important commercialisation success factor.

Product revenues have the greatest influence on the profitability of the plant in the biorefinery studied in this paper. Therefore, a commercialisation strategy for the platform chemicals must pay special attention to primarily serving the high-quality and thus high-priced customer segment and not the low-priced mass market. The platform chemicals produced are usually needed for intermediate products. This requires a high degree of flexibility and application know-how. Therefore, the revenues generated should be invested in application-oriented research efforts that focus not only on the further technical development of the plant, but also on possible new market applications. The increase in product revenue also has the greatest impact on the return on investment (ROI) and the payback period (PP). With the basic assumptions made, the annualised ROI for the reference case is 24.6% and the PP 4.4 years (process variant 22.7% and 7.8 years, respectively). If revenues can be increased by 20%, the annualised ROI in the reference case is 49.5% and the PP is 2.2 years (process variant 47.2% and 3.75 years, respectively).

All six biorefinery products investigated in this work deliver a positive financial contribution to the result. For further extractable substances, however, it would have to be examined whether this statement is still valid, especially if further process steps have to be installed for this purpose. CO_2 credits were not included in the economic analysis because this requires government authorisation to participate in emissions trading. These alone would only marginally improve profitability, because the costs for primary energy account for just 10% of the total operating costs in the reference case examined and merely 4% in the process variant.

In the eco-efficiency analysis, both variants examined with the assumptions made perform almost equally on the economic side. However, the process variant with biomass incineration plant and thermal self-supply performs significantly better ecologically than the reference case. In the case of an industrial symbiosis, participation in emissions trading and/or rising material and energy prices, the process variant benefits from economic performance and then moves into the best area of the eco-efficiency portfolio compared to the reference case. The present eco-efficiency analysis should be seen as a basis from which to review the eco-efficiency portfolio for concrete implementation cases within the framework of an industrial symbiosis. For this purpose, other environmental impacts such as particulate matter may have to be added.

5.2. Commercialisation Aspects

Against the background of the information gathered from the literature review and the economic analysis carried out, it seems advisable to realise the demonstration operation as well as the commercialization of the biorefinery in an open innovation environment. The possible separation of biomass pre-treatment and extraction of different platform chemicals from the condensate facilitates the participation of different types of actors in the open innovation network.

The literature research has shown that traditional players find it difficult to adapt their more or less proven traditional way of thinking to the new conditions of a circular economy and constantly changing market conditions. This is especially true when these players have served large market volumes at low prices, but rising energy prices and current climate policy could change this behaviour in the future. Nevertheless, it is important for the network and commercialisation of the biorefinery to involve innovative, knowledge gap closing and application-oriented actors especially from the fields of speciality and fine chemicals, the pharmaceutical, cosmetic and bioeconomy sectors as shown in Figure 9. The network's external recognition must be strong enough to attract and embrace further innovative actors, e.g., from the utility sector. This also means that the open innovation system must be designed in such a way that it can develop continuously. If robust industrial symbiosis can be developed within the network, this will have a positive impact on the profitability of the biorefinery, as infrastructure and personnel can be shared.

In order to trigger the necessary investments, it is necessary either to obtain state funding or to attract financially strong actors. This will also raise the question of how the profits will be distributed fairly in the network and who will take over the project management at which point in time. Against the background of these challenges, it makes sense to take a closer look at existing open innovation platforms from other industrial sectors, such as ARENA 2036 [65], for the mobility and production of the future, or to join a platform from the bioeconomy sector, such as TALENT4BBI [66].

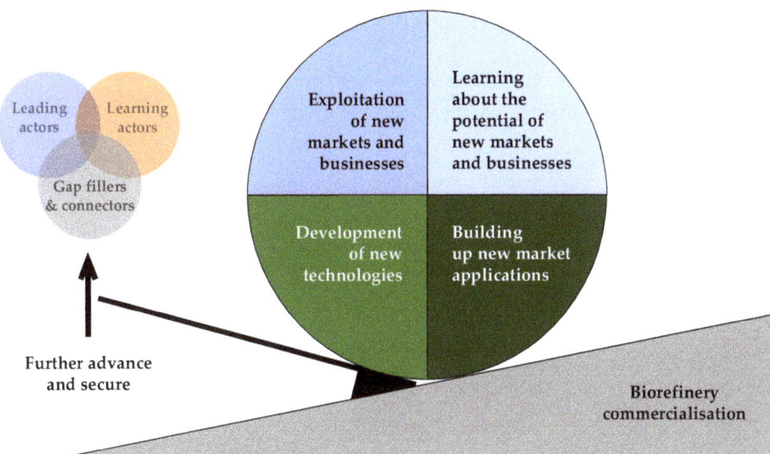

Figure 9. Biorefinery commercialisation proposal.

6. Conclusions

The economic analysis of the investigated novel biorefinery shows that it can be operated economically under the current conditions and the assumptions made. Because the plant is small compared to the conventional production capacities, the infrastructure and personnel costs are disproportionately high. However, the plant size can be increased if enough feedstock is available. Moreover, this dilemma can be solved if the biorefinery is used as part of a production network and the above-mentioned costs only have to be charged in part. Selling the surplus heat has little impact on profitability but makes it independent from the purchase of natural gas, which is more environmentally friendly and interesting for a sustainable industrial symbiosis. Focusing on niche markets that are willing to pay a higher price for high-quality chemicals is crucial for the profitability of the biorefinery operation. The plant size, the structure and the flexibility of the biorefinery in terms of the biomass feedstock, the mode of operation and the generated products are advantageous for an industrial symbiosis. The different, flexibly generated products make the biorefinery interesting for a variety of different industries. The plant is at the development stage between laboratory and demonstration scale and has yet to be established in the circular economy environment. It would therefore make sense to carry out further development and commercialisation within the context of an open innovation platform. Such a platform would allow different actors to participate in the project, and hence, valuable synergies could be leveraged. In summary, the flexibility of the plant, its low complexity and the focus on niche markets are a good basis for the sustainability, competitiveness and resilience of the novel biorefinery.

The investigated biorefinery is ecologically and economically very well suited to support the achievement of the 2030 target for the reduction in greenhouse gas emissions as well as the target of climate neutrality by 2050 within the framework of the European Green Deal. It is also appropriate for realising the shift from linear economy to circular economy value chains.

The successful upscaling and implementation of the developed biorefinery concept will depend on the ability to treat different types of feedstock/residues, the automatic and robust operation of the different processes (with reduced staff intervention) and the market demand of the obtained products. Therefore, further works will focus on the continuous separation of chemicals, the flexible extraction of different valorisation products and the upscaling of the SHS-based drying/torrefaction with different biomass residues.

Supplementary Materials: The following are available online at https://www.mdpi.com/article/10.3390/su14042338/s1, Direct investment costs for Reference Case and Process Variant.

Author Contributions: Conceptualization, B.R., P.K.-M. and A.D.; methodology, P.K.-M.; software, B.R.; validation, B.R.; formal analysis, B.R. and P.K.-M.; investigation, B.R.; resources, P.K.-M.; data curation, B.R.; writing—original draft preparation, B.R. and P.K.-M.; writing—review and editing, P.K.-M. and A.D.; visualization, B.R. and P.K.-M.; supervision, P.K.-M.; project administration, P.K.-M.; funding acquisition, P.K.-M. and A.D. All authors have read and agreed to the published version of the manuscript.

Funding: This research was funded by the FEDERAL MINISTRY OF EDUCATION AND RESEARCH (BUNDESMINISTERIUM FÜR BILDUNG UND FORSCHUNG), grant number 031B0664.

Institutional Review Board Statement: Not applicable.

Informed Consent Statement: Not applicable.

Data Availability Statement: Data are contained within the article or Supplementary Material.

Conflicts of Interest: The authors declare no conflict of interest.

References

1. D'Amato, D.; Veijonaho, S.; Toppinen, A. Towards sustainability? Forest-based circular bioeconomy business models in Finnish SMEs. *For. Policy Econ.* **2018**, *110*, 101848. [CrossRef]
2. Vu, H.P.; Nguyen, L.N.; Vu, M.T.; Johir, M.A.H.; McLaughlan, R.; Nghiem, L.D. A comprehensive review on the framework to valorise lignocellulosic biomass as biorefinery feedstocks. *Sci. Total. Environ.* **2020**, *743*, 140630. [CrossRef] [PubMed]
3. Cheah, W.Y.; Sankaran, R.; Show, P.L.; Ibrahim, T.N.B.T.; Chew, K.W.; Culaba, A.; Chang, J.-S. Pretreatment methods for lignocellulosic biofuels production: Current advances, challenges and future prospects. *Biofuel Res. J.* **2020**, *7*, 1115–1127. [CrossRef]
4. Hilz, X. Assessing Areas of Concern in the Commercialisation Process of Biorefineries: An Importance Performance Analysis. Master's Thesis, Universität Graz, Graz, Austria, 2021.
5. Reim, W.; Sjödin, D.; Parida, V.; Rova, U.; Christakopoulos, P. Bio-economy based business models for the forest sector—A systematic literature review. In *Proccedings of International Scientific Conference "Rural Development 2017", Kaunas, Lithuania, 22 December 2017*; Aleksandras Stulginskis University: Kaunas, Lithuania, 2017. [CrossRef]
6. Domínguez-Robles, J.; Cárcamo-Martínez, Á.; Stewart, S.A.; Donnelly, R.F.; Larrañeta, E.; Borrega, M. Lignin for pharmaceutical and biomedical applications—Could this become a reality? *Sustain. Chem. Pharm.* **2020**, *18*, 100320. [CrossRef]
7. Kircher, M. *Bioökonomie im Selbststudium: Unternehmensstrategie und Wirtschaftlichkeit*; Springer: Berlin/Heidelberg, Germany, 2020; ISBN 978-3-662-61004-6.
8. Orive, M.; Cebrián, M.; Amayra, J.; Zufía, J.; Bald, C. Techno-economic assessment of a biorefinery plant for extracted olive pomace valorization. *Process Saf. Environ. Prot.* **2021**, *147*, 924–931. [CrossRef]
9. Hansen, T.; Coenen, L. Unpacking resource mobilisation by incumbents for biorefineries: The role of micro-level factors for technological innovation system weaknesses. *Technol. Anal. Strat. Manag.* **2016**, *29*, 500–513. [CrossRef]
10. Wang, Y. *Development of a Quantitative Risk Analysis Approach to Evaluate the Economic Performance of an Industrial-Scale Biorefinery*; University of British Columbia: Vancouver, BC, Canada, 2018.
11. Elaradi, M.B.; Zanjani, M.K.; Nourelfath, M. Integrated forest biorefinery network design under demand uncertainty: A case study on canadian pulp & paper industry. *Int. J. Prod. Res.* **2021**, *59*, 1–19. [CrossRef]
12. Chai, L.; Saffron, C.M. Comparing pelletization and torrefaction depots: Optimization of depot capacity and biomass moisture to determine the minimum production cost. *Appl. Energy* **2016**, *163*, 387–395. [CrossRef]
13. Chandel, A.K.; Garlapati, V.K.; Singh, A.K.; Antunes, F.A.F.; Da Silva, S.S. The path forward for lignocellulose biorefineries: Bottlenecks, solutions, and perspective on commercialization. *Bioresour. Technol.* **2018**, *264*, 370–381. [CrossRef]
14. Kaserer, W. Fasern für fast alles im Leben. In *CSR und Klimawandel*; Sihn-Weber, A., Fischler, F., Eds.; Springer: Berlin/Heidelberg, Germany, 2020; pp. 381–393; ISBN 978-3-662-59747-7.
15. Chen, W.-H.; Lin, B.-J.; Lin, Y.-Y.; Chu, Y.-S.; Ubando, A.T.; Show, P.L.; Ong, H.C.; Chang, J.-S.; Ho, S.-H.; Culaba, A.B.; et al. Progress in biomass torrefaction: Principles, applications and challenges. *Prog. Energy Combust. Sci.* **2020**, *82*, 100887. [CrossRef]
16. Zetterholm, J. Evaluation of Emerging Forest-Industry Integrated Biorefineries: Exploring Strategies for Robust Performance in Face of Future Uncertainties. Ph.D. Thesis, Lulea University of Technology, Lulea, Sweden, 2021.
17. Lager, T.; Blanco, S.; Frishammar, J. Managing R&D and innovation in the process industries. *R&D Manag.* **2013**, *43*, 189–195. [CrossRef]
18. Mossberg, J.; Frishammar, J.; Söderholm, P.; Hellsmark, H. Managerial and organizational challenges encountered in the development of sustainable technology: Analysis of Swedish biorefinery pilot and demonstration plants. *J. Clean. Prod.* **2020**, *276*, 124150. [CrossRef]

19. Tsvetanova, L.; Carraresi, L.; Wustmans, M.; Bröring, S. Actors' strategic goals in emerging technological innovation systems: Evidence from the biorefinery sector in Germany. *Technol. Anal. Strat. Manag.* **2021**, *33*, 1–14. [CrossRef]

20. Kasnitz, L. Building a Biorefinery Business: If it does not Fit, Make it Fit—Strategies for Successful Commercialization. Master's Thesis, Lund University, Lund, Sweden, 2017.

21. Musiolik, J.; Markard, J.; Hekkert, M.; Furrer, B. Creating innovation systems: How resource constellations affect the strategies of system builders. *Technol. Forecast. Soc. Chang.* **2020**, *153*, 119209. [CrossRef]

22. Budzianowski, W.M. High-value low-volume bioproducts coupled to bioenergies with potential to enhance business development of sustainable biorefineries. *Renew. Sustain. Energy Rev.* **2017**, *70*, 793–804. [CrossRef]

23. Odegard, I.; Croezen, H.; Bergsma, G. Cascading of Biomass: 13 Solutions for a Sustainable Bio-Based Economy. Making Better Choices for Use of Biomass Residues, by-Products and Wastes. Available online: https://ce.nl/wp-content/uploads/2021/03/CE_Delft_2665_Cascading_of_Biomass_def_1348490086.pdf (accessed on 12 November 2021).

24. Donner, M.; Gohier, R.; de Vries, H. A new circular business model typology for creating value from agro-waste. *Sci. Total Environ.* **2020**, *716*, 137065. [CrossRef]

25. Ahlström, J. Shaping Future Opportunities for Biomass Gasification: The Role of Integration. Ph.D. Thesis, Chalmers University of Technology, Gothenburg, Sweden, 2020.

26. Hassan, S.S.; Williams, G.A.; Jaiswal, A.K. Lignocellulosic Biorefineries in Europe: Current State and Prospects. *Trends Biotechnol.* **2019**, *37*, 231–234. [CrossRef]

27. Gabriella, N. A Delphi Study of Factors Affecting Forest Biorefinery Development in the Pulp and Paper Industry: The Case of Bio-Based Products. Master's Thesis, Universität Graz, Graz, Austria, 2016.

28. Arora, A.; Banerjee, J.; Vijayaraghavan, R.; MacFarlane, D.; Patti, A. Process design and techno-economic analysis of an integrated mango processing waste biorefinery. *Ind. Crop. Prod.* **2018**, *116*, 24–34. [CrossRef]

29. Nitzsche, R.; Budzinski, M.; Gröngröft, A.; Majer, S. Bewertungsansätze bei der Optimierung von Bioraffinerie-Konzepten. In *Bioenergie. Vielseitig, Sicher, Wirtschaftlich, Sauber?!* Nelles, M., Ed.; DBFZ Deutsches Biomasseforschungszentrum Gemeinnützige GmbH: Leipzig, Germany, 2014; pp. 57–68. ISBN 2199-9384.

30. Wilén, C.; Sipilae, K.; Tuomi, S.; Hiltunen, I.; Lindfors, C.; Sipilä, E.; Saarenpää, T.-L.; Raiko, M. Wood Torrefaction: Market Prospects and Integration with the Forest and Energy Industry. Available online: https://www.vttresearch.com/sites/default/files/pdf/technology/2014/T163.pdf (accessed on 11 November 2021).

31. Hagberg, M.B.; Pettersson, K.; Ahlgren, E.O. Bioenergy futures in Sweden—Modeling integration scenarios for biofuel production. *Energy* **2016**, *109*, 1026–1039. [CrossRef]

32. Ahlström, J.M.; Pettersson, K.; Wetterlund, E.; Harvey, S. Value chains for integrated production of liquefied bio-SNG at sawmill sites—Techno-economic and carbon footprint evaluation. *Appl. Energy* **2017**, *206*, 1590–1608. [CrossRef]

33. Arvidsson, M.; Heyne, S.; Morandin, M.; Harvey, S. Integration Opportunities for Substitute Natural Gas (SNG) Production in an Industrial Process Plant. Available online: https://www.aidic.it/cet/12/29/056.pdf (accessed on 11 November 2021).

34. Budzianowski, W.M.; Postawa, K. Total Chain Integration of sustainable biorefinery systems. *Appl. Energy* **2016**, *184*, 1432–1446. [CrossRef]

35. Hellsmark, H.; Hansen, T. A new dawn for (oil) incumbents within the bioeconomy? Trade-offs and lessons for policy. *Energy Policy* **2020**, *145*, 111763. [CrossRef]

36. Cesena, E.A.M.; Mutale, J.; Rivas-Davalos, F. Real options theory applied to electricity generation projects: A review. *Renew. Sustain. Energy Rev.* **2013**, *19*, 573–581. [CrossRef]

37. Horst, S.; Höfer, J.; Kleine-Möllhoff, P.; Wennagel, F.; Wiech, N.; Pfost, M.; Atmaca, B.; Gries, J.; Epple, R. Kostenkalkulation im Anlagenbau: Modell zur Bewertung der Konkurrenzfähigkeit im Entwicklungsstadium. *Chem. Ing. Tech.* **2020**, *92*, 1033–1043. [CrossRef]

38. Humphreys, K.K. *Project and Cost Engineers Handbook*, 4th ed.; Marcel Dekker: New York, NY, USA, 2004. [CrossRef]

39. Albaroudi, H.; Atayi, K.; Baleka, K.; Osae, E.; Pushparakan, S.; Taylor, A. *Acetic Acid Process Plant Design*; University of Hull: Hull, UK, 2016.

40. Prinzing, P.; Rödl, R.; Aichert, D. Investitionskosten-Schätzung für Chemieanlagen. *Chem. Ing. Tech.* **1985**, *57*, 8–14. [CrossRef]

41. Roy, B.; Kleine-Möllhoff, P.; Dalibard, A. Superheated Steam Torrefaction of Biomass Residues with Valorisation of Platform Chemicals—Part 1: Ecological Assessment. *Sustainability* **2022**, *14*, 1212. [CrossRef]

42. Castedello, M.; Schöninger, S. Cost of Capital Study 2020: Global Economy—Search for Orientation? Available online: https://hub.kpmg.de/kapitalkostenstudie-2020?utm_campaign=Kapitalkostenstudie%202020&utm_source=AEM (accessed on 29 September 2021).

43. Norder Band, A.G. Legierungszuschlag.info: Legierungszuschlag für Werkstoff 1.4301. Available online: https://legierungszuschlag.info/wkst/4301 (accessed on 6 July 2021).

44. Janisch, A. Kostenfaktoren im Stahlbau: Was Kostet ein kg Stahlbau 2021? Available online: https://jactio.com/kostenfaktoren-im-stahlbau/ (accessed on 4 November 2021).

45. CheCalc. Shortcut Heat Exchanger Sizing. Estimates LMTD, Exchanger Surface Area, Number of Tubes, Shell Diameter and Number of Shell in Series. Available online: https://checalc.com/calc/ShortExch.html (accessed on 26 January 2022).

46. Eltrop, L. Leitfaden Feste Biobrennstoffe, Gülzow-Prüzen. 2014. Available online: http://www.fnr.de/fileadmin/allgemein/pdf/broschueren/leitfadenfestebiobrennstoffe_web.pdf (accessed on 1 October 2021).

47. Horsch, J. *Kostenrechnung: Klassische und neue Methoden in der Unternehmenspraxis*, 3rd ed.; Springer Fachmedien Wiesbaden: Wiesbaden, Germany, 2018; ISBN 978-3-658-20029-9.
48. Ingenieure, V.D. *Methods for Evaluation of Waste Treatment Processes: Part 2, Samples Calculations*; VDI 3925 Part 2; Beuth Verlag: Berlin, Germany, 2018.
49. Saling, P.; Kicherer, A.; Dittrich-Krämer, B.; Wittlinger, R.; Zombik, W.; Schmidt, I.; Schrott, W.; Schmidt, S. Eco-efficiency analysis by basf: The method. *Int. J. Life Cycle Assess.* **2002**, *7*, 203–218. [CrossRef]
50. Rüdenauer, I.; Grießhammer, R.; Bunke, D.; Gensch, C. Integrated Environmental and Economic Assessment of Products and Processes. *J. Ind. Ecol.* **2005**, *9*, 105–116. [CrossRef]
51. Statista. Annual Electricity Prices (Including Electricity Tax) for Industrial Businesses in Germany from 1998 to 2021: (In Euro Cents per Kilowatt Hour). Available online: https://www.statista.com/statistics/1050448/industrial-electricity-prices-including-tax-germany/ (accessed on 18 January 2022).
52. Statista. Gaspreise* für Gewerbe- und Industriekunden in Deutschland in den Jahren 2011 bis 2021: (In Euro-Cent pro Kilowattstunde). Available online: https://de.statista.com/statistik/daten/studie/168528/umfrage/gaspreise-fuer-gewerbe-und-industriekunden-seit-2006/#:~{}:text=Zum%201.,2%2C95%20Cent%20pro%20Kilowattstunde (accessed on 18 January 2022).
53. Centrale Agrar-Rohstoff Marketing- und Energie-Netzwerk. Marktpreise Pellets. Available online: https://www.carmen-ev.de/service/marktueberblick/marktpreise-energieholz/marktpreise-pellets/ (accessed on 10 January 2022).
54. *AGFW I Der Effizienzverband für Wärme, Kälte und KWK e.V*; Statistik Fernwärme: Preisübersicht, Frankfurt am Main, Germany, 2020; Available online: https://www.agfw.de/index.php?eID=tx_securedownloads&p=345&u=0&g=0&t=1640739723&hash=a7afb9b043c6373adbd36969041eb8fe0ef7f864&file=/fileadmin/user_upload/Wirtschaft_u_Markt/markt_und_preise/Preisbildung-_Anpassung/2020_AGFW_Preisuebersicht_Webexemplar.pdf (accessed on 1 October 2021).
55. Breitkopf, A. Durchschnittlicher Preis für Methanol auf dem Europäischen Markt in den Jahren von 2012 bis 2021. Available online: https://de.statista.com/statistik/daten/studie/730823/umfrage/durchschnittlicher-preis-fuer-methanol-auf-dem-europaeischen-markt/ (accessed on 7 June 2021).
56. Roth, C. Natriumacetat Trihydrat, 25 kg, $\geq$99%, Ph. Eur., USP. Available online: https://www.carlroth.com/de/de/von-a-bis-z/natriumacetat-trihydrat/p/3856.5 (accessed on 7 June 2021).
57. Roth, C. Natriumformiat, 1 kg, $\geq$99%, p.a., ACS. Available online: https://www.carlroth.com/de/de/natriumsalze-na/natriumformiat/p/4404.3 (accessed on 7 June 2021).
58. Aldrich, S. Furfural for Synthesis: 98-01-1. Available online: https://www.sigmaaldrich.com/DE/en/product/mm/804012 (accessed on 26 January 2022).
59. Aldrich, S. Furfural, $\geq$98%, FCC, FG: W248908-25KG-K. Available online: https://www.sigmaaldrich.com/DE/en/product/aldrich/w248908 (accessed on 26 January 2022).
60. Aldrich, S. Furfural, Natural, $\geq$98%, FCC, FG: W248924-10KG-K. Available online: https://www.sigmaaldrich.com/DE/en/product/aldrich/w248924?gclid=Cj0KCQjwqp-LBhDQARIsAO0a6aIH_8uFG_r4gPO_Df_vJMdpAFRMJvyEiDR5Cfq0wiiAdlot-Vq77ioaAlkREALw_wcB (accessed on 26 January 2022).
61. Scott, G. *Marktpreise Plattformchemikalien*; AVA Biochem: Zug, Switzerland, 2021.
62. Zinke, H.; El-Chichakli, B.; Dieckhoff, P.; Wydra, S.; Hüsing, B. Bioökonomie für Die Industrienation: Ausgangslage für Biobasierte Innovationen in Deutschland Verbessern, Berlin. 2016. Available online: https://www.biooekonomierat.de/fileadmin/Publikationen/berichte/Hintergrundpapier_ISA_Vero__ffentlichung_2.pdf (accessed on 17 September 2021).
63. Carus, M.; Raschka, A.; Piotrowski, S. *Entwicklung von Förderinstrumenten für die Stoffliche Nutzung von Nachwachsenden Rohstoffen in Deutschland (Kurzfassung): Volumen, Struktur, Substitutionspotenziale, Konkurrenzsituation und Besonderheiten der Stofflichen Nutzung Sowie Entwicklung von Förderinstrumenten, Mai 2010 = The Development of Instruments to Support the Material Use of Renewable Raw Materials in Germany (Summary)*, 2., Geringfügig Überarb. Aufl., Juli 2010; Nova-Institut für Politische und Ökologische Innovation GmbH: Hürth, Germany, 2010; ISBN 9783981202731.
64. Rakotovao, M.; Gobert, J.; Brullot, S. Bioraffineries rurales: La question de l'ancrage territorial. *GeoCantemir* **2017**, *44*, 85–100. [CrossRef]
65. Froeschle, P.; ARENA2036: The Research Campus. The Innovation Platform for Mobility and Production of the Future. Available online: https://www.arena2036.de/en/ (accessed on 17 November 2021).
66. The European Bioeconomy Network. TALENT4BBI: Training Future Leaders 4 the European Bio-Based Industries. Available online: https://eubionet.eu/category/projects/open-innovation-platforms-and-facilities/ (accessed on 17 November 2021).

 *sustainability*

MDPI

Article

Resource Pressure of Carpets: Guiding Their Circular Design

Virginia Lama [1,2,3] , Serena Righi [1] , Brit Maike Quandt [2] , Roland Hischier [3] and Harald Desing [3,*]

1   CIRSA—Centro Interdipartimentale di Ricerca per le Scienze Ambientali, Alma Mater Studiorum—University of Bologna, Via dell'Agricoltura 5, 48123 Ravenna, Italy; virginia.lama2@unibo.it (V.L.); serena.righi2@unibo.it (S.R.)
2   Tisca Tischhauser AG, Sonnenbergstrasse 1, 9055 Bühler, Switzerland; m.quandt@tisca.com
3   Empa, Technology and Society Laboratory, Lerchenfeldstrasse 5, 9014 St. Gallen, Switzerland; roland.hischier@empa.ch
*   Correspondence: harald.desing@empa.ch

Abstract: When designing a product, many decisions are made that determine the environmental impacts that the product will eventually exert on our planet. Therefore, it is paramount to have considered the environmental performance already in the design phase. In this contribution, we showcase the application of the recently developed resource pressure (RP) method to assess the environmental sustainability of various carpet design alternatives. This method consists of qualitative guidelines and a quantitative indicator. With the Earth's carrying capacity as a reference, the product system is evaluated in relation to its consumption of primary resources and the final generation of waste. Several scenarios are developed by following the design guidelines provided by this method. Those scenarios aim at identifying the most promising circular strategies for reducing the products' resource pressure. To assess the validity of the RP method, the results are compared to a simplified LCA study. This comparison showed a close correlation for most of the considered impact categories. It confirms that the RP method can effectively predict environmental impacts across a wide range of impact categories, reducing the amount of necessary data and simplifying the calculations. It can therefore support designers in considering the environmental effects easily, from the beginning of the design process onward. Moreover, the simplicity of this method makes it attractive for application by practitioners who are not themselves experts in environmental assessments.

Keywords: circular economy; product design; carpet

 **check for updates**

Citation: Lama, V.; Righi, S.; Quandt, B.M.; Hischier, R.; Desing, H. Resource Pressure of Carpets: Guiding Their Circular Design. *Sustainability* **2022**, *14*, 2530. https://doi.org/10.3390/su14052530

Academic Editor: Ana Ramos

Received: 26 January 2022
Accepted: 18 February 2022
Published: 22 February 2022

Publisher's Note: MDPI stays neutral with regard to jurisdictional claims in published maps and institutional affiliations.

Copyright: © 2022 by the authors. Licensee MDPI, Basel, Switzerland. This article is an open access article distributed under the terms and conditions of the Creative Commons Attribution (CC BY) license (https:// creativecommons.org/licenses/by/ 4.0/).

1. Introduction

Today's environmental burdens are mostly caused by the extraction and usage of natural resources [1]. The safe limits for many vital Earth system processes have already been crossed [2,3], leading to the current climate crisis and biodiversity loss. Therefore, it is essential to develop solutions by which to preserve our primary resources.

The overall sustainability of a product, and, thus, the potential resource utilization efficiency, is determined to a large extent by its conception and design [4]. The European Commission report on the strategy for plastics in a circular economy (CE) shows that over 80% of the environmental impacts related to products are determined during the design stage [5]. Applying circular strategies to product design has the potential to significantly reduce the environmental impacts on our planet [6,7]. However, CE potentials for most sectors are still unclear, hindering investments into new circular business models and waste management infrastructures [8]. Moreover, the current structure, composition, and assembly of complex products such as carpets prevent closing these material cycles and, thus, pose challenges in reaching circularity. The carpet sector, however, poses great potential for circular innovation as Europe generates 1.6 million tons of post-consumer carpet waste every year, 60% of which is landfilled, 37–39% is incinerated, and only 1–3% is recycled [9].

Today, design guidance for making circular products and services environmentally sustainable is available either in the form of guidelines (e.g., [10,11]) or are based on simplified LCA approaches (e.g., [12]). Neither of the types of approaches is widely adopted in industry, as they either depend on specific knowledge or rely on time-consuming and complex procedures [7,13]. LCA is usually applied ex post, analyzing the environmental impacts of the finished products, and it has little influence on design decisions [13,14]. Furthermore, LCA was designed for analyzing the linear economy and there is a need to develop approaches for measuring circularity in a standardized way [15]. Circularity metrics, in contrast, describe the circularity of material flows only, missing their environmental (and social) implications [15–17]. Few studies compare the results obtained with design methods with ex post LCAs (e.g., [18,19]); however, such an assessment would be required for documenting the effectiveness of design methods.

For easy use at the company level, a scientifically sound but easy-to-apply method is required. In this contribution, we have applied the recently developed resource pressure (RP) method [7] to assess the environmental sustainability of different carpet designs and circular strategies. The RP method comprises qualitative guidelines and a quantitative indicator, with the aim of minimizing the consumption of limited resources. Respecting planetary boundaries [3], primary resource consumption is limited by the environmental impacts caused during extraction, processing, and disposal [20]. Product systems induce resource consumption both directly, through the consumption of primary materials, and indirectly, through the generation of final waste. Circularity strategies can reduce both the required primary material and the generated final losses. The RP method allows researchers to quantify on a case-by-case basis the effectiveness of circularity strategies on reducing primary resource consumption, as well as environmental impacts.

Today's carpet industry uses a wide range of fibers [21] for the creation of a complex and multi-component structure [22]. The basic structure for tufted and woven carpets are shown in Figure 1 and include: pile yarn (loop or cut pile), primary backing, a bonding agent and a secondary backing for the former (a), and pile yarn (loop or cut pile), binding chain, stuffer (filling chain), weft yarn, and a bonding agent for the latter (b). The spectrum of materials that can be used is large: it ranges from natural fibers, like cotton (CO) and wool (WO), to synthetic fibers like polyamide-6 (PA6) or polyamide-66 (PA66), polyester (PES), polypropylene (PP), and acrylics. Therefore, numerous possible material combinations within one product are employed [23].

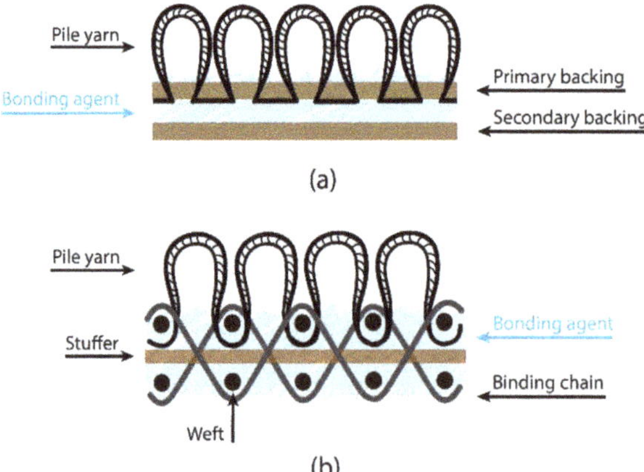

Figure 1. The basic structure of carpet for a tufted model (**a**) and a woven model (**b**).

PA6 and wool are primarily used as pile yarn for tufted and woven carpets. The bonding agent links and secures the face fibers to the primary backing (tufting) or the supporting yarn construction (weaving). For tufted products, the secondary backing gives further stability to the carpet structure.

Additional layers, such as foam backing, although not investigated here, provide further features, such as thermal and acoustic insulation [24].

Depending on the carpet's intended application, each of the components has a specific function and must, therefore, satisfy a set of requirements. Residential and commercial buildings require vastly different specifications [25]. Customer expectations influence both the visual design and the construction. While, e.g., wear resistance is more relevant in the commercial sector, comfort plays a large role in residential orders [25,26]. Other notable properties are insulation, wear resistance, acoustics, moisture resistance, color-fastness, light-fastness, and reflection [27]. An overall classification and rating of carpets, based on their features, can be found in the European standard for textile floor coverings, EN 1307:2014 [28].

2. Materials and Methods

The RP method is a tool for supporting design decisions, based on the utilization of resources, and its detailed description can be found in the recently published paper by Desing et al. [7]. The RP method quantifies the pressure exerted by the ecological resource budgets (ERB) [20,29,30] on the amount of a resource that is necessary to produce a product with a specific design. ERBs measure environmental impacts in relation to Earth system boundaries, originating from primary resource extraction and end-of-life (EoL) treatment. ERBs can be calculated either based on the ecological resource availability (ERA) method [20], if the absolute environmental performance of a product with regard to a defined resource consumption pattern is of concern, or based on the ecological resource potential (ERP) method [30], if the aim is to reduce the environmental impacts of a new design. Since this study focuses on design improvements, ERBs were obtained using the ERP method.

The RP method provides the designer with a tool allowing, at the same time, the reduction of the pressure on primary resources and the maximization of the utility of materials. As shown in Figure 2, the RP of a product depends on ERBs, together with product design parameters such as its mass ($m_{product}$), manufacturing losses (γ_m), product lifetime (t_L), primary material content (α'), recyclability (η_r) and cascadability (η_c). Those factors represent the essential elements for the calculation of the RP, according to Equation (1).

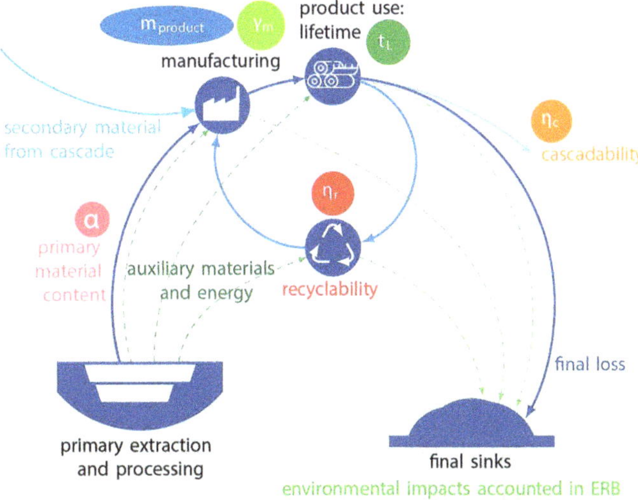

Figure 2. Overview of the resource pressure design method, reproduced and adapted from [4].

Thus far, the RP method has not been tested on complex products, such as carpets. Therefore, the present study applies the RP method to the development and evaluation of various carpet designs and compares their results with a simplified life-cycle assessment (LCA). LCA is a well-established but ex post methodology for the assessment of environmental performance. With the on-hand comparison, the potential benefits and drawbacks of both approaches are highlighted within the context of CE. The term "simplified" refers to the fact that the LCA was established, mainly using averaged data from the database *ecoinvent v3.6*, instead of modeling the processes with case-specific information. All calculations for RP and the simplified LCA were carried out using Microsoft® Excel®.

$$\tau = \frac{1}{2} \times \frac{m_{product}}{ERB} \times \frac{1}{t_L} \times (1 + \gamma_m) \times (1 + \alpha'(1 - \eta_r) - \eta_r - \eta_c) \tag{1}$$

3. Case Study

In collaboration with Tisca Tischhauser AG, a Swiss textile company, we assessed the environmental sustainability of different carpet designs. The various design scenarios are based on different circular strategies and demonstrate the application of the RP method.

Tisca Tischhauser AG is a full-service provider of high-quality textiles for indoor and outdoor use. The product range includes textile floor coverings, curtains, upholstery and decorative fabrics, as well as sports turf. Among others, woven and tufted carpets are produced, which were selected for this case study. For both types of carpets, different design scenarios are developed and evaluated in the present study. The two types differ significantly, both in the technique with which they are produced and also in their basic structure (see Figure 1). Due to this wide range of possibilities, it was possible to investigate different scenarios for both product groups. Environmental impacts and potential improvements in circularity and sustainability were considered separately for the two carpet types. Six to ten scenarios were developed for woven (W1–6) and tufted (T1–10) carpets, respectively. A simplified LCA was carried out for a subset of the selected scenarios and compared to the results of the RP method. The overall goal of the study was to determine the influence of design changes within each carpet type, based on the RP results.

In order to select a reference product for each of the two case studies, a screening process was carried out, comparing the available products from Tisca in each respective category. The carpet design scenarios were then developed in both categories by following the design guidelines provided in the RP method itself (Table 1) and converting those guidelines into technically feasible design choices for the respective carpet type. The latter process was supported by the expertise of the company's personnel.

For both methods (i.e., RP and LCA), the functional unit (FU) is equal to 500 m^2 of commercial floor covering, having a lifespan of 10 years. The system boundaries for the LCA are defined as "cradle-to-grave", referring to the carpet product system only, thus using the cut-off approach for secondary resources (Figure 3). This means that secondary material inputs enter the product system burden-free, i.e., without the environmental burdens related to their primary material production. Impacts for secondary materials are counted only in terms of their processing for the specific application. On the other hand, primary material inputs are evaluated, including the environmental impacts of raw material extraction. Environmental impacts for EoL treatment are considered for any processes that involve the carpet product system only; thus, this represents either incineration or recycling. Cascading and the cascaded product system are not included in the simplified carpet LCA.

Table 1. Carpet design scenarios (rows) for the woven (W1–6) and tufted (T1–10) types developed following the RP design guidelines (ticked columns).

	Choose Materials with Large ERB	Mass in Product ↓	Primary Material ↓	Recyclability ↑	Cascading ↑
Mass ↓ (W1 andT1)		✓			
Primary material ↓ (W2 and T2)			✓		
Mass ↓ + primary material ↓+ RE (W3 and T3)		✓	✓		
Material choice with large ERB (W4 and T4–5)	✓				
Recyclability ↑ (T6)				✓	
Recyclable carpet design (W5 and T7)					
Design + recyclability ↑ + primary material (T8)				✓	
Design + recyclability ↑ + secondary material (W6 and T9)				✓	✓
Cascading ↑ (T10)					✓

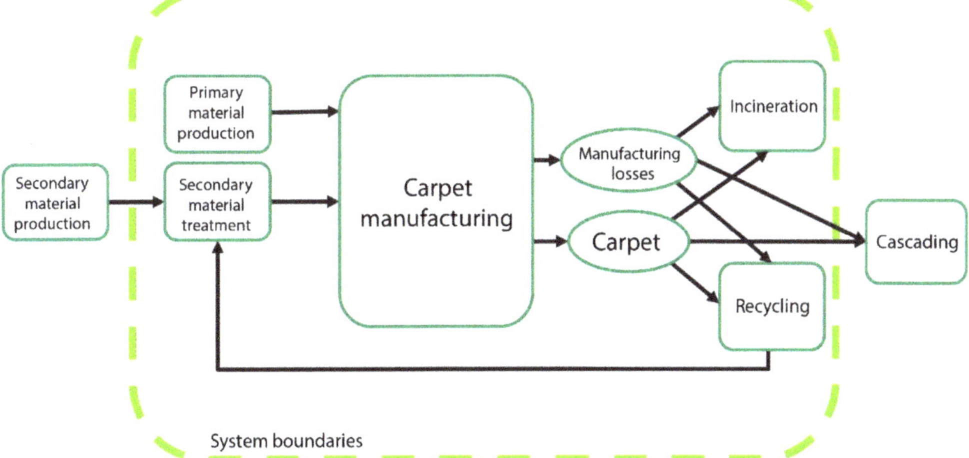

Figure 3. Carpet product system LCA: the green dashed box represents the system boundaries for the LCA study; therefore, the environmental impacts were only evaluated for the unit processes included within the system boundaries. Elements in ovals represent the product outputs of the carpet's manufacturing process and do not contribute to any environmental impacts themselves.

4. Results

4.1. Screening Process

The screening process across the entire range of the company's products yielded the following general tendencies for carpet design characteristics, regarding resource pressure:

- Pile mass ↑: RP value ↑.
- RP (wool carpets) > RP (synthetic pile material).
- RP (PA66 pile material) > RP (PA6 pile material).
- RP (Commercial sector) < RP (Residential sector).

From the full range of products, one woven (W0) and one tufted (T0) carpet, both for commercial application and made from PA6 as a pile material, with average RP values, were selected as the respective reference products. The spread in terms of RP across the different products ranged from 22% lower than the selected product and 68% higher for the woven, 32% lower, and 35% higher for tufted carpets, respectively. Choosing an average product allows for a scenario involving a lighter product, as well as other designs that might lead to improvements in the carpet's sustainability.

4.2. RP and LCA Results

The RP method aims at giving designers guidance on potential strategies to improve the sustainability of their products. Compared to the reference product, every design scenario can then be evaluated in terms of reduced resource pressure. Figure 4 shows the relative resource pressure of all considered design scenarios for woven and tufted carpets. It is essential to note that the numerical value of RP itself is irrelevant for the comparison of different design scenarios. It is only when the RP is calculated with ERA budgets that its value represents the fraction of global sustainable resource availability consumed by the production of a FU. The RP serves here as a relative score, comparing and ranking the design alternatives. Thus, it helps in identifying the best- and worst-case scenarios. We compared the RP for woven and tufted carpet scenarios separately, due to their different structures and properties, as they exhibit different potentials for design change.

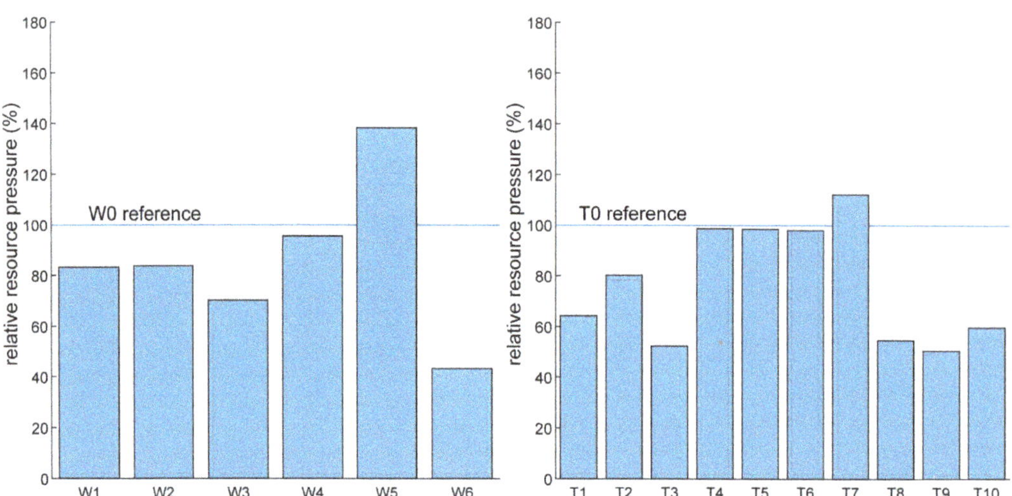

Figure 4. Relative resource pressure for woven (**left**) and tufted (**right**) carpets and the corresponding design scenarios.

The RP results show that scenarios involving the choice of a material with large ERBs (i.e., W4, T4 and T5) do not significantly improve the RP result compared to the reference scenario. This is because the carpet's structural elements that were subjected to this material substitution represent just a marginal part of the overall carpet, being indeed an insignificant weight-fraction of the overall carpet. Material substitutions only result in a large benefit when their share in the product and/or the difference in ERPs is large. If all the structural components of the carpet, including the pile material, were converted to materials with a larger ERP, greater improvement in the RP result could be obtained. However, material substitutions in the design are limited by the functional requirements, applications, and aesthetics, which are of crucial importance for success in the market. Moreover, the implementation of circular EoL strategies like recycling or cascading might require the use of a specific material and, therefore, represent a further limitation in the spectrum of potential material to be used.

On the other hand, reducing the mass in the product has a major impact on lowering the RP of the reference product. These scenarios (W1, T1) show much lower RP compared to the reference scenarios, as well as to the previously mentioned ones (i.e., W4 and T4–5). Carpet mass reduction with respect to the reference product is assumed to be according to the feasibility of this operation within the two types of carpets, i.e., the grade of reduction is different between woven (11% mass reduction) and tufted (21%) carpet design scenarios, due to the substantial difference in their structure. In both cases, the mass reduction is determined by the difference between the reference product and the lightest available product in the company's portfolio. Therefore, it is not surprising that the difference between the RP result of the tufted carpet compared to its reference is of a greater magnitude than in the case of the woven carpet, but this is rather a function of which reference product was chosen.

While reducing mass leads to an intuitive reduction in RP, reducing the primary material content requires a more detailed interpretation. In these scenarios (i.e., W2 and T2), the environmental impacts related to the production of primary material are avoided by using secondary material, while the mass of the product remains unchanged. Furthermore, additional impacts related to the energy involved in the processing of secondary material are considered. For this reason, the RP value for these scenarios, even if it is advantageous with respect to the reference products, does not contribute to better performance when compared to the mass-reduction scenarios. It must be noted that there is no significant difference between W1 and W2, given the limited mass reduction that could be attained for woven carpets. Combining the mass reduction with the reduction of primary material, and assuming 100% renewable energy for the required processing of secondary material (W3 and T3), results in the lowest RP among the scenarios W0–3 and T0–3.

For recycling scenarios, a chemical recycling process is considered. Material can only be recycled when it fulfills strict requirements regarding its composition and contamination. The multi-material composition of the selected reference products does not allow the chemical recycling of the entire carpet but only of a small portion, in the case of the tufted carpet (T6), which leads to a slight improvement in the RP value compared to its reference. Consequently, a recyclable carpet was designed for the woven and tufted types of carpet that fulfill the requirements of the chemical recycling process. The structure of the so-called "recyclable carpet design" (i.e., W5 and T7) had to be changed significantly from its reference product. The RP of this recyclable carpet design with its different composition, which does not get recycled, is significantly higher than the reference product. Only when they are actually recycled (W6, T8, and T9) do they perform best.

As recycling 100% of the carpet is physically impossible, primary input material is still necessary, although in reduced quantities. Replacing this remaining primary material input with secondary material from other product systems (e.g., PET bottles) reduces the RP further (T9 in comparison to T8).

Results regarding the product cascading (T10), e.g., by using EoL carpets for insulation purposes, can reduce the RP, but these results need to be carefully interpreted, as will be explained further in the discussion section.

To validate the RP, a comparison between RP and LCA results was undertaken and is shown in Figure 5. RP, as a single-score indicator, provides clear guidance on which design scenario is the best to implement. LCA, on the other hand, provides information on a wide range of impact categories, providing a more comprehensive understanding of the environmental impacts of each scenario. This, however, adds complexity to the interpretation of the results and prevents a straightforward and unambiguous ranking. By using these two different approaches for the assessment of the environmental impacts of a product, we aimed to find out how far the "simpler" RP method could predict the more comprehensive results of an LCA. The prediction accuracy (covariance) of impacts across multiple LCA impact categories with RP is further reported in Table 2.

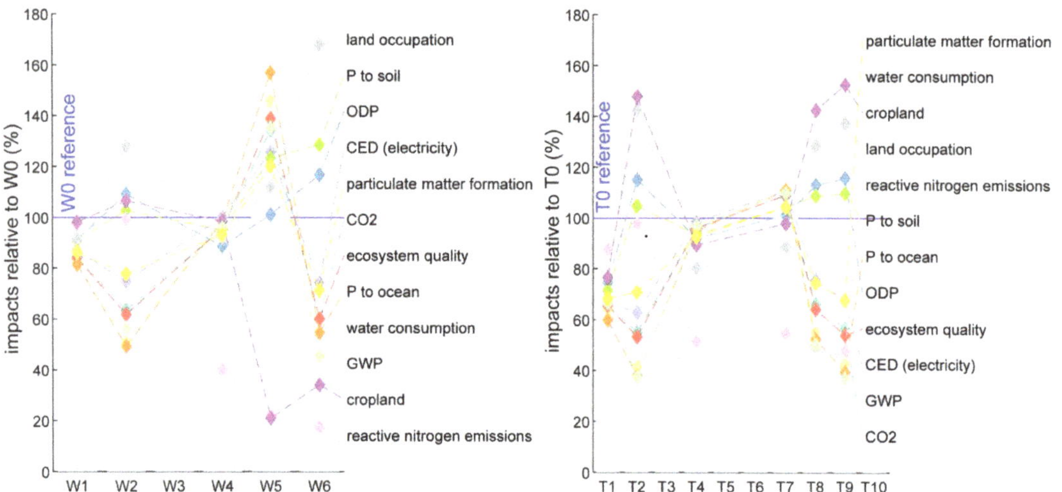

Figure 5. RP and LCA results for the woven (**left**) and tufted (**right**) carpet scenarios. ODP = ozone depletion potential, CED = cumulative energy demand, GWP = global warming potential, P = phosphorus, CO_2 = carbon dioxide.

Table 2. Pearson covariance among RP and LCA impact categories for both woven and tufted CDSs. Results are distinguished into three categories, based on the type of correlation: green refers to a strong positive correlation (>0.8), orange to a weak but still positive correlation (0 < x < 0.8), and red to none or a negative correlation (<0). ODP = ozone depletion potential, CED = cumulative energy demand, GWP = global warming potential, P = phosphorus, CO_2 = carbon dioxide.

	Woven CDSs	Tufted CDSs
CO_2	0.92	0.88
GWP	0.92	0.87
CED (electricity)	0.90	0.85
Ecosystem quality	0.91	0.88
Water consumption	0.94	0.86
Particulate matter formation	0.94	0.91
P to ocean	0.87	0.87
P to soil	−0.32	0.00
Reactive nitrogen emissions	0.20	0.19
ODP	−0.61	−0.23
Land occupation	−0.73	−0.54
Cropland use	0.13	−0.55

Figure 5 shows that RP (bars) corresponds with many, but not all impact categories (lines). Furthermore, some scenarios show diverging results for different impact categories, making their interpretation more complex.

For both woven and tufted carpets, the highest difference across impact categories is shown by scenarios involving the use of secondary material and recycling, which both include the recycling process's energy-related impacts. This means that this type of process-related impact requires particular attention in scenario modeling and the subsequent result interpretation operations through the two methodologies.

The correlation between the RP and the LCA results for the different impact categories can be expressed using Pearson covariance. Its results (summarized in Table 2) show a close correlation among RP and CO_2 emissions, GWP, CED, ecosystem quality, water consumption, particulate matter formation, and the addition of phosphorus (P) to the ocean. All these impact categories identify W1, W2, W6, and T1, T2, T9 scenarios as the most promising strategies by which to improve the carpet's sustainability.

In contrast, those results related to the other impact categories, i.e., P to the soil, reactive nitrogen (N) emissions, ODP, land occupation, and cropland use, show inversely correlated or uncorrelated results. For example, the LCA results for CO_2 emissions, GWP, CED, ecosystem quality, water consumption, particulate matter formation, and the addition of P to the ocean of scenarios W6 and T8–9 follow the RP results; however, for the impact categories of P to the soil, ODP, land occupation and cropland, impacts are even worse than the reference products (W0 and T0). This is because, in the ERP calculation, the most pressing environmental category becomes limiting for each resource, which, in most cases, is CO_2. As such, any impact category correlated with the limiting boundary can be predicted well, while others are not. This is, however, not a problem, as the most limiting boundary category is identified in the ERP procedure.

5. Discussion and Outlook

This study shows an application of the resource pressure method to a complex product, with a carpet consisting of several materials. The method itself is applied in a straightforward manner, with the data needed mostly covered by key performance indicators (KPIs) that are typically monitored at the company. For instance, a rejection rate in manufacturing gives additional information on the average material cost of a produced product apart from its material weight. The main challenge of the application of the method was the initial data collection and compilation inside and outside the company, as the method requires data in a new format. Nevertheless, after having defined the relevant parameters and their interconnectedness, the method can be used by product designers and technical specialists with minimal training.

It is true, however, that the results obtained through the RP method showed, in some cases, a significant divergence from the results of the LCA study. In addition, LCA results for different impact categories also differ, making it difficult to give absolute preference to one design alternative over the others. In addition, LCA has been performed only for all impact categories included in the ERP method [30], acknowledging that other impact categories may be relevant as well. A weighting procedure for the LCA impact categories could be used to obtain guidance for the prioritization of design scenarios. However, such a weighting procedure would require defining how relevant the impact categories are for this specific case study. An impact category can be of more or less relevance to the subject under investigation. Weighting is the subject of an ongoing debate among LCA practitioners because of the subjectivity that this operation involves. In contrast, the RP method provides a single score result, despite considering multiple environmental impact categories for the calculation of ERBs. The most limiting boundary and, therefore, the most relevant impact category is limiting the ERB of each material that comprises the product. Throughout this procedure, all impact categories can be taken into account and the one most relevant with respect to Earth system boundaries is automatically selected. The limiting boundary for all the materials comprising the carpet is CO_2 emissions, and the RP and LCA results for the related impact categories are strongly correlated (>0.85, see Table 2). This way, even if not all indicators point in the same direction, the RP result should be followed, keeping in mind that there are other impact categories that may suggest something different. Thus, an added, simplified LCA might help with choosing between two similar RP scenarios.

The scope of this RP-based study is still limited to a scenario of cradle-to-gate plus the additional EoL impacts. Environmental impacts from the use phase (e.g., through cleaning or wear) have been neglected in the present calculations. Furthermore, as we have selected the so-called "cut-off" system model, impacts occurring in other but related product systems (e.g., the product system using cascaded material from the carpet) are excluded. Therefore, future studies should try to include these systems as well.

The study is further limited to a chosen set of scenarios. Further scenarios, which could be of relevance for carpets, are possibilities for the cascading use of waste carpets, life extensions, or the use of alternative materials (e.g., bio-based polymers [31]). Cascadability, for instance, is not only a function of the material's properties and the product requirements but also of the available market. Thus, only a fraction of the material can be considered for cascading, the fraction for which an actual and large enough market exists [7]. Reported alternative applications for carpet waste include equestrian surface materials [32], low-cost composite tooling materials [33], eco-efficient lightweight concrete [34], noise barriers for highway and infrastructure applications [35], and injection-molded thermoplastics [32]. Studies investigating the potential application of cascaded carpet in structural composites have been extensively reviewed by Sotayo et al. [36], showing a vast range of alternative processes for carpet waste to be diverted from EoL treatment. Using carpets as an alternative raw material in the production of acoustic panels, for example, is already feasible [35,37–39]. Cascading, however, should be evaluated together with the up- and downstream applications. Detailed studies need to be carried out, comparing the resulting environmental impacts of the treatment of carpets to produce such a finished product, and verifying whether this would represent an advantage over the current state of the art. Another interesting aspect that should be taken into account is the end of life of the "secondary product". If the acoustic panels could be recycled at the end of their life, thus allowing the recovery of at least part of the raw materials to produce new products, this would be a mechanism perfectly in line with the circular economy. On the other hand, if these products were not recyclable, this would, in a way, be prolonging the lifespan of the materials and, thus, delaying but not preventing the final loss.

The resource pressure method can serve as an absolute environmental indicator [7], assessing the absolute environmental sustainability of an item with regard to Earth system boundaries. However, resource budgets need to be calculated alongside the ERA method [20], as these are absolutely sustainable resource budgets under a specified allocation approach. To be useful in guiding design decisions [7,20], "desirable" resource budgets need to be defined based on a set of societal values, guaranteeing the fulfillment of basic needs for a decent life for all. Using these desirable ERA budgets will allow the setting of targets for the reduction of primary resource consumption for society and regarding specific products [40].

6. Conclusions

The circular design of carpets can reduce resource pressure significantly. However, a design for recycling alone leads to a higher RP; only when an item is recycled can it yield substantial improvements. This requires companies, consumers, and countries to ensure that circular strategies are implemented to their full capacity. It is not only new business models that can help but also regulations and take-back schemes. The organization of reverse logistics and ownership questions need to be addressed. More transparent supply chains can enable better recycling [41].

The RP method has proved to be applicable to complex products and effective in guiding design decisions. In comparison with LCA, RP correlates with many impact categories;

however, this is not the case with all categories, thus requiring a careful interpretation of the obtained results.

The RP method still showed significant potential in providing design guidance in the context of CE, successfully integrating all the relevant design parameters in a very simple and straightforward structure. The RP method constitutes an initial step in promoting the transition toward CE, enabling designers to include environmental considerations in the product design process. The present work has shown that recycling and using secondary material input are the most promising strategies to improve carpet sustainability. Further evaluations, also including the economic and social aspects related to CE, need to be carried out in order to assess the feasibility of bringing the identified circular strategies into practice. From a purely methodological perspective, the RP method showed several advantages in supporting the decision-making process. Unlike LCA, it avoids the need to investigate which impact categories are relevant or not in the context of the case study, still taking into account the most relevant impact categories in relation to the boundaries. Therefore, the RP method proved to be more suitable in the context of product design from a CE perspective. However, it could still be associated with an LCA study to obtain more detailed information across a wider range of impact categories. Future progression in the RP method through more comprehensive data, tools for estimating the necessary parameters, and potential adaptations to special cases may result in its establishment as an independent method, therefore providing scientific guidance in the design phase to a wider spectrum of users.

Author Contributions: Conceptualization, H.D. and B.M.Q.; methodology, H.D.; validation, H.D. and R.H.; formal analysis, V.L.; investigation, V.L. and H.D.; resources, B.M.Q. and R.H.; data curation, V.L. and B.M.Q.; writing—original draft preparation, V.L.; writing—review and editing, all authors; visualization, H.D. and V.L.; supervision, H.D. and B.M.Q.; project administration, B.M.Q. and R.H.; funding acquisition, R.H. and V.L. All authors have read and agreed to the published version of the manuscript.

Funding: V.L. was funded by the EIT Climate-KIC Master Label Program and Tisca Tischhauser AG. H.D. was funded by the Swiss National Science Foundation (SNSF) in the framework of the project "LACE—Laboratory for an Applied Circular Economy" (grant number 407340_172471) as part of the National Research Program "Sustainable Economy: resource-friendly, future-oriented, innovative" (NRP 73).

Institutional Review Board Statement: Not applicable.

Informed Consent Statement: Not applicable.

Data Availability Statement: The data presented in this study may be available on request from the corresponding author. The data are not publicly available due to their confidential character.

Acknowledgments: V.L. acknowledges Nick Tischhauser, Volkmar Reichmann, Annett Waibel, Oliver Mildner, Pascal Weber and Tina Korthing for the support in providing all the technical data necessary to carry out the study presented in this paper.

Conflicts of Interest: M.Q., as an employee of Tisca, has collected data, edited the manuscript, and vetoed the inclusion of data if that data pertains to innovation that is yet to be published and marketed. No data was changed or misrepresented. V.L. is similarly bound by an NDA. All other authors declare no conflict of interest.

References

1. United Nations Environment Programme—UNEP. *Global Resource Outlook*; UNEP: Nairobi, Kenya, 2019.
2. Rockström, J.; Steffen, W.; Noone, K.; Lambin, E.; Lenton, T.M.; Scheffer, M.; Folke, C.; Schellnhuber, H.J.; De Wit, C.A.; Hughes, T.; et al. Planetary Boundaries: Exploring the Safe Operating Space for Humanity. *Ecol. Soc.* **2009**, *14*, 1–33. [CrossRef]
3. Steffen, W.; Richardson, K.; Rockström, J.; Cornell, S.E.; Fetzer, I.; Bennett, E.M.; Biggs, R.; Carpenter, S.R.; de Vries, W.; de Wit, C.A.; et al. Planetary Boundaries: Guiding Human Development on a Changing Planet. *Science* **2015**, *347*, 1259855. [CrossRef]
4. Reuter, M.A.; van Schaik, A.; Gutzmer, J.; Bartie, N.; Abadías-Llamas, A. Challenges of the Circular Economy: A Material, Metallurgical, and Product Design Perspective. *Annu. Rev. Mater. Res.* **2019**, *49*, 253–274. [CrossRef]

5. European Commission. A European Strategy for Plastics in a Circular Economy. 2018. Available online: https://eur-lex.europa. eu/legal-content/EN/TXT/?qid=1516265440535&uri=COM:2018:28:FIN (accessed on 18 December 2020).

6. Desing, H.; Brunner, D.; Takacs, F.; Nahrath, S.; Frankenberger, K.; Hischier, R. A Circular Economy within the Planetary Boundaries: Towards a Resource-Based, Systemic Approach. *Resour. Conserv. Recycl.* **2020**, *155*, 104673. [CrossRef]

7. Desing, H.; Braun, G.; Hischier, R. Resource Pressure—A Circular Design Method. *Resour. Conserv. Recycl.* **2021**, *164*, 105179. [CrossRef]

8. Wilts, H. *Germany on the Road to a Circular Economy?* Friedrich Ebert Stiftung: Bonn, Germany, 2016.

9. Deutsche Umwelthilfe, e.V. *Swept under the Carpet: The Big Waste Problem of the Carpet Industry in Germany*; Environemntal Action Germany: Radolfzell, Germany, 2017.

10. Bovea, M.D.; Pérez-Belis, V. Identifying Design Guidelines to Meet the Circular Economy Principles: A Case Study on Electric and Electronic Equipment. *J. Environ. Manag.* **2018**, *228*, 483–494. [CrossRef]

11. Toxopeus, M.E.; van den Hout, N.B.; van Diepen, B.G.D. Supporting Product Development with a Practical Tool for Applying the Strategy of Resource Circulation. *Procedia CIRP* **2018**, *69*, 680–685. [CrossRef]

12. Broeren, M.L.M.; Molenveld, K.; van den Oever, M.J.A.; Patel, M.K.; Worrell, E.; Shen, L. Early-Stage Sustainability Assessment to Assist with Material Selection: A Case Study for Biobased Printer Panels. *J. Clean. Prod.* **2016**, *135*, 30–41. [CrossRef]

13. Brundage, M.P.; Bernstein, W.Z.; Hoffenson, S.; Chang, Q.; Nishi, H.; Kliks, T.; Morris, K.C. Analyzing Environmental Sustainability Methods for Use Earlier in the Product Lifecycle. *J. Clean. Prod.* **2018**, *187*, 877–892. [CrossRef]

14. Desing, H. *Product and Service Design for a Sustainable Circular Economy*; ETH Zürich: Zürich, Switzerland, 2021.

15. Rigamonti, L.; Mancini, E. Life Cycle Assessment and Circularity Indicators. *Int. J. Life Cycle Assess.* **2021**, *26*, 1937–1942. [CrossRef]

16. Haupt, M.; Hellweg, S. Measuring the Environmental Sustainability of a Circular Economy. *Environ. Sustain. Indic.* **2019**, *1–2*, 100005. [CrossRef]

17. Corona, B.; Shen, L.; Reike, D.; Rosales Carreón, J.; Worrell, E. Towards Sustainable Development through the Circular Economy— A Review and Critical Assessment on Current Circularity Metrics. *Resour. Conserv. Recycl.* **2019**, *151*, 104498. [CrossRef]

18. Schöggl, J.-P.; Baumgartner, R.J.; Hofer, D. Improving Sustainability Performance in Early Phases of Product Design: A Checklist for Sustainable Product Development Tested in the Automotive Industry. *J. Clean. Prod.* **2017**, *140*, 1602–1617. [CrossRef]

19. Suppipat, S.; Teachavorasinskun, K.; Hu, A.H. Challenges of Applying Simplified LCA Tools in Sustainable Design Pedagogy. *Sustainability* **2021**, *13*, 2406. [CrossRef]

20. Desing, H.; Braun, G.; Hischier, R. Ecological Resource Availability: A Method to Estimate Resource Budgets for a Sustainable Economy. *Glob. Sustain.* **2020**, *3*, e31. [CrossRef]

21. Chaudhuri, S.K. 2—Structure and properties of carpet fibres and yarns. In *Advances in Carpet Manufacture*, 2nd ed.; Goswami, K.K., Ed.; The Textile Institute Book Series; Woodhead Publishing: Sawston, UK, 2018; pp. 17–34.

22. Wang, Y.; Zhang, Y.; Polk, M.; Kumar, S.; Muzzy, J. Recycling of Carpet and Textile Fibers. In *Plastics and the Environment*; Andrady, A.L., Ed.; John Wiley & Sons: New York, NY, USA, 2005; pp. 697–725. ISBN 9780471721550.

23. The Carpet and Rug Institute (CRI). *The Carpet Primer*; The Carpet and Rug Institute, Inc.: Dalton, GA, USA, 2012.

24. Ege Carpets. *Carpet Handbook*; Ege Carpets. egecarpets, Print.: Herning, Denmark, 2018.

25. Whitefoot, D. 1—Carpet types and requirements. In *Advances in Carpet Manufacture*; Goswami, K.K., Ed.; Woodhead Publishing Series in Textiles; Woodhead Publishing: Sawston, UK, 2009; pp. 1–18. ISBN 978-1-84569-333-6.

26. Baranwal, B. 17—Classification of carpets. In *Advances in Carpet Manufacture*, 2nd ed.; Goswami, K.K., Ed.; The Textile Institute Book Series; Woodhead Publishing: Sawston, UK, 2018; pp. 467–483.

27. Ege Carpets. *The Architect's Guide to Choosing the Right Carpet*; Ege Carpets Online Guide: Herning, Denmark, 2019.

28. *EN 1307:2014*; Textile Floor Coverings Classification. European Committee for Standardization (CEN): Brussels, Belgium, 2014.

29. Desing, H.; Widmer, R.; Beloin-Saint-Pierre, D.; Hischier, R.; Wäger, P. Powering a Sustainable and Circular Economy—An Engineering Approach to Estimating Renewable Energy Potentials within Earth System Boundaries. *Energies* **2019**, *12*, 4723. [CrossRef]

30. Desing, H.; Braun, G.; Hischier, R. Ecological Resource Potential. *MethodsX* **2020**, *7*, 101151. [CrossRef]

31. Ivanović, T.; Hischier, R.; Som, C. Bio-Based Polyester Fiber Substitutes: From GWP to a More Comprehensive Environmental Analysis. *Appl. Sci.* **2021**, *11*, 2993. [CrossRef]

32. Jain, A.; Pandey, G.; Singh, A.K.; Rajagopalan, V.; Vaidyanathan, R.; Singh, R.P. Fabrication of Structural Composites from Waste Carpet. *Adv. Polym. Technol.* **2012**, *31*, 380–389. [CrossRef]

33. Mishra, K.; Das, S.; Vaidyanathan, R. The Use of Recycled Carpet in Low-Cost Composite Tooling Materials. *Recycling* **2019**, *4*, 12. [CrossRef]

34. Fashandi, H.; Pakravan, H.R.; Latifi, M. Application of Modified Carpet Waste Cuttings for Production of Eco-Efficient Lightweight Concrete. *Constr. Build. Mater.* **2019**, *198*, 629–637. [CrossRef]

35. Mishra, K.; Vaidyanathan, R.K. Application of Recycled Carpet Composite as a Potential Noise Barrier in Infrastructure Applications. *Recycling* **2019**, *4*, 9. [CrossRef]

36. Sotayo, A.; Green, S.; Turvey, G. Carpet Recycling: A Review of Recycled Carpets for Structural Composites. *Environ. Technol. Innov.* **2015**, *3*, 97–107. [CrossRef]

37. Lakshminarayanan, K. *Scaling Up of Manufacturing Processes of Recycled Based Carpet Composites*; Oklahoma State University: Stillwater, OK, USA, 2011.
38. Asdrubali, F. Survey on The Acoustical Properties of New Sustainable Materials for Noise Control. In *Proceedings of Euronoise*; European Acoustics Association: Tampere, Finland, 2014.
39. Pan, G.; Zhao, Y.; Xu, H.; Ma, B.; Yang, Y. Acoustical and Mechanical Properties of Thermoplastic Composites from Discarded Carpets. *Compos. Part B Eng.* **2016**, *99*, 98–105. [CrossRef]
40. Desing, H.; Braun, G.; Hischier, R. The Resource Reduction Index—Evaluating Product Design's Contribution to a Sustainable Circular Economy. In Proceedings of the European Roundtable on Sustainable Consumption and Production 2021, Graz, Austria, 8–10 September 2021.
41. Braun, G.; Som, C.; Schmutz, M.; Hischier, R. Environmental Consequences of Closing the Textile Loop—Life Cycle Assessment of a Circular Polyester Jacket. *Appl. Sci.* **2021**, *11*, 2964. [CrossRef]

 *sustainability*

MDPI

Article

The Mechanism of Forming the Strategic Potential of an Enterprise in a Circular Economy

Aleksandra Kuzior [1,*], Olena Arefieva [2], Zarina Poberezhna [2,*] and Oleksiy Ihumentsev [2]

[1] Faculty of Organization and Management, Silesian University of Technology, 26 Roosevelt Str., 41-800 Zabrze, Poland

[2] Faculty of Economics and Business Administration, National Aviation University, 1, Liubomyra Huzara Ave., 030580 Kyiv, Ukraine; lena-2009-19@ukr.net (O.A.); aleks7536@gmail.com (O.I.)

* Correspondence: aleksandra.kuzior@polsl.pl (A.K.); zarina_www@ukr.net (Z.P.)

Abstract: In the framework of this study, significant features of the formation of the strategic potential of the enterprise in a circular economy are identified. The characteristics and elements of the strategic potential of the enterprise, which can ensure its integrity and continuity of operations, are highlighted. The authors conducted and analyzed a theoretical review of the concept of the "circular economy" and its impact on business and resource conservation and environmental protection. The conditions for the transition to a circular economy at the macro level are formed. The key stages of ensuring the strategic potential of the enterprise, taking into account the internal and external environmental factors, are highlighted. The authors forecast the volume and dynamics of waste until 2027 using the Cobb–Douglas function. The mechanism of the formation of the strategic potential of the enterprise in the conditions of a circular economy is offered. This mechanism provides for the potential compliance with the strategic goals of the enterprise, as well as the rationality and balance of structural elements. Assessing the compliance of strategic potential with the developed strategy allows decisions to be made on the implementation of measures to meet the objectives of the enterprise, or to search for opportunities and reserves to improve its level. A set of measures aimed at the effective implementation of the proposed mechanism and the results of resource-efficient production is developed.

Keywords: potential; strategic potential; circular economy; mechanism of formation of enterprise strategic potential; strategic management

 check for
updates

Citation: Kuzior, A.; Arefieva, O.; Poberezhna, Z.; Ihumentsev, O. The Mechanism of Forming the Strategic Potential of an Enterprise in a Circular Economy. *Sustainability* **2022**, *14*, 3258. https://doi.org/10.3390/su14063258

Academic Editors: Ana Ramos and Donato Morea

Received: 25 December 2021
Accepted: 5 March 2022
Published: 10 March 2022

Publisher's Note: MDPI stays neutral with regard to jurisdictional claims in published maps and institutional affiliations.

Copyright: © 2022 by the authors. Licensee MDPI, Basel, Switzerland. This article is an open access article distributed under the terms and conditions of the Creative Commons Attribution (CC BY) license (https://creativecommons.org/licenses/by/4.0/).

1. Introduction

The strategic priority of the sustainable development of the state is to form a qualitatively new model of the national economy. It should be based on the symbiosis of the circular and the ecological economy, promoting the use of local resources to meet the needs of the economy and the formation of closed material and resource cycles. Circularity is one of the forms of dynamic socioeconomic system development at different levels of management.

According to Ansoff, the strategy of the enterprise, which relates to defining the goals and objectives of the organization and ensuring that its relations with the external environment meet its internal capabilities, will allow economic stability in the market [1].

The substantive aspect of the theory of strategic management emphasizes that enterprises operating in the same external environment develop differently, and have different successes depending on the content of the strategy and strategic potential implemented [2]. Therefore, an effective business management strategy is the key to success.

The definition of strategic potential is based on a systematic approach to considering the conditions and results of the functioning of the enterprise as an open system. According to this approach, the enterprise is considered to be a system of resources, the interaction of which determines the achievement of the results. The potential capabilities of the enterprise in the effective use of resources characterize the strategic potential of the enterprise.

When forming the strategic potential of the enterprise, it is necessary to proceed from the fact that its structure is a certain interdependent set of local potentials, i.e., the potential of each type of resource that ensures the most effective long-term goals and strategic directions of enterprise development. For each strategic direction, a common vision is formed, which reflects its degree of attractiveness to the company compared to others. Further comparison will make it possible to choose the most attractive strategic direction and most effective type of enterprise strategy, taking into account the state and trends of the strategic potential [3].

In the context of limited natural resources, the peculiarity of the circular economy as an objective basis for the functioning of the economic environment is that the waste products are not garbage, but rather are useful resources for the output of an innovative product, created by business entities involved in innovative cooperation. The symbiosis of an ecological, circular economy and strategic potential needs the development of the mechanism of interaction between the subjects of the circular economy and the tools to ensure this interaction. Therefore, in our opinion, the formation of the mechanism to ensure the enterprise strategic potential should focus not on the individual components of the system, but on the principles of its construction and features of its management, allowing the building of a rational strategy to find ways to ensure adaptability and flexibility in a transformational period.

2. Literature Review

The modern market environment is characterized by the instability and unpredictability of events, insufficiently effective economic legislation, and a lack of developed infrastructure. These, among other factors, make it impossible for companies to function normally and lead to the need to formulate alternative strategies.

Scientists from various research centers, such as Ansoff [1], Kozenkov [2], Strickland, Thompson [3], Shershneva [4], and others, have covered the problems of strategic management in their research. Strategic management is more often discussed in the context of modern technologies supporting decision-making processes [5–13]. However, despite the achievements of the scientists, many theoretical and practical issues in this field of knowledge are still insufficiently solved, so the study of the issues of strategic potential management is important and necessary.

Research on the formation of a circular economy has been conducted by many scientists [14–29]. From the point of view of the implementation of the circular economy, the innovative potential of enterprises is also significant [30–32]. The analysis of the results of research on the circular economy allowed us to determine the lack of a unified approach to the formation of interactions between the circular and ecological economies, which determined the relevance and practical significance of the chosen field of research. The generalization of the results of scientific research on the problems of the circular economy made it possible to conclude that the circular economy model is a system that, firstly, is based on the renewal of resources, and secondly, in which the input external flows in the form of resources do not have a negative impact on society and the environment.

According to Bogatskaya, the strategic potential of an enterprise is a totality of the available resources and capabilities (abilities) for the development and implementation of the enterprise strategy [33]. However, according to the authors, the strategic potential of the enterprise is characterized not only by the available resources, but also by the potential capabilities of the enterprise to improve the efficiency of their use. Thus, Gedroyts characterizes strategic potential as the economic capabilities of an organization that can be used to achieve strategic goals [34]. This is why managing the process of forming the strategic potential of an enterprise under the conditions of instability and unpredictability of events is extremely important, because it allows the analysis of the influence of external and internal environmental factors, and provides opportunities for the elimination of potential threats in a circular economy.

Numerous scientific publications are devoted to the methodological application of various aspects of the cyclic development of economic processes. The idea is not new. It first appeared in the considerations of Kenneth E. Boulding in 1966 [35]. The term "circular economy" was used in the study by Kneese in 1988 [36]. In the program document of the concept of sustainable development (Agenda 21, 1992), the idea of recycling was promoted [37]. The importance of scientific advice in achieving the goals of sustainable development was emphasized [38]. The circular economy concept was presented by the European Union in 2014. A new plan in this area was outlined by the EU in 2020, outlining a vision of a climate-neutral and competitive economy based on a circular economy [39]. In this article, the circular economy will be presented in symbiosis with the ecological economy and the strategic potential of the company. Table 1 shows the views of the main scholars who have had a great influence on the characterization of the concept of the circular economy in the context of strategic potential.

Table 1. Basic interpretations of the concept of a "circular economy".

Author	Definition of "Circular Economy"	Source
Pakhomova, Rikhter, and Vetrova (2017)	Recovery and closed economy. Characteristic minimization of consumption of primary raw materials and amount of processed resources, with simultaneous reduction of areas occupied by corresponding landfills and unorganized landfills	[40]
Geissdoerfer (2017)	Regenerating system in which resource costs, emissions, and energy losses are minimized by closing and reducing material and energy cycles	[41]
Korhonen, Nuur, and Feldmann (2018)	Sustainable Development Initiative, which aims to reduce linear material and production flows in production-based and consumption-based systems, use of material cycles, and renewable and cascading energy flows	[42]
Lieder and Rashid (2016)	Solve problems related to waste generation, resource scarcity, and sustainable economic benefits	[43]
Pilyugina (2016)	An economy that increases people's wellbeing and ensures social justice, while reducing risks to the environment	[44]

An analysis of the scientific publications on the study of circular economics shows a deep understanding of the existing problems by scientists and the gradual formation of theoretical and methodological approaches to their solution. Based on the generalization of the views on this category, we can identify the following areas of interpretation of the circular economy: restrictions to the level of democratic freedoms; limiting population growth; limiting the level of individual consumption; overcoming economic inequality at the country and global levels; radical increase in investment in resource recovery; and prohibition of the implementation of large technogenic and dangerous projects.

A prerequisite for the formation of the strategic potential of the enterprise in a circular economy is the search for new tools that can ensure harmony between economic growth and environmental sustainability to ensure economic and environmental security and, consequently, to reduce the harmful effects on the environment. This is why the formation of the strategic potential of the enterprise should be based on the proper use of resources, which consists of the recycling of almost any commodity, which will ensure the further implementation of innovative and investment processes and ensure zero-waste production. In contrast to the traditional economy, the circular model is the most successful way to conserve resources and materials, so it is able to overcome the potential threats from environmental pollution, can implement environmental policy, and can reduce the preventive costs of the enterprise.

Thus, according to the authors, the main elements of the strategic potential of the enterprise consist of the following elements: (1) the quantity and quality of resources in enterprises (production and property potential), namely, the number of employees, fixed and non-productive assets or inventories, financial and intangible resources, patents, licenses, information, and technology; (2) the educational and qualification characteristics of the personnel of the enterprise and their ability to create certain kinds of products or services (labor potential); (3) marketing potential, systematic research market, consumers, and competitors; the production of high quality products to meet the needs of consumers; increasing the level of competitiveness among enterprises in the market; the application of modern tools in the practice activity of the enterprise; acquaintance of consumers with products or services; the development and maintenance of corporate style and image enterprises in the market; (4) financial potential: ensuring the appropriate level of financial stability, liquidity, and profitability of the enterprise; (5) information potential: the ability of the enterprise to generate, transform and use information resources; (6) innovation potential: the use of modern forms and methods of organization and business process management; renewal of technical and technological basis of production; (7) organizational and managerial potential.

3. Materials and Methods

To ensure the economic sustainability of the enterprise, the system should cover several elements, among which is the monitoring of the sustainability of the business processes of the enterprise under the influence of the external and internal environment, in the context of investment and security aspects, as well as the development of effective measures to prevent the risks of unstable activity [45].

It must be recognized that economic development directly depends on saving the use of resources, as it will reduce the dependence on raw materials through the continuous processing of goods and materials.

Based on a systematic approach, the mechanism of the formation of the strategic potential of the enterprise is evaluated as the total value of the potential of the enterprise, and the value of its elements. The total strategic potential of the enterprise is not calculated as a simple sum of its constituent elements, but as an integral indicator through the heterogeneity and sometimes incomparable estimates of individual elements, as well as the need to consider the synergistic effect of their interaction (Formula (1)):

$$SPE = \sum_{i=1}^{n} \int (p_1, p_2, p_3, \dots p_n) \tag{1}$$

$$\text{condition } SPE \rightarrow p_n + 1,$$

where SPE—strategic potential of the enterprise, and $p_1, p_2, p_3, \dots, p_n$—elements of the strategic potential of the enterprise [4].

It is the process of the formation of the strategic potential of the enterprise that allows the effective use and careful saving of limited resources.

The concept of a circular economy involves the efficient use of resources and a reduction of the negative impact on the environment from the production and consumption of goods and services on all cycles, that is, from the extraction of raw materials to the final use. The formation of the strategic potential of the enterprise in a circular economy will reduce the negative impact on the environment by reducing the use of resources in production and, as a result, creating a cleaner and safer environment; reducing production costs by reducing the quantity of primary resources used; and creating new markets and, consequently, new jobs, which contributes to the overall level of prosperity.

The mechanism of the formation of strategic potential of the enterprise in the conditions of a circular economy, shown in Figure 2, is based on the selection and creation of its unique combinations of resources and distinctive competencies, taking into account the results of the search for new opportunities. This is achieved through the research of enterprise and market potential, which creates the pre-conditions for the readiness of the enterprise for market changes, increasing the range of alternatives for enterprise development.

4. Results and Analysis

Taking into consideration the basic structural constraints outlined above, the circular economy on the principles of strategic enterprise potential should be much more efficient and technologically advanced, but the main criterion of its efficiency will no longer be the size of the profit (Figure 1).

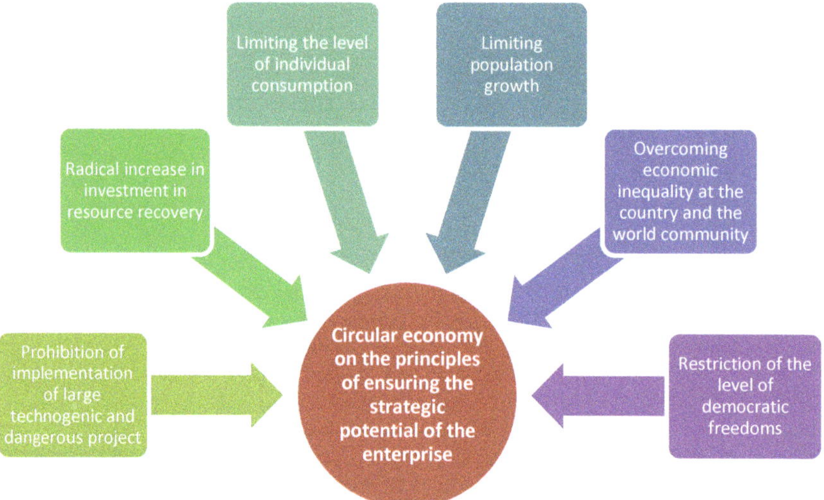

Figure 1. Task limits and forming conditions for transition to circular economy on the macro-level. Source: own study.

Depending on the specifics of their activities, enterprises can set different strategic goals in the area of their development, such as maintaining and strengthening the efficiency of the distribution and use of resources and energy in production chains, initiating new niches for business, changing the approach to services, transforming products into services, and creating exchange platforms, among others. Integrating the principles of a circular economy into the enterprise operations can create a long-term competitive advantage. For this purpose, it is recommended to follow a systematic approach, including the planning and analysis of external and internal environments, the development of strategic alternatives, and the formation and implementation of mission and strategy. Each of the abovementioned stages has some peculiarities in the conditions of the implementation of the circular economy.

Different aspects of revealing the relationship between the organizational behavior of the enterprise and its development have elements of corporate culture, which influence the very process of its management, reflected in the work of Kwilinski [46]. Thus, the features of corporate culture also influence the vitality of the enterprise, and as a consequence, can affect its level of profit. Moreover, one of the ways of forming the strategic potential of the enterprise and providing a higher level of competitiveness is creative human capital, as well as knowledge and research results. Their effective implementation in enterprises will contribute to the successful economic development of the country [15].

When constructing the mechanism of the formation of the strategic potential of the enterprise in the conditions of a circular economy, the system approach is applied, which can provide self-organization and self-reproducibility as a separate process, and the enterprise as a whole in the context of the national economy.

The mechanism of the formation of the strategic potential of the enterprise in the conditions of a circular economy, shown in Figure 2, is based on the selection and creation of its unique combinations of resources and distinctive competencies, taking into account the results of the search for new opportunities. This is achieved through the research of enterprise and market potential, which creates the pre-conditions for the readiness of the enterprise for market changes, increasing the range of alternatives for enterprise development.

The main feature of the proposed mechanism is the formation of strategic potential in a circular economy, because the principles of cyclicality and conservation of resources ensure the profitability of the enterprise by reducing costs and resources. The main stages of the formation of the strategic potential of the enterprise, within the limits of the application of this mechanism, are the result. The main components of strategic potential and the circular economy are determined under the influence of the factors and methods of application. As a result, this mechanism provides resource-efficient and clean production, environmental protection, waste-free production, and the minimization of consumer resources.

Prospects for further research are associated with the development of the adaptability and flexibility of the strategic potential of the enterprise.

The positive effect of the implementation of this mechanism of the formation of the strategic potential of the enterprise will be expressed in:

- Promoting the creation of new productions and differentiation of products that will have a high level of competitiveness at the international level;
- Increasing the innovativeness of both the enterprise and its products;
- Orientation of efforts to achieve full and productive employment of the population;
- Preservation of the effective creation of new workplaces and improvements to the quality of the labor force.

Another significant part of the circular economy is the issue of the formation of an environmentally effective waste management policy.

Figure 3 shows general analytical information regarding waste generation and management indicators for the period 2011–2019. According to the data, the amount of waste sent to landfills tends to decrease, although this positive trend is due to a negative factor—a decrease in the pace of development of the real sector of the economy.

As can be seen from Figure 3, the largest indicator has exactly the same formation of waste as its utilization and disposal, which indicates a significant problem in this aspect and the pollution of the environment.

In 2014, this indicator increased sharply to 0.084%, which is directly related to the beginning of military operations and subsequent territorial changes in the country, which affected the nationwide volume of waste accumulation, and continued to grow until 2018, amounting to 0.09%. In 2019, the region's share in the total volume of waste accumulated in Ukraine decreased slightly (to 0.078%) and generally remains insignificant.

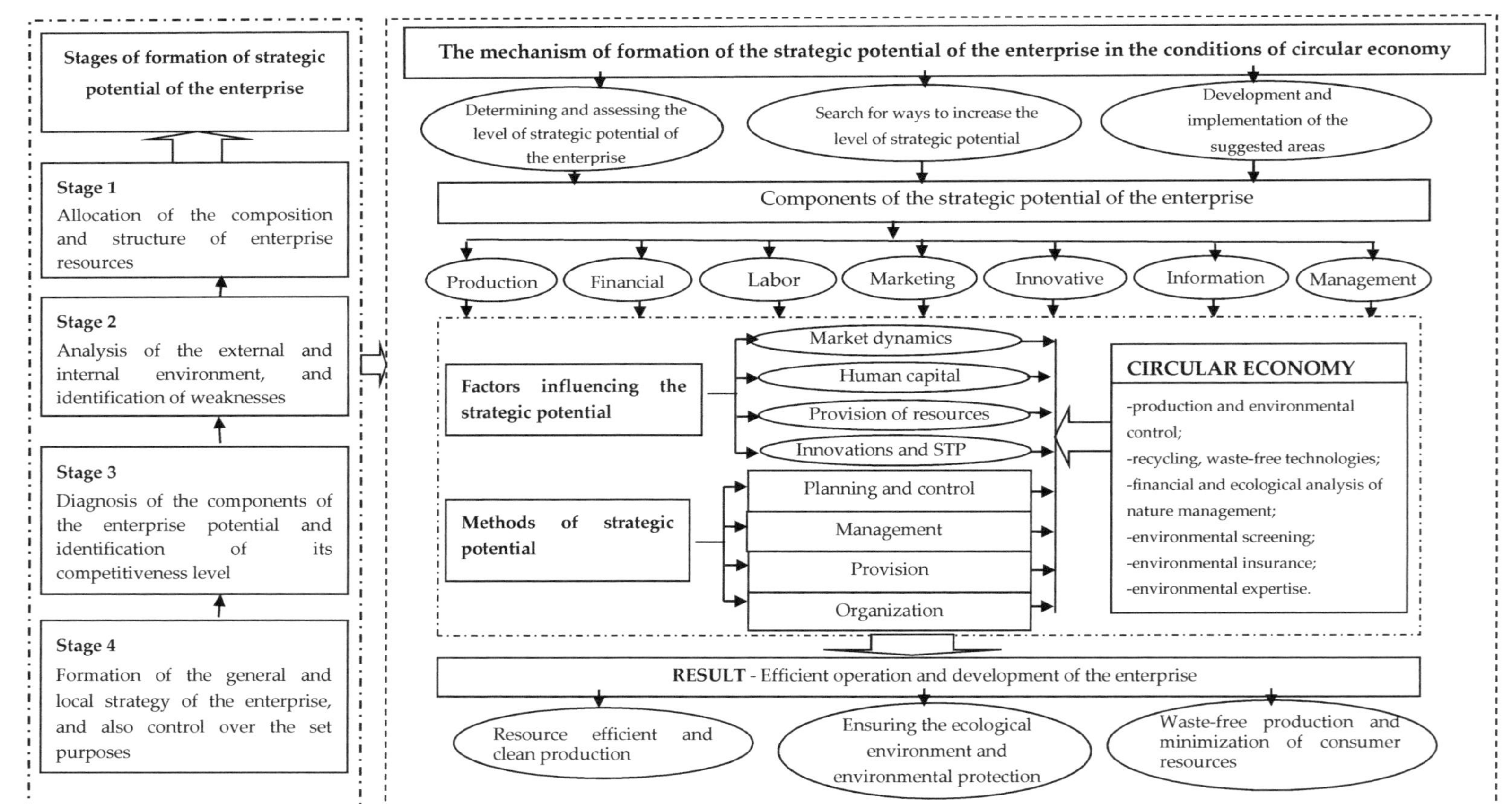

Figure 2. Mechanism of the formation of strategic potential of the enterprise in the conditions of a circular economy. Source: own study.

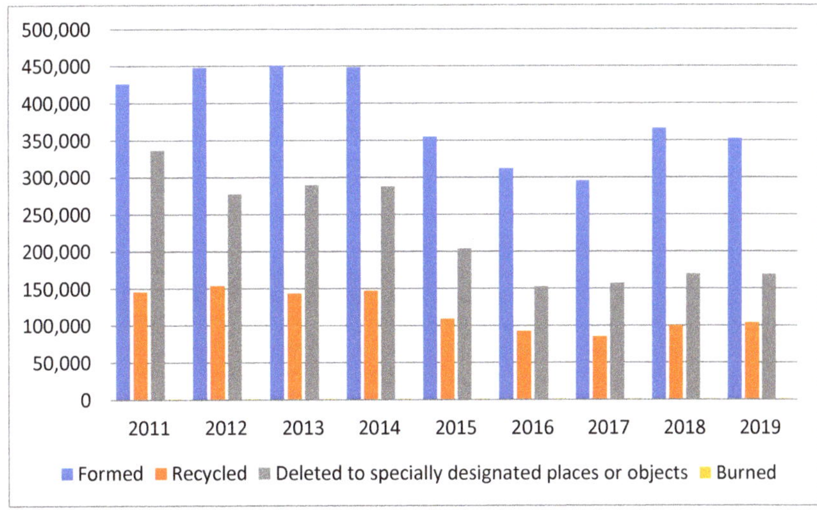

Figure 3. Indicators of waste generation and handling for 2011–2019 (thousand tons). Source: State Statistics Service of Ukraine [47].

According to the global experience, countries such as Germany, the USA, China, Japan, and Great Britain are cited as references regarding methods and means of non-adverse impacts of the lack of waste disposal. Germany is one of the world's leading countries in terms of the amount of recycled waste, where 66% of waste is recycled. The government of this country has obliged manufacturers to label goods according to specific waste categories. The introduction of the collateral value of packaging, along with multicolored containers for different types of waste, contributes to the recycling processes. Campaigning has also played an important role. For Germans, it is a civic obligation to promote waste sorting [48].

The USA has introduced separate garbage disposal (waste sorted by the owner is taken away free of charge). As in Germany, there is a deposit value of packaging. Innovative technologies for recycling and waste disposal should be noted [48].

China has a large number of recycling plants. At this stage, the government promotes recycling by introducing fees for separate waste. The government plans to impose fines for unsorted garbage. Recycling is also promoted by special garbage collectors who buy garbage from ordinary people and resell it to special institutions. Campaigning has been implemented to a lesser extent [48].

In Japan, the problem of recycling is especially urgent given the size of the country. Large-scale propaganda is not necessary due to the peculiarities of religion (Shintoism) and the worldview of the Japanese. It is quite usual to classify waste into four categories: incinerable, non-incinerable, recyclable, and large-sized. Penalties for violating recycling rules can be imposed onto the entire housing cooperative. Ultra-modern technology is used for recycling and incineration. Japan is another world-leading country in terms of recycling and waste disposal [48].

For a long time, Great Britain lagged behind other countries in waste management. However, in recent years, the situation has improved significantly. A system of waste sorting has been introduced. Regarding the advocacy work, the government has chosen a slightly different strategy. All violations of waste sorting rules are punished with substantial fines. Even the excessive weight of waste is considered to be a violation [48].

To ensure the minimization of the appearance and destruction of unprocessed waste and to prolong the duration of the exploitation of products, the main aim is the effective use of available resources by forming the strategic potential of the enterprise in a circular economy.

The characteristic features of strategic potential are the reflection of the past, i.e., a set of properties accumulated by the system during its establishment, and conditioning the possibility of its functioning and development; determination of the level of practical application and use of available opportunities; and orientation towards development (for the future), i.e., it can be stated that strategic potential is the basis for developing an effective strategy and ensuring an appropriate level of enterprise development. The mechanism of the formation of the strategic potential of an enterprise is determined by:

- The volume and quality of its available resources (the number of employed workers, basic production and non-production funds or inventories, financial and intangible resources and patents, licenses, information, and technology);
- The ability of managers and other categories of personnel to create certain types of products, qualification, and motivation potential;
- The ability of management to optimally use the available resources of the enterprise;
- Information capabilities, i.e., the capabilities of an enterprise to generate and transform information resources for use in production, commercial, and managerial activities;
- Innovation capabilities of the enterprise to update the technical and technological basis of production, the transition to new competitive products, the use of modern forms and methods of organization, and management of economic processes;
- Financial capabilities of fundraising [49].

To study the relationship of the above processes, it is advisable to use the Cobb–Douglas production function (Formula (2)) [4]:

$$Q = AL^{\alpha}K^{\beta} \tag{2}$$

In our cost-effectiveness model, we evaluate the effectiveness of applying additional waste processing steps. Of course, additional waste processing steps are a separate component of the circular economy. First, the economic efficiency of increasing resource profitability is the most important consideration, with the second being the economic impact of other circular economy measures. Third, the introduction of additional stages of waste processing, in the vast majority of cases, does not require extremely complex technical solutions.

When creating a model of economic efficiency, we rely on several principles, which are listed below:

1. We operate with three categories of objects of the production cycle: resources (R), wastes (W), and products (Q). We assume that resources are directly proportional to capital.
2. To get rid of conventions with units of measurement, we take into account objects of all three categories in their monetary expression.
3. We assume that during processing, the waste is converted into resources of the same branch of production.
4. For each branch, there is a certain idealized impossible situation in which waste is not generated at all. In this case, the volume of output will only be determined by the applicable resources and the type of production itself. For our model, we slightly modify the Cobb–Douglas function. First, since it is the change in resources and the corresponding change in output that we are considering, we can assume that L = constant. Moreover, we have already noted that resources are directly proportional to capital. Therefore, we will use the Cobb–Douglas function in the following form (Formula (3)):

$$Q = 1.5438 \times C^{1.4919252} \times R^{0.473957048} \tag{3}$$

where C—certain coefficient of proportionality [4].

In the resulting model, there is an increasing effect of scale, because the sum of α and β exceeds 1 (equal to 1.9659). This means that if resources (R) and waste (W) increase in a certain proportion, the amount of waste will increase in a larger proportion. It was found that, on average, during the analyzed period there was an annual increase in the amount of enterprise waste by 10% and a decrease in the amount of resources by 3%.

Given these assumptions, let us build a forecast of the waste volume for 2022–2027 (Figure 4).

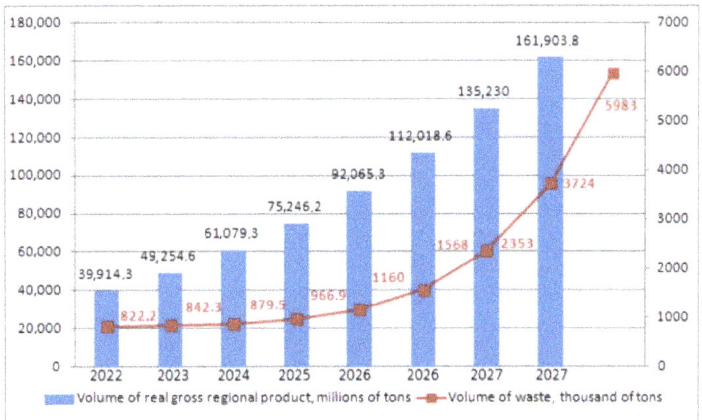

Figure 4. Outlook of the volume of waste, t/million UAH, 2022–2027. Source: State Statistics Service of Ukraine [47].

This is why, with the increase of the load on the environment and the growth of global environmental problems, attention on the formation of the strategic potential of the enterprise with an emphasis on the environmental component is increasing. The introduction of the components of strategic potential, such as production, finance, labor, marketing, innovation, and information, is, in most developed countries, an effective tool for solving both environmental and economic problems.

5. Conclusions

The conducted study, which included the application of conceptual foundations to the formation of the mechanism of the strategic potential of the enterprise in a circular economy, allows us to suggest a number of priority measures that should be implemented in order to ensure this process:

- Multiplication of competitive advantages by improving the quality of products in accordance with the needs of consumers;
- Saving of resources in the process of production at the expense of mastering innovations, realization of the economy of scale, scientific and technical innovations, and improvement of the management system;
- Saving resources and increasing customer loyalty by improving the quality of products;
- Increasing the competitive position of the enterprise due to the introduction of flexible communications and supply chains, organization of service, and warranty service.

The implementation of the suggested measures will allow enterprises to ensure the strategic potential, forming a set of social interrelations, which in a circular economy will contribute to the implementation of the mission of the enterprise, increasing the competitiveness of its activities and ensuring stable development in the transformation of the economic environment of activity.

The authors demonstrated that the processing and conservation of resources allows the company to save production costs and maximize profits. This is why the questions raised regarding the formation of the mechanism of the strategic potential of the enterprise in the conditions of a circular economy will reduce the influence on the ecological environment [28].

The experience of developed countries shows the possibility of achieving significant economic, environmental, and social effects through the formation of the waste industry and its transformation into an integral element of the socioeconomic infrastructure of the economy. At the same time, the current unsatisfactory state of waste management in the rural areas of Ukraine encourages the development of effective mechanisms of interaction of different parts of society to address the issues of sustainable development [43].

Author Contributions: Conceptualization, A.K. and O.A.; methodology, A.K., O.A., Z.P. and O.I.; validation, A.K. and O.A.; formal analysis, A.K., O.A., Z.P. and O.I.; investigation, A.K., O.A., Z.P. and O.I.; resources, A.K., O.A., Z.P. and O.I.; data curation, O.A.; writing—original draft preparation, A.K., O.A., Z.P. and O.I.; writing—review and editing, A.K. and O.A.; visualization, A.K., O.A., Z.P. and O.I.; supervision, A.K. and O.A.; project administration, A.K.; funding acquisition, A.K. All authors have read and agreed to the published version of the manuscript.

Funding: The research received funding under the research subsidy of the Faculty of Organization and Management of the Silesian University of Technology for the year 2022 (13/990/BK_22/0170).

Institutional Review Board Statement: Not applicable.

Informed Consent Statement: Not applicable.

Data Availability Statement: Not applicable.

Conflicts of Interest: The authors declare no conflict of interest.

References

1. Ansoff, Y. *Novaya Korporatyvnaya Stratehyya. [New Corporate Strategy]*; Pyter Publications: St. Petersburg, Russia, 1999; 416p.
2. Kozenkov, D.Y.E. Osnovni vymohy do formuvannya systemy stratehichnoho upravlinnya pidpryyemstvom. *Akad. Ohlyad.* **2011**, *1*, 83–88.
3. Tompson, A.; Stryklend, D. Stratehycheskyy medezhment. In *Yskusstvo Razrabotky y Realyzatsyy*; Yunyty: Berlin, Germany, 1998; 576p.
4. Shershn'ova, Z.Y.E. Stratehichne Upravlinnya. Baryery Stratehichnoho Planuvannya. Available online: http://studentbooks.com.ua/content/view/642/42/1/3/ (accessed on 10 December 2021).
5. Mirza, J. Supporting strategic management decisions: The application of digital twin system. *Strateg. Dir.* **2021**, *37*, 7–9. [CrossRef]
6. Schaefer, J.L.; Siluk, J.C.M.; Carvalho, P.S.D. An MCDM-based approach to evaluate the performance objectives for strategic management and development of Energy Cloud. *J. Clean. Prod.* **2021**, *320*, 128853. [CrossRef]
7. Dong, A. ERP and Artificial Intelligence based Smart Financial Information System Data Analysis Framework. In Proceedings of the 6th International Conference on In-ventive Computation Technologies, ICICT, Coimbatore, India, 20–22 January 2021; pp. 845–848. [CrossRef]
8. Prokopenko, O.; Kichuk, Y.; Ptashchenko, O.; Yurko, I.; Cherkashyna, M. Logistics Concepts to Optimise Business Processes. *Stud. Appl. Econ.* **2021**, *39*, 4712. [CrossRef]
9. Kwilinski, A.; Kuzior, A. Cognitive Technologies in the Management and Formation of Directions of the Priority Development of Industrial Enterprises. *Manag. Syst. Prod. Eng.* **2020**, *28*, 133–138. [CrossRef]
10. Olena, S.; Tetyana, V. Neuro-genetic hybrid system for management of organizational develop-ment measures. *CEUR Workshop Proc.* **2020**, *2732*, 411–422.
11. Temich, S.; Pollak, A.; Kucharczyk, J.; Ptasiński, W.; Mężyk, A.; Gąsiorek, D. Prediction of energy con-sumption in the industry 4.0 platform—Solutions overview. *J. Theor. Appl. Mech.* **2021**, *59*, 455–468. [CrossRef]
12. Baryshnikova, N.; Kiriliuk, O.; Klimecka-Tatar, D. Enterprises' strategies transformation in the real sector of the economy in the context of the COVID-19 pandemic. *Prod. Eng. Arch.* **2021**, *27*, 8–15. [CrossRef]
13. Elia, G.; Margherita, A. A conceptual framework for the cognitive enterprise: Pillars, maturity, value drivers. *Technol. Anal. Strateg. Manag.* **2021**, 1–13. [CrossRef]
14. Berezin, O.V. Zavdannya ta mekhanizm optymizatsiyi struktury potentsialu pidpryyemstva. Visnyk Natsional'noho universytetu vodnoho hospodarstva ta pryrodokorystuvannya. *Ekonomika. Chastyna II Zb. Nauk. Pr.* **2007**, *4*, 20–28.
15. Prokhorova, V.; Iarmosh, O.; Shcherbyna, I.; Kashaba, O.; Slastianykova, K. Innovativeness of the creative economy as a component of the Ukrainian and the world sustainable development strategy. *IOP Conf. Ser. Earth Environ. Sci.* **2021**, *628*, 12035.
16. Chukhno, A.A. Modernizatsiya ekonomiky ta ekonomichna teoriya. *Ekon. Ukrayiny* **2012**, *10*, 24–33.
17. Nguyen, H.; Stuchtey, M.; Zils, M. Remaking the industrial economy. *McKinsey Quaterly*, 2014. Available online: https://www.mckinsey.com/business--functions/sustainability--and--resource--productivity/our--insights/remaking--the--industrial--economy(accessed on 14 December 2021).

18. Briguglio, M.; Llorente-González, L.J.; Meilak, C.; Pereira, Á.; Spiteri, J.; Vence, X. Born or grown: Enablers and barriers to circular business in europe. *Sustainability* **2021**, *13*, 13670. [CrossRef]
19. van Engelenhoven, T.; Kassahun, A.; Tekinerdogan, B. Circular business processes in the state-of-the-practice: A survey study. *Sustainability* **2021**, *13*, 13307. [CrossRef]
20. Centobelli, P.; Cerchione, R.; Esposito, E.; Passaro, R.; Shashi. Determinants of the transition towards circular economy in SMEs: A sustainable supply chain management perspective. *Int. J. Prod. Econ.* **2021**, *242*, 108297. [CrossRef]
21. Acerbi, F.; Taisch, M. Information Flows Supporting Circular Economy Adoption in the Manufacturing Sector. *IFIP Adv. Inf. Commun. Technol.* **2020**, *592*, 703–710. [CrossRef]
22. Sawe, F.B.; Kumar, A.; Garza-Reyes, J.A.; Agrawal, R. Assessing people-driven factors for circular economy practices in small and medium-sized enterprise supply chains: Business strategies and environmental perspectives. *Bus. Strategy Environ.* **2021**, *30*, 2951–2965. [CrossRef]
23. Morea, D.; Fortunati, S.; Martiniello, L. Circular economy and corporate social responsibility: Towards an integrated strategic approach in the multinational cosmetics industry. *J. Clean. Prod.* **2021**, *315*, 128232. [CrossRef]
24. Smol, M.; Marcinek, P.; Koda, E. Drivers and barriers for a circular economy (Ce) implementation in poland—A case study of raw materials recovery sector. *Energies* **2021**, *14*, 2219. [CrossRef]
25. Chakrabarty, A.; Nandi, S. Electronic waste vulnerability: Circular economy as a strategic solution. *Clean Technol. Environ. Policy* **2021**, *23*, 429–443. [CrossRef]
26. Adami, L.; Schiavon, M. From circular economy to circular ecology: A review on the solution of environmental problems through circular waste management approaches. *Sustainability* **2021**, *13*, 925. [CrossRef]
27. Järvenpää, A.-M.; Kunttu, I.; Jussila, J.; Mäntyneva, M. Data-Driven Decision-Making in Circular Economy SMEs in Finland. In *Research and Innovation Forum 2021*; Springer Proceedings in Complexity; Springer: Cham, Switzerland, 2021; pp. 371–382. [CrossRef]
28. Shashi, C.P.; Cerchione, R.; Mittal, A. Managing sustainability in luxury industry to pursue circular economy strategies. *Bus. Strategy Environ.* **2021**, *30*, 432–462. [CrossRef]
29. Zaloznova, Y.; Kwilinski, A.; Trushkina, N. Reverse logistics in a system of the circular economy: Theoretical aspect. *Econ. Her. Donbas* **2018**, *4*, 29–37.
30. Grebski, M.E.; Grebski, W. Psychology of Creativity and Innovation in Engineering and Business Curriculum. *Multidiscip. Asp. Prod. Eng.—MAPE* **2021**, *4*, 387–394. [CrossRef]
31. Grebski, M. Mobility of the Workforce and Its Influence on Innovativeness (Comparative Analysis of the United States and Poland). *Prod. Eng. Arch.* **2021**, *27*, 272–276. [CrossRef]
32. Kuzior, A.; Zozulak, J. Adaptation of the Idea of Phronesis in Contemporary Approach to Innovation. *Manag. Syst. Prod. Eng.* **2019**, *27*, 84–87. [CrossRef]
33. Bohats'ka, N.M. Stratehichnyy Potentsial Pidpryyemstva. Available online: www.rusnauka.com/33_DWS_2010/Economics/ (accessed on 10 October 2021).
34. Hedroyts, H.Y.U. Vyznachennya sutnosti ponyattya «stratehichne upravlinnya». Ekonomichni nauky/10. *Ekon. Pidpryyemstva* **2012**, *2*, 22–23.
35. Boulding, K.E. The Economics of Knowledge and the Knowledge of Economics. *Am. Econ. Rev.* **1966**, *56*, 1–13. Available online: http://www.jstor.org/stable/1821262 (accessed on 14 December 2021).
36. Kneese, A.V. The Economics of Natural Resources. *Popul. Dev. Rev.* **1988**, *14*, 281–309. [CrossRef]
37. *Agenda 21: Earth Summit: The United Nations Programme of Action from Rio*; United Nations: New York, NY, USA, 1993.
38. Kuzior, A. Polish and German Experiences in Planning and Implementation of Sustainable Development]. *Probl. Ekorozw. Probl. Sustain. Dev.* **2010**, *5*, 81–89. Available online: http://ekorozwoj.pol.lublin.pl/no9/h.pdf (accessed on 15 December 2021).
39. A New Circular Economy Action Plan. For a Cleaner and More Competitive Europe. Communication from the Commission to the European Parliament, the Council, the European Economic and Social Committee and the Committee of the Regions, Brussels. 2020. Available online: https://eur-lex.europa.eu/legal-content/EN/TXT/?uri=COM:2020:98:FIN&WT.mc_id=Twitter (accessed on 14 December 2021).
40. Pakhomova, N.V.; Rikhter, K.K.; Vetrova, M.A. Transition to circular economy and closedloop supply chains as driver of sustainable development. *Vestn. St. Peterbg. Univ. Ekon.* **2017**, *33*, 244–268. [CrossRef]
41. Geissdoerfer, M. The Circular Economy: A new sustainability paradigm. *J. Clean. Prod.* **2017**, *10*, 757–768. [CrossRef]
42. Korhonen, J.; Nuur, C.; Feldmann, A. Circular economy as an essentially contested concept. *J. Clean. Prod.* **2018**, *175*, 117–125. [CrossRef]
43. Lieder, M.; Rashid, A. Towards circular economy implementation: A comprehensive review in context of manufacturing industry. *J. Clean. Prod.* **2016**, *1*, 36–51. [CrossRef]
44. Pilyugina, M. The circular economy model as a new approach to sustainable development. *Stroit.—Form. Sredy Zhiznedeyatelnosti* **2016**, *31*, 148155.
45. Arefieva, O.; Piletska, S.; Khaustova, V.; Poberezhna, Z.; Zyz, D. Monitoring the economic stability of the company's business processes as a prerequisite for sustainable development: Investmentand security aspects. *IOP Conf. Ser. Earth Environ. Sci.* **2021**, *628*, 12042. [CrossRef]

46. Kwilinski, A. Trends of development of the information economy of Ukraine in the context of ensuring the communicative component of industrial enterprises. *Econ. Manag.* **2018**, *1*, 64–70.
47. Ofitsiynyy Veb-Sayt Derzhavnoyi Sluzhby Statystyky Ukrayiny. Available online: http://www.ukrstat.gov.ua/ (accessed on 10 November 2021).
48. Mikhno, I.S. Metody finansuvannya utylizatsiyi vidkhodiv. Svitovyy dosvid [Methods of the waste utilization financing. World Experience]. *Ekon. Finans. Menedzhment Aktual'ni Pyt. Nauk. i Prakt.* **2015**, *2*, 68–78.
49. Pashchenko, O.P. Potentsial pidpryyemstva u systemi stratehichnoho upravlinnya rozvytkom. Naukovyy visnyk Khersons'koho derzhavnoho universytetu. *Seriya Ekon. Nauk.* **2014**, *8*, 77–80.

 *sustainability*

Article

Transforming Linear Production Chains into Circular Value Extended Systems

Carlos Scheel * and Bernardo Bello 

EGADE Business School, Tecnologico de Monterrey, San Pedro Garza García 66269, Mexico;
bernardo.bello.92@gmail.com
* Correspondence: cscheel@tec.mx

Abstract: Different schools of thought, theories, and concepts have been developed to diminish the social and environmental impact that the take–make–dispose linear economic model has produced. Such is the case of industrial ecology (IE) and circular economy (CE). However, the principles and guidelines in IE literature are focused more on resource efficiency without considering the social externalities. In the same sense, CE literature has not brought clear guidance about how to circularize linear businesses and is mainly focused on recycling strategies, which could be the least profitable and attractive option among the circular business models (CBM). Based on the sustainable wealth creation through disruptive innovation and enabling technologies (SWIT) framework and the business model framework, we have developed a roadmap to transform linear value chains into an industrial ecology cluster of zero-waste chains and enabling institutions called a circular value extended system (CVES), which is able to exploit non-usual business opportunities of waste and residue revaluation. This systemic approach opens the possibilities of creating a socially inclusive, environmentally resilient, and economically viable system of capital. A case study is presented to clarify the design process and application of the framework. Our contribution entails guidelines to transform linear value chains into a cluster of circular economy systems capable of producing sustainable increasing returns to benefit multiple regional stakeholders.

Keywords: circular economy; zero-residues industrial ecology; multiple non-usual businesses; circular business clusters; circular value extended system

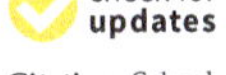 check for updates

Citation: Scheel, C.; Bello, B. Transforming Linear Production Chains into Circular Value Extended Systems. *Sustainability* **2022**, *14*, 3726. https://doi.org/10.3390/su14073726

Academic Editor: Ana Ramos

Received: 23 February 2022
Accepted: 15 March 2022
Published: 22 March 2022

Publisher's Note: MDPI stays neutral with regard to jurisdictional claims in published maps and institutional affiliations.

Copyright: © 2022 by the authors. Licensee MDPI, Basel, Switzerland. This article is an open access article distributed under the terms and conditions of the Creative Commons Attribution (CC BY) license (https://creativecommons.org/licenses/by/4.0/).

1. Introduction

"Sustainability is a "vision", something most of us share, but does not give advice on how to act."-Schnitzer and Ulgiati, 2007

Resources for industrial production are becoming scarcer and more expensive to extract, mainly for developing countries. The Global Footprint Network [1] has reported that humans use as many ecological resources as if we lived on 1.75 Earths. In response, some industries and nations have begun transforming their linear production chains into industrial ecosystems capable of reducing the consumption of virgin materials and creating symbiotic linkages of waste and residue sharing [2,3].

These industrial ecosystems are often called eco-industrial parks (EIP) [3]. Examples of EIP entail the cases of Kalundborg, Denmark [4], Guigang, China [5], Kawasaki, Japan [6], Jyvaskyla, Finland [7], Las Gaviotas, Colombia [8], among others. The research on EIP has focused on optimizing energy and materials within the EIP to reduce the environmental impact compared to multiple stand-alone linear production chains.

Nevertheless, the design and efficiency of EIPs still have some gaps in developing a holistic zero-waste (zero-emissions) economy. The optimization of energy and materials exchange within EIPs does not assess the real impact of upstream and downstream supply chains outside the industrial network [9,10]. Therefore, there is a need to expand the scope of industrial ecology (IE) principles so that business leaders can analyze the impact that the linear value chains have on the environment and on the community where they are embedded.

In this sense, circular economy (CE) scales up the concepts of industrial metabolism and optimization to an economy-wide system to establish a new model of economic development, production, distribution, and recovery of products [11]. However, the implementation of CE is mainly focused on recycling and product-as-a-service systems (PSS) and not on how to transform linear business models into a cluster of multiple business models in practice [12,13]. Moreover, CE literature is mostly centered on environmental aspects and marginally addresses the social and institutional implications [14].

Consequently, this paper aims to bring guidelines on implementing CE building blocks to create self-sustained wealth that produces value-added to society, the environment, and the firm by transforming linear value chains into a cluster of non-usual businesses. We seek to facilitate the design of a mechanism capable of transforming linear production chains into a viable ecological cluster of firms, institutions, and supporting stakeholders to deliver sustainable increasing returns (SIR) for all the participants through the leverage of business opportunities found on waste, obsolete products, new capabilities and residue revalorization [13].

The development of the model is based on sustainable wealth creation through disruptive innovation and enabling technologies (SWIT) framework [13], which entails concepts from blue economy [15], industrial ecology [16], principles from the circular economy approach [17], as well as systems dynamic modeling of complex phenomena [18,19]. The value added from the SWIT framework to the CE discussion is its systemic approach, which points out the need for a holistic innovation considering all key stakeholders instead of product innovation or the participation of just isolated stakeholders in the transformation of sustainable production processes. The model we propose has been called a circular value extended system (CVES), and its aim is the generation of a balanced value proposition [20] among the economic, social, and environmental spheres. For this, we have extended the contribution from Osterwalder et al. [21] to a triple value proposition to facilitate practitioners to distinguish non-usual business models that may be part of industrial ecology business clusters, able to create sustainable value for all stakeholders. To widen the understanding of our proposal, we share the methodology applied in an action research case to show the advantages and differentials of the approach.

This article is organized as follows: Section 2 reviews the basis of industrial ecology, circular economy, and sustainable wealth generation through disruptive innovation and enabling technologies (SWIT) framework, as well as the business model literature. Section 3 describes the action research methodology undertaken to apply the framework in our business case. Section 4 describes the economic, social, and environmental context in Bartica, Guyana, where the framework has been proposed for implementation. Section 5 illustrates the roadmap for transforming a linear chain into a CVES with a stepwise exemplification for our business case in the gold mining industry in Bartica. Section 6 entails the discussion of the management implications and the advantages of the CVES for the creation and assembly of CBMs. Finally, Section 7 involves the relevant conclusions from this work, including limitations and further research.

2. Literature Review

2.1. Industrial Ecology and Performance of Eco-Industrial Parks

The IE literature was born as a response to the conflict between industrial activities and ecological systems [22] when traditional and reactive approaches such as "end-of-pipe" pollution control methods were ineffective to treat industrial waste [16]. The concept of industrial symbiosis [3], built on the notion of symbiotic biological relationships in nature, plays a central role in implementing IE practices. Each of the firms involved in industrial symbiosis systems seek to deliver extended value for their customers but coexist in harmony with natural ecosystems [23].

The IE literature research of industrial parks as ecosystems further evolved into the concept of eco-industrial parks (EIPs) [24,25]. EIPs are arrangements of firms that comply with the concept of industrial symbiosis through the exchange of resources, such as raw materials, waste, water, energy, and information [2,3]. The industrial symbiosis concept has been widely implemented to design and manage EIP by researchers, specialized consultants, and personnel from economic and environmental development agencies in industrialized and developing countries [23]. Based on results of previous research that considered warming potential (GWP), water acidification, biodiversity degradation, CO_2 emissions, among other measurements [4–6], EIP performance has shown that EIPs have effectively reduced the environmental impact from the industry.

Nevertheless, the design and the measures of EIP usually overlook the consideration of downstream and upstream chains and their environmental impact, as well as other externalities and relationships that may have relevant implications on the original design of EIPs [9,10]. Sokka, Pakarinen, and Melanen [26] highlighted the importance of these considerations in a case study of the EIP based on a pulp and paper plant in Kymi, Finland, where they find that 70% of direct emissions from the EIP activity comes from the raw material production, which is not an activity considered within the EIP. Though industrial symbiosis in EIP can provide environmental benefits compared to value chains in isolation, the evaluation focused just on the EIP network can yield a different picture from the environmental impact when upstream processes are included in the study as well [26]. Therefore, radical innovations and a broader holistic vision are needed to approach a truly sustainable business model for wealth creation [13,27].

2.2. Circular Economy and Circular Business Models

Similar to the IE stream, CE was born as an alternative to the take-make-dispose economic model [17,28], and it can be traced back to the environmental economy [29] and the ecological economy literature [30]. These pillars of CE pointed out that the environment has different roles: it provides resources, a life support system, and a sink for emissions and waste. However, there is no price or market of these environmental goods (environmental services) even though they have a clear value for humanity, provoking that deterioration of the environment due to industrial activity can increase without limits if no regulations and specific actions are enacted [29,30]. Scaling up the concepts of industrial symbiosis and optimization from IE, the circular economy literature seeks to establish a new model of development, production, distribution, and recovery of products to revalorize the emissions and waste to decouple economic growth and resource use [11,13,31].

Circular business models (CBM) are a new kind of business model (BM) where the value creation is grounded on keeping the economic value embedded into products after their use and exploiting it for new types of market offerings [12]. Though the 3R's principles (reduction, reduce and recycle) [11] or the ReSOLVE (regenerate, share, optimize, loop, virtualize, exchange) framework from the Ellen MacArthur Foundation [17] are the base for CBMs, most of them are focused on recycling and product-as-a-service-system (PSS) [11,12]. This focus on recycling could be an obstacle for further implementing CE initiatives because it could be the least sustainable solution based on resource efficiency and profitability [1]. Additionally, these business model frameworks are just business-oriented, while circular economy needs a holistic vision and planning, which considers all stakeholders' involvement within the community as discussed in the SWIT framework [13,32].

Moreover, another gap in the CE literature is the lack of research regarding how to practically transform linear BM into a circular one [12]. Sometimes, CE theoretical concepts are given as a suggestion to firms and politicians, without any explanation or guideline on implementing such practices [12]. Just evoking a concept such as CE in management systems is unlikely to lead to the radical innovations needed for sustainable business models [27]. Therefore, the CE research must move forward the development of CBM to more inclusive, collaborative, and non-usual initiatives [33] that can deliver value not just for firms but also for citizens, institutions, and the environment.

Additionally, the discussion about CBM centers involves cleaner production processes but omits the rebound effects that material and energy efficiency might generate [34]. The rebound effects literature acknowledges that energy and material efficiency can reduce the environmental impact, but if the consumption of the goods rises beyond a certain amount, this benefit can be surpassed by the net impact [35,36]. CBM is not the exception to this phenomenon [37], and thus, we discussed the implications of rebound effects in our model.

2.3. Sustainable Wealth Creation through Disruptive Innovation and Technologies (SWIT) Framework

The sustainable wealth creation through disruptive innovation and enabling technologies (SWIT) framework [13] is a systemic approach for regional sustainable economic growth based primarily on circular economy (CE) and industrial ecology (IE) perspectives [3] using a system thinking approach [18,19] as an integral and collaborative concept. This approach focuses on how the unit of analysis interacts with its other constituent parts. Thus, instead of fragmenting the system configuration into different elements and analyzing them individually, it accounts for the more significant number of interactions and its feedback effects and synergies among external stakeholders [13,32].

The SWIT systemic framework provides the conditions for the interaction of procedures at three levels of complexity: product-residues chain level, stakeholders' synergy cluster level, and a regional level that accounts for external institutions [13,32]. The product-residue level entails the circularization of linear production chains by revaluating and transforming residues and waste into closed-loop chains of multiple processes, called zero-value residue industrial ecology systems (ZRIES). The cluster-level or circular value extended system (CVES) articulates the synergies of multiple ZRIES business units within a region, with a common goal for its inhabitants. Finally, the sharing value system (SVS) at the regional level provides the necessary conditions (resources, technologies, policies, infrastructure, institutions), resource allocation management, and inclusive governance to achieve an effective generation of sustainable wealth for all stakeholders [13].

The SWIT framework attempts to overcome the lack of design in the academic literature of economic growth with a systemic vision [13]. The objective of this paper is to extend the implementation of this framework addressing the lack of guidelines for the circularization of linear BM [12]. Based on the SWIT framework, this paper contributes with guidelines based on the CVES concept to transform linear value chains into valuable circular value systems. In contrast to EIPs, we seek to address the downstream and upstream chains and their environmental impact and the relevant externalities and relationships for the design of EIPs [9,10,26].

The purpose of the CVES cluster is to achieve effective and efficient clean production and waste revaluation that creates economic rents, social inclusiveness, and quality of life growth for the population by creating sustainable increasing returns while increasing natural resilience [13,32]. To achieve this, the proposed design of the CVES includes all stakeholders involved in the development of the value proposition and not just the business-oriented participants.

2.4. Triple Value Proposition Design

Our proposal extends the conventional business model framework from Osterwalder and Pigneur [20] to the inclusion of diverse stakeholders. Osterwalder and Pigneur [20] defined a BM as the description of the rationale of how an organization creates, delivers, and captures value. Conventional business models conceptualize value generation from a "one-shot" approach; instead, we proposed a model that is focused on creating value by the iteration of several cycles in the circular system through time, where various stakeholders participate, creating multiple value benefits for all.

Osterwalder et al. [21] have broken down the value proposition and the customer segment from the business model canvas for further analysis. The value proposition map (Figure 1) breaks the value proposition down into products and services, pain relievers, and gain creators for the customer [21]. The customer segment profile (Figure 1) breaks the customer segment down into its jobs, pains, and gains [21]. Although these guidelines are helpful for conventional BMs, we extend the value proposition map and the customer segment profile [21] for the development of the CVES extended value proposition.

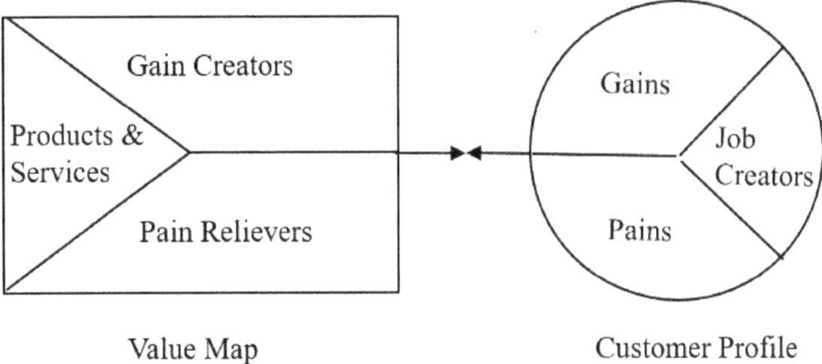

Figure 1. Value Proposition Map and the Customer Segment Profile adapted from Osterwalder et al. [21].

Mainly, we consider that the CVES serves three "customers" based on the triple-bottom-line of sustainability: the society, the environment, and the industrial firms. Thus, the circular approach requires the engagement of three customers, forming a triple value proposition (TVP). In Section 5, we present the designing steps for the CVES, which entails developing the TVP.

3. Methodology: Validating the CVES with Action Research

The CVES has been designed to articulate the synergies of the cluster of multiple ZRIES chains [13]. The present work aims to describe how to implement the CVES model to transform linear value chains into a cluster of multiple initiatives. The CVES model has been validated by previous research [13,31,32], which include cases such as the transformation of residues from the food industry [38], the exploitation of the residues from the palm oil production [13], and the transformation of a complete community into a cluster of businesses [32].

The method we have employed is a participatory action research approach in which diagnosing and planning are performed in collaboration between researchers and key stakeholders from the community [39]. The action research methodology is ideal for validating the framework because it is collaborative, generates theory grounded in action by evaluating the applicability of existing theories, and the relationships amongst stakeholders are relevant. These elements are essential for the case study we address, in which we depend on the balance of the three main stakeholders (environment, society, and industry) and supporting institutions (ABIIGS: academy, banks, infrastructure, industry leaders, government, and citizens of the local society).

Motivated by Guyana's Department of Environment of British Guyana and the UNEP, the town of Bartica was selected to develop the CVES mechanism to balance economic development, social welfare, and environmental resilience. In the following section, we present the action research case of Bartica, providing an overview of the social, economic, and environmental state of the town.

4. Case Study: Bartica, British Guyana

This case study was performed between 2017–2018, motivated by the Green State Development Strategy (GSDS): Vision 2040 from Guyana's policy development led by Guyana's Department of Environment. The United Nations Environmental Program (UNEP) Regional Director of Latin America and the Caribbean, and the GSDS Coordination Desk, look for the engagement of the private sector in the second session of "Green Conversation" under the theme From Green Towns to Green Cities.

We considered implementing the SWIT framework [13] as a development plan for Bartica since the mining industry is the main economic driver in Guyana, but the benefits for its population are not being reflected in the economic development of the country. Additionally, the mining industry has had extraordinary impacts on the environment, such as deforestation, topsoil removal, and contamination of watercourses. Thus, developing a CVES focused on "circularizing" the gold mining industry in this region might be an enabler for economic growth, environmental impact reduction, and improvement of the population's life quality.

Guyana is well endowed with natural resources, where gold and bauxite have played an essential role in the local economy. Artisan, small and medium-scale mines (ASMS) have dominated Guyana's mining industry. Under the Mining Act 1989, the Guyana Geology and Mine Commission (GGMC) receives over 50 percent of its income through royalties and fees. Furthermore, the GGMC serves as a regulatory entity that also provides training to miners on environmentally friendly practices.

Nevertheless, growth in Guyana's gross domestic product (GDP) has been highly volatile over the past decades due to geopolitical events, natural disasters, and global community price fluctuations. In 2015 and 2018, very significant oil reserves were discovered, putting Guyana at a critical point in its history, providing its citizens with an opportunity to change its development path. The GSDS: Vision 2040 plan seeks to develop sustained economic growth that is low-carbon, climate-resilient, and promotes social cohesion, good governance, and careful management of finite resources.

After several unstructured interviews with local authorities and the people from Bartica, observation, and document review, we identified the key activities and relevant residues and wastes from the gold mining industry in Bartica. The following section describes the roadmap and the steps undertaken to design a CVES based on the gold mining industry.

5. Research Results: Roadmap for the Development of the CVES

This section describes stepwise the design and development of a CVES. The steps (Figure 2) are based on previous case studies the authors have developed [13,32,38], experts think tanks, document review from Guyana, and unstructured interviews with local leaders from Bartica. Additionally, we framed this roadmap with the literature from IE, CE, SWIT, and business models (BM) frameworks. The exemplification of the steps is performed with the actions undertaken in the action research case developed in Bartica, British Guyana.

5.1. Identification and Deployment of the Key Linear Value Chains

The original concept of the value chain has evolved with the idea of shared value [40], highlighting that the core of economic value creation must simultaneously fulfill the society's aspirations to achieve durable economic success. One way to create shared value is to develop local support groups for the firm assembled as a cluster 40]. The first step we proposed for creating a circular cluster is to trace the *extended value map* [13]. This structure allows identifying valuable activities of an industry chain, the processes involved, the raw material required, the machinery, and identifying the residues generated that could be used as inputs by decomposer firms [13].

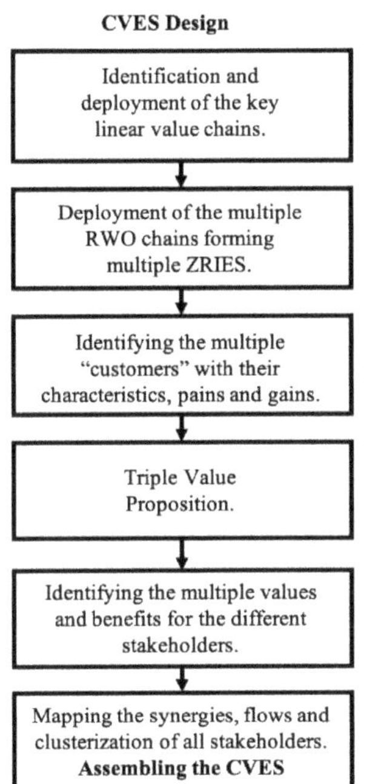

Figure 2. Roadmap for the CVES cluster design.

Data gathering at this phase is performed through primary and secondary sources (documents, interviews, observation), used in eco-industrial parks (EIP) designing processes to optimize the exchange of raw materials, water, waste, and energy [2,3]. The identification of residues, waste, and obsolete products (RWO) can be carried out by experts on the supply chain or other different practices. As performed in IE literature, the waste from the value chain stages can be studied by material flow analysis (MFA) [41], substance flow analysis (SFA) [42], physical input–output accounting, and life cycle assessment (LCA) [43,44].

At this stage, the objective is to develop the most detailed value chain where all the production processes, residues, and wastes are depicted. All stages of the value chain should be identified to propose systemic and holistic solutions [45]. In Figure 3, we have represented the extended value map for the gold production in Guyana based on the available documentation of the mining industry [46,47]. The RWO we have identified in Guyana's mining industry is the tailings generated from excavation, the vegetal material that was in the original landscape, the toxic slurry from the purification treatment of gold, the molds that are used to cast the metal, the metal scrap and the tires from obsolete machinery that are left on the open field and on the city.

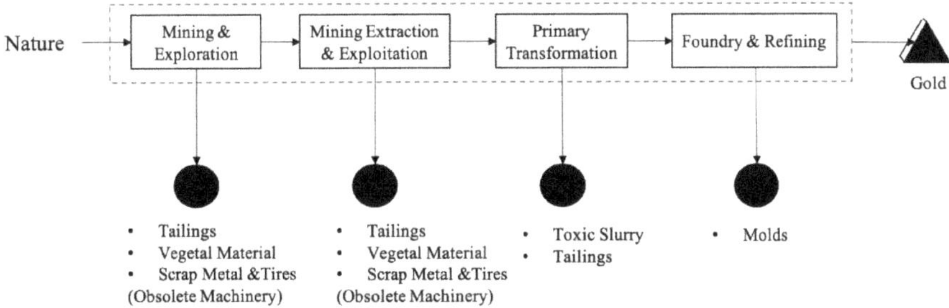

Figure 3. Extended value map for the natural resource extraction for gold production.

5.2. Deployment of the Multiple RWO Chains Forming Multiple ZRIES

The SWIT framework proposes that the RWO should be transformed through technological procedures into sequences of processes capable of decomposing and producing valuable products that generate new residue-nutrient chains for other processes [31]. The revalorization of RWO can start with scavengers and decomposers firms in the industrial system. Scavenger firms can collect materials, dismantle and make them accessible for other companies, and decomposer firms enable recycling by breaking down complex molecules into simpler nutrients for other processes [38,48]. Moreover, dematerialization or virtualization [17,49,50] is also essential for the ZRIES (zero residues/waste) chains design and the creation of sustainable wealth. Information systems facilitate the connection between RWO generators and transformation processes by exchanging information to coordinate different actors [50].

It is important to recognize that added processes and products generate more residues or other types of waste than the original linear value chain. Therefore, this is an iterative process between identifying processes and residues until the chains have zero-residues or harmless residues [13]. The objective of each new process is to reduce the RWO and create additional value for the stakeholders. In Figure 4, we have listed and deployed different processes that can be generated by decomposer and scavenger firms to transform the RWO from the extended value chain from the gold mining industry into valuable products.

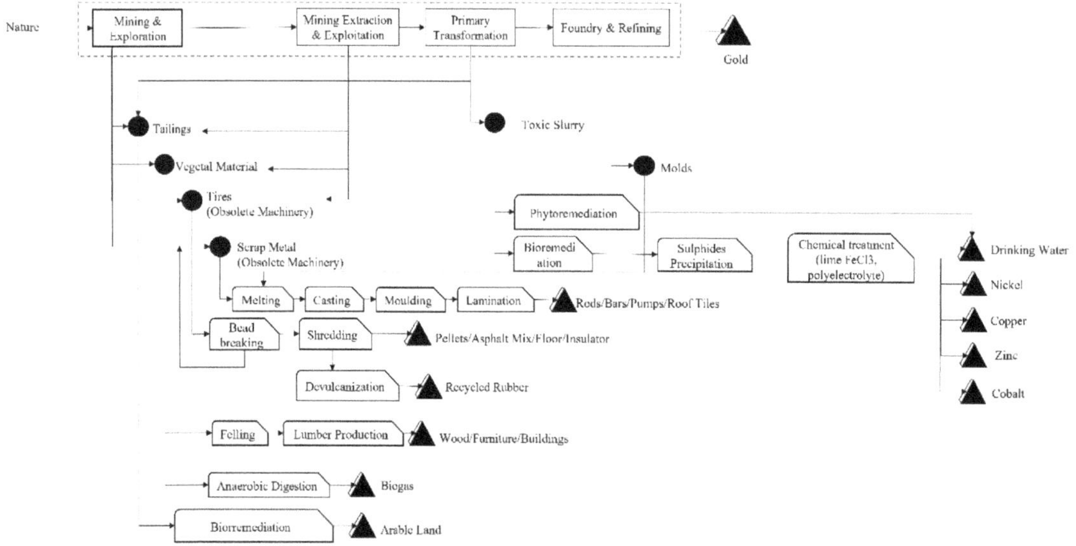

Figure 4. Processes and valuable products forming multiple ZRIES for the gold mining industry in Bartica, Guyana.

These ZRIES chains have the potential not only to restore the environment but to generate value through different products, which at a regular linear value chain are part of the waste and never recover. In Table 1, we can observe the waste generated by the linear model of the gold mining activity. Furthermore, Table 1 describes the potential new products that can be generated by the gold mining industry by rethinking the system with ZRIES chains.

Table 1. Linear gold mining scenario vs. ZRIES system potential products.

The End-of-Life Scenario of Linear Gold Mining Industry Model	The Potential of Multiple ZRIES Chains.
Abandoned scrap metal from drilling machinery, bulldozers and transportation vehicles.	Each ton of recycling steel recovered for steel production uses approximately 32% less energy and generates 20% CO_2 emissions [51].
Abandoned tires from machinery and transportation vehicles with components such as metals, stabilizers, plasticizers and colorants that leach into the soil causing death to local species and bacteria in the soil [52].	There are different processes to produce new products or energy with less environmental impact such as pyrolysis, devulcanization and grinding [53,54].
Accumulation of tailings with different risks to the surrounding population's health due to volatile particles and, in the presence of sulfides, the heavy metals can be leached and contaminate soil and bodies of water nearby [55].	Remediation of soil by extracting or reducing the concentration of heavy metals makes the soil available for plant growth. There are also different processes but the most promising one is bioremediation [55].
The bodies of water are subject to mercury contamination and the leach of heavy metals from the mining activity [56,57].	Remediation costs can be reduced with phytoremediation and bioremediation techniques. Furthermore, some valuable products such as metals sulfides can be obtained [56,57].
The vegetation (plants, foliage) that is not any kind of timber plant is taken apart from the mining zone and treated as waste.	In combination with other wastes, the vegetation can be converted into biogas through anaerobic digestion [58].

5.3. Identifying the Multiple "Customers" with Their Characteristics, Pains and Gains

The value creation of our model is not centered on just one customer segment; we propose that the value creation is delivered to three "customers" based on the spheres of the triple-bottom-line (TBL): the society, the environment, and the industrial firms. Thus, this step involves an in-depth analysis of the value that the three "customers" receive from the gold extended linear value chain (Table 2).

Table 2. Triple Customer Segment Profile for the gold mining industry in Bartica, Guyana.

	"Customers"		
	Environment	**Society (Local)**	**Firms**
(1) What does this client need?	Complete its natural cycles. Natural degradation. Remediation. Conservation of regional natural resources	Good public health. Useful land for housing, economic activities, or amusement. Good revenues and prosperity Clean and drinking water.	Profit rising. Return of investment. Avoid fines. Good reputation
(2) How this product or service is beneficial to the client?	It does not generate any benefit for the local environment	Is benefited from the taxes paid by international extracting firms. It is a source of jobs.	It is a product with high demand and impacts several industries.
(3) How does this product or service negatively affect the client?	The pollution intoxicates and degrades the ecosystem. Impacts the landscape.	Pollution of water by residues. Diseases due to waste. Unsuitable land for living and recreational activities.	Bad reputation among consumers and society due to pollution. Negative international standards of competitiveness.

Osterwalder et al. [21] customer segment profile involves three parts: (1) customer jobs, the description of what the (2) the gain creators, which is how the product or service creates customer benefits; and (3) the pain relievers, which refers to how the product or service alleviate "pains" or discomforts to the client. Based on this model, we propose three questions that should be addressed, taking into account the three customers:

1.   What does this client need?
2.   How this product or service is beneficial to the client?
3.   How does this product or service negatively affect the client?

The customers and these questions can be arranged as shown in Table 2 and Figure 5. All intersections must be filled to expand the scope of benefits and pains delivered by the product or service. This process should be validated through surveys and interviews performed with the local community. This stage highlights the problems that the environment, society, and firms can face due to the activity of the linear value chain.

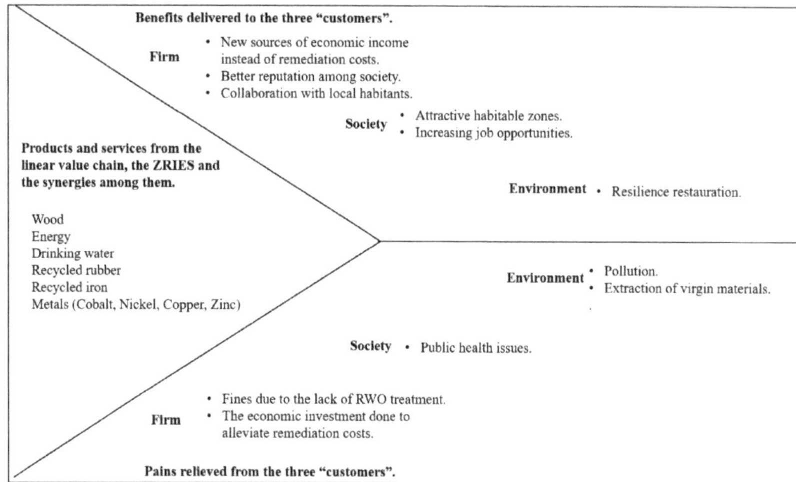

Figure 5. Triple Value Proposition. (Adapted from Osterwalder et al. [21]). Based on some of the main "products" result of the gold mining industry.

5.4. Triple Value Proposition

The development of the triple customer segment profile and the ZRIES description points out the problems that the customers face and multiple business opportunities that can be addressed [38]. These opportunities can form a cascade of initiatives that create an increasing value for the three types of "customers".

Based on the BM framework [21], the triple impact value map also entails three parts: (1) products and services delivered by the CVES, (2) the benefits to the three "customers" that the CVES delivers, and (3) the pains that the CVES relieves (Figure 5).

The TVP is a tool that helps to set the goals that need to be achieved for the value creation for the three customers. Consequently, the extended value chain and the initiatives that identify a business opportunity from the RWO must comply with the TVP. In other words, *value creation for one customer does not compromise the prosperity of other customers.* Thus, the creation of value must come from the synergy and balance between more than one customer.

5.5. Identifying the Multiple Values and Benefits for the Different Stakeholders

Following the SWIT principles [13], the processes and technologies capable of creating maximum value from the transformation of residues or waste must have been identified at this point (Section 5.2). However, it is still needed to connect the different ZRIES chains by identifying synergies among them and involving the stakeholders in the community that have a close interaction with the linear value chain and can also be interested in the business opportunities that have been identified. Chertow [3] described this as stakeholder processes, which entail the construction of committees. These committees consist of covenants and codes to rule the interaction among the stakeholders [59].

With the stakeholders identified and involved in the creation of the CVES, we finish with the TVP emerging statements, which have a similar purpose as mission statements [60,61]; they give clarity about the synergistic participation of the three customers; they highlight the importance and the purpose of each element of the cluster, and they guide the strategic plan for the stakeholders involved. Due to the systemic nature and dynamics of the CVES and the participants involved, different TVP emerging statements are used to describe the initiatives that were designed to be simultaneously working (Table 3).

Table 3. TVP emerging statements describing the value creation for the three customers.

TVP Statements
The activity of the new business opportunities generated from RWO (e.g., tailings, toxic slurry, tires, scrap metals) as raw materials originate new jobs (e.g., welders, handymen). The new jobs reduce the income inequality in the region and increase the social welfare of Bartica's inhabitants. This can raise the participation of the population with the cluster.
The identification of new business opportunities transforms RWO into beneficial raw materials (e.g., recycled metals and recycled rubber). This facilitates the production of valuable products in the market, generating new economic flows. For the firms, it represents more revenues or utilities. Therefore, the firms participating in the cluster can increase their activity within the cluster or attract more participants.
The material reprocessing due to the cluster activity can diminish the extraction of virgin materials (e.g., more raw material for tires or extraction of virgin metals). The reduction of extraction activities avoids the pollution of resources such as water, air and soil. Meanwhile, the environment's resilience increases, finally raising the availability of resources for a new cycle of the CVES cluster.
The material reprocessing generates improved public health for the local inhabitants (e.g., respiratory diseases), increasing their social welfare. This can raise the participation of the local population with the cluster.
The material reprocessing generates the reduction of remediation costs (e.g., water purification) and, in consequence, the increase of economic utilities. Finally, the rise of utilities can foster the engagement of the firm with the cluster.

5.6. Mapping the Synergies, Flows, and Clusterization of All Stakeholders, to Assemble the CVES Cluster

Finally, to better picture the CVES model we are creating, we developed a system dynamic model representing the TVP emerging statements. This model and the TVP emerging statements are made simultaneously to clarify and balance the relationship of each element in the cluster. The system dynamic model represents the linear value chain, the new business opportunities, and its relationship among the three customers. This model facilitates the monitoring of the CVES as a whole and for each of its elements. Based on the systemic approach [18,19], this model can have balance and reinforcement loops. The former tries to bring things to the desired state, and the latter produces both growth and decay; that is, the compound change in one direction with even more change (Figure 6).

In summary, the CVES map shows the dynamics among *"flows, feedbacks, synergies and stakeholders"*; creating reinforced (R) loops or balanced (B) loops, where the whole cluster has a general *purpose*: create sustainable wealth. For instance, the economic "customer" has new economic flows and fewer remediation costs to address; the environmental "customer" receives additional value with less pollution and more resilience; the social "customer" has an increased economic benefit through new jobs, improved public health, and social equality. The main balance loop in the model is due to rebound effects. We further discuss rebound effects in our model in the following section.

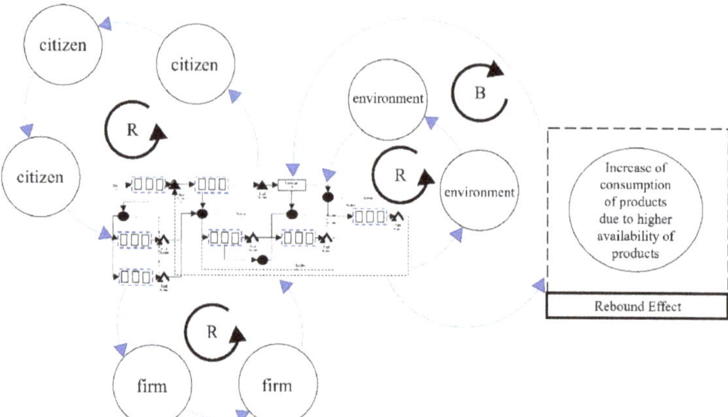

Figure 6. System Dynamics interaction of the CVES (multiple ZRIES–zero emission–chains) and the three key stakeholders.

Based on the goals and outputs of the initial report "Providing cross-sectorial solutions to support Bartica Green Town" provided by UNEP and the information gathering provided by the local team (local officials, civil society representatives from commerce chambers, businessmen, policymakers, NGOs, community leaders, among others), we have proposed to develop "Bartica's sustainable ecosystem", which is the implementation of the CVES cluster and all required capabilities and conditions "to decouple the economic growth of the environmental damage and of the social inequality" of the region, creating sustainable wealth for all citizens, its industries and its natural resources (Figure 7).

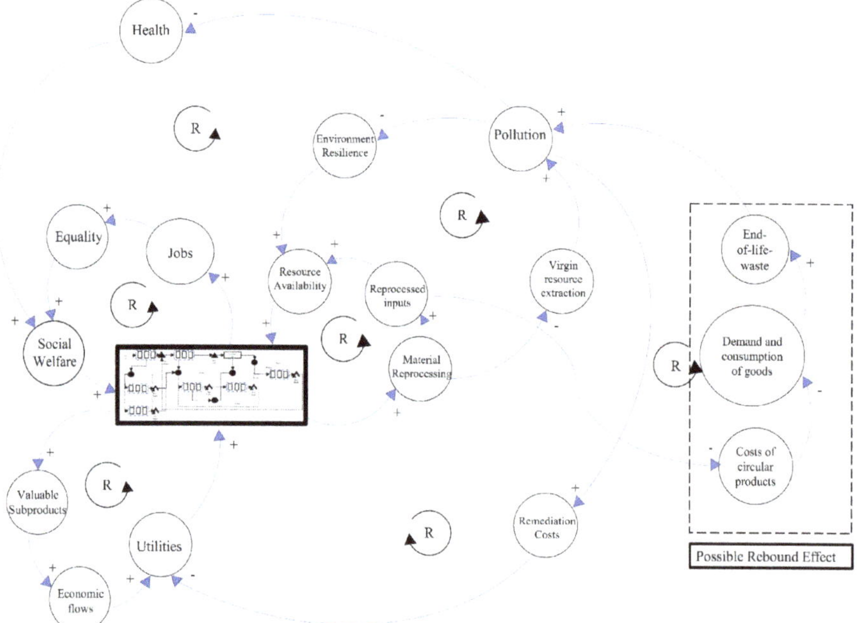

Figure 7. Bartica's sustainable ecosystem (CVES) integrates all linear chains created from the core industry of the town, the gold industry. This map shows the interaction of the system dynamics approach applied to the multiple stakeholders and businesses.

6. Discussion

Linear BMs have diminished the availability of natural resources and the planet's resilience. Although "greener" practices have attempted to alleviate the social and environmental impact of conventional BMs through more efficient processes, we have not diminished the real problems, and it is expected that they even could increase. Increasing the efficiency of linear BMs is ineffective because they do not focus on extending the life of products, the take-back of obsolete products, and holistic management of used materials to transform once again into nutrients for a new cycle of production [62].

However, going from the theory to the implementation of CE principles is not an easy task and involves a deep knowledge of external conditions that affect all regional stakeholders. Guidelines about CE business creation are scarce and focused on a few business models such as PSS, C2C loops, inverse logistics, and recycling activities [12]. With the development of the CVES, we attempt to move research towards this goal. The CVES is a "systemic and holistic approach" [13] that can guide designers and business strategists on how to implement sustainability practices to create self-sustained wealth that creates extended value to the society, the environment, and the firm by transforming linear value chains into a cluster of non-usual business, helping on the transition towards a circular economy.

We have found three main differential characteristics of the CVES in contrast to the conventional value chain. First, it is a dynamic process and not a static chain; the unit processes are not planned to be performed one after another; the cycles of energy, material, and information are performed simultaneously. Secondly, the CVES is not just driven by the activity economic actors; it needs the collaboration of multiple actors to address the shared value delivered to the social, environmental, and economic spheres. Finally, the sum of all initiatives and their synergies in the CVES can produce what we call "sustainable increasing returns" (SIR) [63]. In other words, it fosters self-managed economic growth, environmental recovery, and equal social development, all together under a holistic incremental value creation.

In the case of the gold mining industry in Bartica, the design and implementation of the CVES imply opening multiple alternatives for the local inhabitants. Bartica, as in several small towns from developing countries, usually depends on the strength of one industry that is mainly based on natural resource extraction, with the impossibility of generating other sources of income. This multiple business approach may bring to the town a green and sustainable alternative. Furthermore, geopolitical opportunities may support this new green deal for the country. When we visited the country, Guyana was in the process of receiving a large amount of funding from royalties from the oil found offshore their territorial limits. This situation may empower this type of initiative to have a more sustainable and resilient town and country.

Rebound Effects Considerations

As we mentioned before, the balance loops in the CVES model are caused by the consequences of rebound effects [37]. In this section, we will discuss how these rebound effects can be anticipated or managed.

First, one assumption is that the circular products could have less quality than linear products and therefore be offered at the lowest prices to compete in the market. However, this first assumption is easy to demonstrate that it should not necessarily be true for all circular products. For instance, some products such as bioplastics or electric vehicles are more expensive than their linear counterparts. Our model attempts to make circular products more competitive by creating a cascade of businesses, while the linear counterparts are produced in isolated linear production chains.

Secondly, to be economically attractive, some products of the CVES should be cheaper to produce than their counterparts. This might generate the rise in consumption of the product, and its net environmental impact mathematically might be higher [37]. At this point, it is important to mention that the businesses involved in the CVES are interdependent, and this limits the production of some goods. For instance, in our Bartica model, scrap metal does not exist without using mining machinery. In consequence, the availability of circular products can not follow the pace of the product demand, which might raise the prices and balance the product demand once again.

Finally, if some of the circular products can follow their demand in the market, we propose that the business that participates in a CVES should create multiple CBMs logic and not just a circular production value chain. In other words, it is not only to concentrate on how the products are produced but on how they are delivered to consumers and how far the businesses extend their responsibility in materials.

7. Conclusions

Due to the ongoing evolution of the CE literature, there is an increasing interest in the topic but also different approaches to CE implementation. In practice, some business seeks to innovate using the CE principles just aiming at reducing the environmental impact and promoting the more efficient use of natural resources (e.g., electric cars, digital platforms, green solutions) [14]. However, reformulating the value proposition of products and services and considering the social implications is marginally addressed [14]. In contrast, our model attempts to balance the economic, environmental, and social spheres. Moreover, the focus is not generating value through more efficient use of resources but the creation of a cascade of multiple businesses, forming a symbiosis that exploits different ways of value generation and becomes "eco-effective" instead (e.g., sharing economy, dematerialization, upcycling) [28].

Our model does not seek to avoid linear value chains but instead is a "systems" oriented approach that focuses on the multiple business opportunities that the linear value chains generate, as nature can create value from waste [64]. Therefore, business designers and strategists who want to apply our model should concentrate on developing a system with different stakeholders, leaving product innovation in second place. Our approach could be helpful for the implementation of "extended producer responsibility (EPR)", where the recovering and remediation costs of transforming waste produced by the linear supply chains can be reduced by reevaluating the residues using the CVES model, and at the end, to produce effective SIRs, for all the three main stakeholders of the natural ecosystem. The reason to consider the CVES as a sustainable increasing returns mechanism is that the success of the feedback loop may attract more participants [63] and, in consequence, increase the value delivered to the three stakeholders (society, environment, industry).

This research, however, has its limitations. First, due to the fragmented and increasing literature of CE, there might be other approaches focused on a different level of analysis or evaluating business models just by their resource efficiency. Secondly, from the action research experience on the field, one limitation identified is that it requires a group of leaders who can manage the tradeoffs between local stakeholders. Additionally, they should have enough technical knowledge to discover non-usual transformation processes (with local resources, machinery, labor capacities) and business opportunities for the cluster of CVES. Thirdly, the business solutions for a specific region might not be replicable to a different geographical context due to the availability of resources, capacities, conditions, characteristics of stakeholders, and mainly to the availability of the proper assembly of ABIIGS institutions (Academy, Banking, Industry, Infrastructure, Government and Society) from a specific region.

However, the steps we have described here can be replicated according to the specific customer segment profile (Table 2) and the local conditions for each new case. Finally, the governance context can foster or hinder the solutions; for instance, the regulations for water quality, food waste collection, or plastic regulations, can change depending on the local context. Nevertheless, this kind of problem can also be solved with the government's involvement as a critical institution in developing the CVES cluster.

The CVES can be improved with further research; multiple new business opportunities can be added accordingly to the expansion of the leading linear production chains. Each step can be further developed or substituted by new methods. Further research can also focus on the personal profiles and capabilities of business and social leaders who have managed to create clusters in CE projects in specific social and political circumstances.

Author Contributions: Conceptualization, C.S.; methodology, B.B.; validation, B.B. and C.S.; formal analysis, B.B.; investigation, B.B.; writing—original draft preparation, C.S. All authors have read and agreed to the published version of the manuscript.

Funding: This research received no external funding.

Institutional Review Board Statement: Not applicable.

Informed Consent Statement: Not applicable.

Data Availability Statement: Not applicable.

Conflicts of Interest: The authors declare no conflict of interest.

References

1. The Global Footprint Network. National Footprint and Biocapacity Accounts. Available online: http://data.footprintnetwork.org/?_ga=2.200497511.871541179.1605312320-1841177995.1605312320#/countryTrends?cn=5001&type=earth (accessed on 1 January 2021).
2. Ayres, R.U.; Ayres, L. *A Handbook of Industrial Ecology*; Edward Elgar Publishing: Cheltenham, UK, 2002.
3. Chertow, M. Industrial symbiosis: Literature and taxonomy. *Annu. Rev. Energy Environ.* **2000**, *25*, 313–337. [CrossRef]
4. Jacobsen, N.B. Industrial symbiosis in Kalundborg, Denmark: A quantitative assessment of economic and environmental aspects. *J. Ind. Ecol.* **2006**, *10*, 239–255. [CrossRef]
5. Zhu, Q.; Lowe, E.A.; Wei, Y.A.; Barnes, D. Industrial symbiosis in China: A case study of the Guitang Group. *J. Ind. Ecol.* **2007**, *11*, 31–42. [CrossRef]
6. Van Berkel, R.; Fujita, T.; Hashimoto, S.; Geng, Y. Industrial and urban symbiosis in Japan: Analysis of the Eco-Town program 1997–2006. *J. Env. Manag.* **2009**, *90*, 1544–1556. [CrossRef]
7. Korhonen, J. Regional industrial ecology: Examples from regional economic systems of forest industry and energy supply in Finland. *J. Env. Manag.* **2001**, *63*, 367–375. [CrossRef]
8. Weisman, A. *Gaviotas: A Village to Reinvent the World*; Chelsea Green Publishing: White River Junction, UK, 2008.
9. Martin, M.; Svensson, N.; Eklund, M. Who gets the benefits? An approach for assessing the environmental performance of industrial symbiosis. *J. Clean. Prod.* **2015**, *98*, 263–271. [CrossRef]
10. Mattila, T.; Lehtoranta, S.; Sokka, L.; Melanen, M.; Nissinen, A. Methodological aspects of applying life cycle assessment to industrial symbioses. *J. Ind. Ecol.* **2012**, *16*, 51–60. [CrossRef]
11. Ghisellini, P.; Cialani, C.; Ulgiati, S. A review on circular economy: The expected transition to a balanced interplay of environmental and economic systems. *J. Clean. Prod.* **2016**, *114*, 11–32. [CrossRef]
12. Rosa, P.; Sassanelli, C.; Terzi, S. Towards circular business models: A systematic literature review on classification frameworks and archetypes. *J. Clean. Prod.* **2019**, *236*, 117696. [CrossRef]
13. Scheel, C. Beyond sustainability. Transforming industrial zero-valued residues into increasing economic returns. *J. Clean. Prod.* **2016**, *131*, 376–386. [CrossRef]
14. Merli, R.; Preziosi, M.; Acampora, A. How do scholars approach the circular economy? A systematic literature review. *J. Clean. Prod.* **2018**, *178*, 703–722. [CrossRef]
15. Pauli, G.A. *The Blue Economy: 10 Years, 100 Innovations, 100 Million Jobs*; Paradigm Publications: Boulder, CO, USA, 2010.
16. Frosch, R.; Gallopoulos, N. Strategies for manufacturing. *Sci. Am.* **1989**, *261*, 144–153. [CrossRef]
17. Ellen MacArthur Foundation. Towards the Circular Economy: Business Rationale for an Accelerated Transition. Available online: https://www.ellenmacarthurfoundation.org/assets/downloads/TCE_Ellen-MacArthur-Foundation_9-Dec-2015.pdf (accessed on 1 January 2021).
18. Richmond, B.; Peterson, S. *An Introduction to Systems Thinking*; High Performance Systems, Incorporated: Lebanon, NH, USA, 2001.
19. Sterman, J.D. *Business Dynamics: Systems Thinking and Modeling for a Complex World*; Irwin/McGraw-Hill: Boston, MA, USA, 2000; Volume 19.
20. Osterwalder, A.; Pigneur, Y. *Business Model Generation: A Handbook for Visionaries, Game Changers, and Challengers*; John Wiley & Sons: Hoboken, NJ, USA, 2010.
21. Osterwalder, A.; Pigneur, Y.; Bernarda, G.; Smith, A. *Value Proposition Design: How to Create Products and Services Customers Want*; John Wiley & Sons: Hoboken, NJ, USA, 2014.
22. Schnitzer, H.; Ulgiati, S. Less bad is not good enough: Approaching zero emissions techniques and systems. *J. Clean. Prod.* **2007**, *15*, 1185–1189. [CrossRef]
23. Neves, A.; Godina, R.; Azevedo, S.G.; Matias, J.C. A comprehensive review of industrial symbiosis. *J. Clean. Prod.* **2020**, *247*, 119113. [CrossRef]
24. Côté, R.; Hall, J. Industrial parks as ecosystems. *J. Clean. Prod.* **1995**, *3*, 41–46. [CrossRef]
25. Lowe, E.A.; Evans, L.K. Industrial ecology and industrial ecosystems. *J. Clean. Prod.* **1995**, *3*, 47–53. [CrossRef]

26. Sokka, L.; Pakarinen, S.; Melanen, M. Industrial symbiosis contributing to more sustainable energy use–an example from the forest industry in Kymenlaakso, Finland. *J. Clean. Prod.* **2011**, *19*, 285–293. [CrossRef]
27. O'Rourke, D.; Connelly, L.; Koshland, C.P. Industrial ecology: A critical review. *Int. J. Env. Pollut.* **1996**, *6*, 89–112. [CrossRef]
28. Webster, K. *The Circular Economy: A Wealth of Flows*; Ellen MacArthur Foundation Publishing: Cowes, UK, 2015.
29. Boulding, K.E. The Economics of the Coming Spaceship Earth. Available online: https://www.laceiba.org.mx/wp-content/uploads/2017/08/Boulding-1996-The-economics-of-the-coming-spaceship-earth.pdf (accessed on 1 January 2021).
30. Pearce, D.W.; Turner, R.K. *Economics of Natural Resources and the Environment*; JHU Press: Baltimore, MD, USA, 1990.
31. Scheel, C.; Aguiñaga, E.; Bello, B. Decoupling economic development from the consumption of finite resources using circular economy. A model for developing countries. *Sustainability* **2020**, *12*, 1291. [CrossRef]
32. Aguiñaga, E.; Henriques, I.; Scheel, C.; Scheel, A. Building resilience: A self-sustainable community approach to the triple bottom line. *J. Clean. Prod.* **2017**, *173*, 186–196. [CrossRef]
33. Kraaijenhagen, C.; van Oppen, C.; Bocken, N. *Circular Business*; Circular Collaboration Publ: Amersfoot, The Netherlands, 2016.
34. Greening, L.A.; Greene, D.L.; Difiglio, C. Energy efficiency and consumption—The rebound effect—A survey. *Energy Policy* **2000**, *28*, 389–401. [CrossRef]
35. Gillingham, K.; Rapson, D.; Wagner, G. The rebound effect and energy efficiency policy. *Rev. Environ. Econ. Policy* **2016**, *10*, 68–88. [CrossRef]
36. Lin, B.; Zhao, H. Technological progress and energy rebound effect in China's textile industry: Evidence and policy implications. *Renew. Sustain. Energy Rev.* **2016**, *60*, 173–181. [CrossRef]
37. Zink, T.; Geyer, R. Circular economy rebound. *J. Ind. Ecol.* **2017**, *21*, 593–602. [CrossRef]
38. Zacarías, M.A.V.; Aguiñaga, E.; Lagunas, E.A. Sustainable entrepreneurship in industrial ecology: The cheese case in Mexico. *Int. J. Trade Glob. Mark.* **2017**, *10*, 19–27.
39. Wen, Z.; Meng, X. Quantitative assessment of industrial symbiosis for the promotion of circular economy: A case study of the printed circuit boards industry in China's Suzhou New District. *J. Clean. Prod.* **2015**, *90*, 211–219. [CrossRef]
40. Porter, M.; Kramer, M. Creating Shared Value: How to reinvent capitalism and unleash a wave of innovation and growth. *Harvard Bus. Rev.* **2017**, *11*, 1–17.
41. Graedel, T.E. Material flow analysis from origin to evolution. *Environ. Sci. Technol.* **2019**, *53*, 12188–12196. [CrossRef]
42. Van der Voet, E. Substance flow analysis methodology. In *A Handbook of Industrial Ecology*; Ayres, R.U., Ayres, L., Eds.; Edward Elgar Publishing: Cheltenham, UK, 2002. [CrossRef]
43. Zhang, L.; He, C.; Yang, A.; Yang, Q.; Han, J. Modeling and implication of coal physical input-output table in China—Based on clean coal concept. *Resour. Conserv. Recycl.* **2018**, *129*, 355–365. [CrossRef]
44. Hauschild, M.Z.; Rosenbaum, R.K.; Olsen, S.I. *Life Cycle Assessment*; Springer: Cham, Switzerland, 2018. [CrossRef]
45. Weimer, L.; Braun, T.; vom Hemdt, A. Design of a systematic value chain for lithium-ion batteries from the raw material perspective. *Resour. Policy* **2019**, *64*, 101473. [CrossRef]
46. Brink, C. Small Mining Operations: Gold in Guyana—Part I: Porknocking on the Puruni River Road. ICMJ's Prospecting and Mining Journal. Available online: https://www.icmj.com/magazine/print-article/gold-in-guyana-part-i-porknocking-on-the-puruni-river-road-2065/ (accessed on 1 January 2021).
47. Clive, Y.T. Too Big to Fail: A Scoping Study of The Small and Medium Scale Gold and Diamond Mining Industry in Guyana. Available online: https://www.yumpu.com/en/document/view/6955171/a-scoping-study-of-the-small-and-medium-scale-gold-and- (accessed on 1 January 2021).
48. Geng, Y.; Côté, R.P. Scavengers and decomposers in an eco-industrial park. *Int. J. Sustain. Dev. World Ecol.* **2002**, *9*, 333–340. [CrossRef]
49. Antikainen, M.; Teuvo, U.; Kivikytö-Reponen, P. Digitalisation as an enabler of circular economy. *Procedia CIRP* **2018**, *73*, 45–49. [CrossRef]
50. Konietzko, J.; Bocken, N.; Hultink, E.J. Online Platforms and the Circular Economy. In *Innovation for Sustainability*; Bocken, N., Ritala, P., Albareda, L., Verburg, R., Eds.; Palgrave Macmillan: Cham, Switzerland, 2019; pp. 435–450. [CrossRef]
51. Johnson, J.; Reck, B.K.; Wang, T.; Graedel, T.E. The energy benefit of stainless steel recycling. *Energy Policy* **2008**, *36*, 181–192. [CrossRef]
52. Landi, D.; Marconi, M.; Meo, I.; Germani, M. Reuse scenarios of tires textile fibers: An environmental evaluation. *Procedia Manuf.* **2018**, *21*, 329–336. [CrossRef]
53. Bockstal, L.; Berchem, T.; Schmetz, Q.; Richel, A. Devulcanisation and reclaiming of tires and rubber by physical and chemical processes: A review. *J. Clean. Prod.* **2019**, *236*, 117574. [CrossRef]
54. Li, X.; Xu, H.; Gao, Y.; Tao, Y. Comparison of end-of-life tire treatment technologies: A Chinese case study. *Waste Manag.* **2010**, *30*, 2235–2246. [CrossRef]
55. Sun, W.; Ji, B.; Khoso, S.A.; Tang, H.; Liu, R.; Wang, L.; Hu, Y. An extensive review on restoration technologies for mining tailings. *Environ. Sci. Pollut. Res.* **2018**, *25*, 33911–33925. [CrossRef]
56. Kosmützky, A.; Krücken, G. Sameness and difference: Analyzing institutional and organizational specificities of universities through mission statements. *Int. Stud. Manag. Organ.* **2015**, *45*, 137–149. [CrossRef]
57. Breznik, K.; Law, K.M. What do mission statements reveal about the values of top universities in the world? *Int. J. Organ. Anal.* **2019**, *108*, 105213. [CrossRef]

58. Bijarchiyan, M.; Sahebi, H.; Mirzamohammadi, S. A sustainable biomass network design model for bioenergy production by anaerobic digestion technology: Using agricultural residues and livestock manure. *Energy Sustain. Soc.* **2020**, *10*, 1–17. [CrossRef]
59. Work and Environment Initiative. *Handbook on Codes, Covenants, Conditions, and Restrictions for Eco-Industrial Parks*; Cornell Center Environment, Cornell University: New York, NY, USA, 1999.
60. Baetz, M.C.; Bart, C.K. Developing mission statements which work. *Long Range Plan.* **1996**, *29*, 526–533. [CrossRef]
61. Campbell, A. Mission statements. *Long Range Plan.* **1997**, *30*, 931–932. [CrossRef]
62. Scheel, C.; Aguiñaga, E. La Economía Circular, una Alternativa a los Límites del Crecimiento Lineal. Available online: https://www.researchgate.net/publication/319839814_Economia_circular_una_alternativa_a_los_limites_del_crecimiento_lineal (accessed on 1 January 2021).
63. Brian, A.W. Increasing returns and the new world of business. *Harvard Bus. Rev.* **1996**, *74*, 100–110.
64. McDonough, W.; Braungart, M. *Cradle to Cradle: Remaking the Way We Make Things*; North Point Press: New York, NY, USA, 2010.

 *sustainability*

Article

Circular Economy Strategies with Social Implications: Findings from a Case Study

Katherine Mansilla-Obando [1,*], Fabiola Jeldes-Delgado [2] and Nataly Guiñez-Cabrera [3]

[1] Faculty of Economics and Business, Universidad Finis Terrae, Santiago 7501015, Chile
[2] School of International Business, CIDEP UV, Faculty of Economic and Administrative Sciences, Universidad de Valparaíso, Valparaíso 2360102, Chile
[3] Department of Business Management, Faculty of Business Sciences, Universidad del Bío-Bío, Chillán 3810189, Chile
*   Correspondence: kmansilla@uft.cl

Abstract: To progress towards sustainable development, more companies are voluntarily committing to move from a linear economy to a circular economy (CE), mitigating resource consumption and waste generation. Despite the commitment of companies, there is a lack of understanding of how stakeholders view reduction, reuse, and recycling (3R), and the social aspects related to them. Stakeholders were asked how they perceive CE strategies, and more specifically, how they perceive that these strategies, observed in the practice of the 3Rs, transcend into social aspects. The objective of this research is to analyse stakeholders' perception of CE strategies using the 3Rs framework and stakeholder theory. Using a qualitative methodology, we conducted a case study for Green Glass, a company that uses glass as an input to manufacture its products. By analysing the content of 20 interviews, 23 videos, and 24 news items related to the company, we found that Green Glass stakeholders perceive the contribution of the 3Rs towards CE and that these have social implications, such as supplier evaluation with social impact, responsibility for the product, and decent work.

Keywords: circular economy; life cycle analysis; social implications; sustainable development; stakeholders; upcycling; 3R strategy

Citation: Mansilla-Obando, K.; Jeldes-Delgado, F.; Guiñez-Cabrera, N. Circular Economy Strategies with Social Implications: Findings from a Case Study. *Sustainability* **2022**, *14*, 13658. https://doi.org/10.3390/su142013658

Academic Editor: Ana Ramos

Received: 3 August 2022
Accepted: 17 October 2022
Published: 21 October 2022

Publisher's Note: MDPI stays neutral with regard to jurisdictional claims in published maps and institutional affiliations.

Copyright: © 2022 by the authors. Licensee MDPI, Basel, Switzerland. This article is an open access article distributed under the terms and conditions of the Creative Commons Attribution (CC BY) license (https://creativecommons.org/licenses/by/4.0/).

1. Introduction

"Until all the glasses in the world are bottled" (Green Glass company)

One of the targets of Sustainable Development Goal (SDG) number 12, imposed by the United Nations (UN) for the Global Compact partner countries in 2015, indicates that by 2030, waste generation should be significantly reduced through prevention, reduction, recycling, and reuse actions [1]. Achieving this goal is a challenge to which we are not oblivious, and the circular economy (CE) supports this implementation as it is a means to achieve sustainable development [2–6].

CE is receiving increasing attention as an alternative to the current linear take–make–dispose system [7]. For their part, companies are transforming their processes from this system to a circular one by considering it as a long-term solution to environmental problems [8–10]. While there are many definitions and interpretations attributed to CE [2,11–13], to the Ellen MacArthur Foundation [14,15], this new economy called CE is defined as "an industrial economy that is restorative or regenerative by intention and design". For its part, for the European Action Plan, the CE is "where the value of products, materials, and resources are retained in the economy for as long as possible, and waste generation is minimized" [16]. This quest to minimize waste is observed, for example, with plastics [8,17,18], food industry [19], electrical appliances [20], or solid waste [21].

For a CE model to be successful, it must contribute to the three dimensions of sustainable development [22]: economic (e.g., by reducing energy use in the production process), environmental (e.g., by reducing waste generation), and social (e.g., by improving employment conditions) [23,24]. In addition, it should consider innovative and cooperative stakeholders that help society achieve sustainability and well-being with low or no material, energy, and environmental costs, which are key elements of CE [8,25–27].

The literature on CE has been prolific, studying its origins and business models (e.g., [12,28]), its determinants and barriers (e.g., [29]), and the methods for implementing strategic models for specific solutions (e.g., [2,30]). However, there is a need to include the discussion of the CE with the stakeholders (e.g., workers or buyers) [26], as well as to examine how the analysis of CE practices can impact social aspects to improve social welfare [31,32]. According to Padilla-Rivera et al. (2020) [31], it is relevant to identify the inclusion of the social dimension in CE for a better understanding of the progress towards the transition to sustainability. In their literature review study, it is concluded that the social aspects related to CE, such as those related to employment, health, security, poverty, and gender, are important because they help understand the negative externalities that arise when moving to a CE.

Thus, this study aims to learn how CE strategies are combined with social implications from the stakeholders' point of view. By doing so, the following research question is addressed: how do stakeholders perceive CE strategies, and specifically, how do they perceive that these strategies, observed in the practice of the 3Rs, transcend in social aspects?

To answer and analyse these questions in depth, the theoretical framework of CE strategies called 3R [33], the stakeholder theory to explain the participation of key stakeholders in CE [34], and the social implications associated with CE are used [31]. A qualitative methodology is used, and the results are obtained through content analysis. For this purpose, this article provides an illustrative example of linking 3R strategies and social aspects as perceived by stakeholders in a company (Green Glass), which has been using a CE model since 2012 by reusing discarded glass bottles.

The interest in glass lies in the fact that it is a universal packaging material that causes adverse effects on the environment when it reaches landfills. This material is mainly used in the beverage sector in producing wine, beer, beverages, water, juices, or preserves; its natural greenish or crystalline colour is 100% recyclable and can be reprocessed an unlimited number of times. This packaging maintains its properties only with a previous washing for new use [35]. This is important because Chile is considered the second largest wine-producing country in Latin America [36], and the stakeholders are responsible for recycling this container, which is the function provided by the Green Glass company.

In Chile, the Ministry of the Environment (MMA) has promoted several programs and laws to promote sustainable and CE practices in the private and public sectors [35]. For example, the #ElijoReciclar program, through which, stakeholders require companies that use glass to make their recycling practices transparent. For this purpose, clean points and green points are placed in different communes of the country, recycling at home, and the deposits of 'green bells' of companies in the glass industry: CristalChile and Cristoro [37]. These companies also seek to provide social actions to corporations such as Coaniquem [38] or animal rehabilitation centers [35]. Another example is the extended producer responsibility law (EPR Law), which indicates that from 2022, the producers and importers must organize, finance, and valorise the collection of their waste, promoting a CE [39]. The stated goal is to enact 65% glass recycling by 2030 [40]. Therefore, the chilean MMA, along with producers, buyers, recyclers, and local city halls, have obligations in the industry. However, hand in hand with innovation in the glass industry, companies such as Green Glass seek to create value with glass together with their stakeholders and work under the paradigm of sustainability.

Therefore, for the analysis of this study, several sources of information are used to understand the case from the point of view of the stakeholders, such as recyclers, buyers, workers, and the company's founding partner. Twenty interviews were conducted with buyers of the Green Glass company, and 23 videos, as well as 24 news items about the Green Glass company available on the web were reviewed.

This article is structured as follows. The following Section 1.1 explains the theoretical framework of the CE: 3R strategies. Section 1.2 presents the stakeholder theory and its relationship with CE. Then, in the following Section 2, the qualitative method applied, the case selection, and the explanation of the industry under study are explained. Section 3 presents the results obtained. Section 4 discusses the findings regarding the framework of CE strategies and related social aspects. Finally, Section 5 presents the study's conclusions, managerial implications, and research limitations.

1.1. Theoretical Framework: Circular Economy Strategies: 3Rs

Circular economy (CE) is studied under the so-called "R" strategies that summarize the main ideas for further circularity [33,41]. The CE strategies: "3R" are defined as reducing, reusing, and recycling, which, when implemented in companies, achieve CE [25,33,42]. Although it is possible to find R imperatives from 4R to 10R, it is still necessary to know about the corporate commitment to 3R [8]. Studies of the 3Rs in Europe, Japan, the USA, Korea, and Vietnam identify CE strategies as initiatives related to waste management policies, where several actors participate [12,25,43,44]. That is why cooperation within and between various stakeholders, such as government, academy, non-governmental organizations, companies, and the public, is important to support and implement these strategies [33].

While [2] indicates that to achieve circularity, the order of the "R" strategies may not be the same for certain products and conditions, for [33], the "Rs" share a hierarchy, having to start with the "R" action of reducing. In this way, the first strategy is to reduce, the second to reuse, and the third to recycle.

Reducing refers to preventing waste generation as a priority [8]. This involves, for example, changing the design of products or packaging to consume the least number of resources [45]; avoiding the use of landfills to reduce greenhouse gas emissions; and managing waste for better circulation of materials [25,44], the work of grassroots recyclers and the awareness of the population for lower consumption.

On the other hand, reusing focuses on operations in which products or components that are not waste are reused for the same purpose, thus reducing the use of virgin material [2,33]. This strategy allows for the reduction of resources, energy, and labour, compared to the production of new products with virgin materials [46], which is very attractive and beneficial [25]. In the case of glass, life cycle analysis (LCA) avoids the emission of harmful substances [46].

Finally, recycling refers to recovering materials that are "waste", which are reprocessed into products, materials, or substances for the original purpose or another purpose [8]. This ensures that materials or products are kept in a reuse cycle; when this is not possible, they are recycled [8]. Therefore, the materials' consistency, purity, and efficiency of the process must be considered, such as that of glass, which has unlimited recycling and reuse [11,47]. Therefore, according to authors, recycling is the least sustainable solution in terms of efficiency and cost-effectiveness compared to previous strategies [48].

1.2. Theoretical Framework: Stakeholders Theory

To address a CE model in which stakeholder participation is crucial [26], we resorted to stakeholder theory [34]. According to [34], organizations are part of a network of relationships with stakeholders who have various interests in the organization. Stakeholders can be workers, suppliers, the government, communities, buyers, and can even, from a non-separatist logic, include owners or shareholder partners [49].

Other studies have used this theory to explain that the adhesion or inhibition of actions of companies towards the adoption of CE strategies responds, for example, to the participation, commitment, cooperation, and pressure from stakeholders on companies [26,50,51].

Based on the objective of this research, stakeholder theory allows key stakeholders to articulate CE strategies: 3R strategies from a logic that fosters the company's commitment to CE [49]. Based on this perspective, we can explain how recyclers, buyers, workers, and the founding partner, participants in the shared strategy, perceive that 3R strategies are present in the CE model of the company in this case study. Likewise, how these CE strategies and tools address and integrate the social aspects of CE [31], for example, the social impact of buyers by making donations; decent work by dignifying the work of waste pickers; and product responsibility that considers communication and labelling that promotes environmental co-creation.

2. Materials and Methods

2.1. Design of the Investigation

Under a qualitative methodology, a case study is used to analyse the perception of CE strategies in depth: 3R in the chilean company "Green Glass". This will allow the analysis of a phenomenon in a natural context, generating interaction between theory and practice [52–57].

2.2. Selection and Identification of the Case: Green Glass Company

The selected case is Green Glass, a chilean company that, since 2012, has been dedicated to transforming glass bottles into glasses. Its headquarters are in Santiago, Chile. Today, this company manufactures 58 thousand glasses per month, reaching 25 countries [58].

This company is a recognized emerging brand because it has a positive impact on society and the environment [59]. It is a member of the World Trade Fair Organization because it carries out fair trade practices, and it also has alliances with non-profit organizations through those that make charitable contributions (for example, the foundation for children with cancer and the reforestation foundation).

The criteria for selecting this company are based on the fact that Green Glass presents an adequate context to study CE in an emerging country in Latin America: Chile, which is characterized by having consumers who are very concerned about the environment and who take advantage of their actions to reduce their waste, in comparison to other countries in the sector [60].

With the above, we refer to the interviews that the founding partner has given in several media, where he expresses the principle of the company, which is to achieve a paradigm shift from the concept of garbage to something that is "useful and beautiful" as it is a "glass" [61]. In Oscar's words (founder of Green Glass): "we are trying so that people can see beyond the glass, and what that glass means" [62]. Likewise, the founding partner makes it clear that his company works to recognize and dignify the work of grassroots recyclers and that they are working to become an ecological company using 100% renewable energy.

In this way, the Green Glass company sees the importance of waste as an essential element to generate value for the environment and consumers, being a tremendous upcycling initiative [63]. In summary, Figure 1 explains the production process of Green Glass, which begins with the collection of glass bottles, and then with the manufacturing process of the glass for subsequent sale. The bottles that come from the garbage are directly collected by the base recyclers, by recycling centres, or are acquired by personnel who work with the waste of the wine companies. Later, the bottles are cleaned, cut, and polished in a treatment cellar. Then, in a creative process, some disruptive theme is selected or related to the culture or current times, for example, Nobel Prize winners of chilean literature, such as Pablo Neruda and Gabriela Mistral, the pride glass; or popular phrases or chilean memes, such as "Tell the truth Rosa", or "Friend, take a risk", which are drawn on the glass, as shown in Figures 2 and 3 with examples of the products. Finally, the glasses are stored and sold through their website as recycled and original glass glasses. In this sale, from Green Glass through special promotions or from the consumer, you can choose to make a donation to a foundation, for example, donate a tree or sterilize a puppy. More than 30 million chilean pesos have already been raised for foundations [64].

Figure 1. Production process of the Green Glass company, adapted from its website.

(**A**) (**B**)

Figure 2. Products offered on the website of the Green Glass company. (**A**) Cultural themes when showing "Chile, a country of poets". (**B**) Relevant topics such as the support of Green Glass and the "Todo Mejora" Foundation for the LGBTIQA+ community. Images obtained from the Green Glass website.

Figure 3. Products offered on the website of the Green Glass company. Popular viral phrases or chilean memes. Images obtained from the Green Glass website.

2.3. Data Collection

Case studies comprise multiple sources and techniques to collect data; in this case, a combination of these primary and secondary sources is used to promote understanding of the phenomenon under study [52,53,56]. The primary sources correspond to 20 buyers interviews and 2 field observations. The secondary sources include 24 news articles, online documentaries, 23 videos related to Green Glass, official communication from the company website, and their social media accounts on TikTok, Instagram and Facebook. These sources were used to confirm the data that emerged during the fieldwork and to complement the narrative of the interviewees, which allowed the data triangulation [65]. Using both sources ensured more excellent reliability of the results, thus reducing the bias of a single observation or the interviews carried out [66,67]. It should be noted that data collection stopped when theoretical saturation was reached [52]. Table 1 shows the number of primary and secondary sources used and Table 2 shows the characterization of the sample of interviewees. Tables A1 and A2 detail the secondary sources used to understand the phenomenon.

Table 1. Data sources used in the Green Glass case.

Source	Quantity
Primary sources	
Buyer interviews	20
Field observations	2
Secondary sources	
News	24
Videos	23

Source: Own elaboration of the authors.

Table 2. Characterization of interviewees: Green Glass buyers.

B[1]	G[2]	A[3]	N[4]	How Did You Hear about Green Glass?	Purchase Reasons
B1	F	37	7	Social networks	Gift-Recycled material
B2	F	33	4	Corporate gift	Gift-Ecofriendly
B3	F	25	2	Social networks	Recycled material
B4	M	35	6	Social networks	Design-Recycled material
B5	M	35	1	Social networks	Design
B6	F	24	1	Social networks	Gift
B7	F	35	2	Gift to friend	Design-Recycled material
B8	F	34	2	Social networks	Gift-Sale-Design
B9	F	26	1	Social networks	Design
B10	F	30	2	Social networks	Gift
B11	F	31	1	I knew the founding partner	Gift
B12	F	34	1	Social networks	Gift
B13	F	38	3	Note in news	Company B-Quality-Gift
B14	M	33	2	Friends with products	Design-Recycled material
B15	F	26	3	Gift search	Gift
B16	F	23	1	Social networks	Donaciones
B17	F	25	1	Note in news	Quality-Recycled material
B18	F	33	3	Social networks	Donations-Recycled material
B19	F	33	5	Interest in CE	Support for chilean SMEs-Recycled material
B20	F	47	1	Social networks	Donations-Recycled material-Marketing

B[1]: buyers' identification. G[2]: gender. F: female. M: male. A[3]: age in years. N[4]: number of times bought in Green Glass. Source: Own elaboration of the authors. Interviews conducted between July and August 2021.

Regarding the semi-structured interviews, these reached the saturation point at number 20; at this point, the interviewees began to reinforce the previous information without adding more new information [68]. The interviews lasted between 60 and 120 min each. These were carried out by the authors and by a research assistant during July and August 2021 by telephone. The questionnaire consisted of 15 questions where they were asked about their background, their purchasing experience, and their perceptions about the relationship between Green Glass and the CE strategies. The buyers and potential interviewees were contacted through the social network Instagram. Those who had bought a product from the Green Glass company at least once were considered viable. Likewise, they were informed of the existence of an information confidentiality report to support anonymity during the interview. Along with these interviews, field observations were made by the authors in the Green Glass company's sales stores for a better understanding of the company's process.

On the other hand, the data from secondary sources were obtained by the three authors, who worked independently in the search, compilation and generation of data to align this information to its further analysis. For example, for the analysis of the videos of the 544.66 min available on the website about the Green Glass company, each author analysed 181.55 min, corresponding to a little more than 3 h of audio-visual material for the analysis of the business. In addition, for the news, the analysis by the author was 8 news items each. This was done during August and September of the year 2021.

2.4. Data Analysis

The data collected were subjected to a careful process of analysis. Before this, it was detected in the information collected, for example, that the interviewees had little knowledge of CE in Green Glass, but after a more profound exploration and debate, the interviewees indicated a negative attitude towards the word sustainability but a favourable attitude toward its underlying dimensions, such as environmental (mentioning the protection of the land, fauna, and the reduction of pollution) and social. To eliminate bias, the data were triangulated with the secondary sources collected and the field observations made. That is why, for the analysis, the indications of the inductive method of [69] were followed based on three stages: open, axial, and selective coding. In the first stage, the three authors independently read and coded the data, sentence by sentence, with discussions and stops in the coding process to arrive at a standard set of codes. In the second stage, the three authors collected the first-order codes that were conceptually similar and relevant. This process builds meaning using an iterative grouping process where the higher order constructs present similarities and relationships between them [52,70]. This leads us to the third stage, in which, according to [71], the topics related to the perceptions of CE by the various actors were added. Here, the results were compared, and the discrepancies found were discussed [72]. In this stage, the creation of topics arose from the interpretation carried out by the researchers, which was carried out actively from the components of the grouped codes identified from the interviews, news and videos [73] and the grounded theory of CE strategies: 3R and stakeholder theory, concerning the related social implications. Based on the theories and through iterative cycles, a coherent account was obtained for the case under study.

3. Results

By exploring the perception of the stakeholders: recyclers, buyers, workers, and the founding partner, regarding the presence of the CE strategies: 3R in the Green Glass company (reduce, reuse and recycle), and their observations about the link with the social aspects associated with CE, the following results were found.

3.1. Reducing Strategy

The stakeholders who perceived the reducing strategy were recyclers, buyers, workers, and the founding partner. This means that these stakeholders recognise that Green Glass prevents the generation of waste, such as glass, which favours its management by reducing pollution and increasing people's awareness. In addition, this strategy is perceived to be connected to social aspects such as: supplier evaluation with social impact, responsibility for the product, and decent work. Specifically, recyclers perceive reducing as the importance of their work in collecting the "rubbish" of others and the social implication of having a decent job, as indicated by one of the recyclers in video 2:

> *"I am proud of what I do. I don't ask anyone for anything, just what I earn. People who litter have different perceptions of us. I feel humiliated, I feel bad, I say, 'Why, why? You're throwing it away, I'm picking it up, I'm cleaning it for you"* (V2, waste picker)

Similarly, the buyers perceive the work of the recyclers in the collection of raw materials for the Green Glass company, relating this to the social implication of decent work for them, as indicated by buyer 4:

> *"The company had a good business style, because they gave work to many people, it was not like big companies, they were independent people, so to speak. Like people who collected bottles, hired them, and they kept collecting bottles so that they would have raw material for the company"* (C4, buyer)

In addition, this stakeholder not only perceives that it is important to recover this waste, but also perceives the contribution that Green Glass makes to different campaigns related to raw materials, such as the Reforestemos foundation, which is socially related to the supplier evaluation with social impact, as indicated by buyer 5. Moreover, buyers emphasize the work they do at the grassroots level and their transparency, being responsible with the product and its impact, as indicated by buyer 3:

> *"Look, what I know is that they have basic recyclers who recover the bottles, then they pass them through the qualification process [...] I also know that, for example, with the purchases one contributes to various campaigns to contribute to reforestation and other aid campaigns. I think it qualifies as a B company"* (C5, buyer)

> *"...] in reality, the work they do at the grassroots, that's what I like the most. Like the process that comes before making the glasses, and they actually, uh, they make it very transparent on their website and the emails that you give, that you get after you sign up and buy. They keep telling you how they do it"* (C3, buyer)

The perception from the workers consists of the collection of bottles from the streets. Although it is perceived by this actor, it is not associated with a social aspect. It is observed in the worker in video 2:

> *"He used to tell me, there are so many bottles lying in the streets, I wish I had a big truck and I could take them all away"* (V2, worker)

In addition, the founding partner perceives this strategy by quantifying the waste collected, and recognizing the work of the recyclers in the process, and the importance of this in the product offered. Therefore, this perception is consistent with the social aspects previously detected: supplier evaluation with social impact, responsibility for the product, and decent work, as the founding partner indicates in video 1:

> *"All of us see that more than a million glass bottles are thrown away, we see that there is a large mass of 60,000 recyclers of people who live off society's waste. Today they are invisible, they are not recognised, they work informally and are like ghosts. We value their work, we work together with them to take people's rubbish and give it back in the form of a glass with that message"* (V1, founding partner)

3.2. Reusing Strategy

The stakeholders who perceived the reusing strategy were the buyers, workers, and the founding partner. These stakeholders recognise that Green Glass transforms "discarded" material into a marketable product, which, in turn, is connected to social aspects such as: supplier evaluation with social impact, responsibility for the product, and decent work. The buyer relates this perceived CE strategy as one that involves the social aspects of supplier evaluation with social impact, as indicated by buyers 15 and 20:

"I mean, I love the idea that they are like recycled bottles" (C15, shopper)

"Well, they are based on sustainability, they turn the bottles into cups and they have different organisations that they are helping, planting more trees for each cup you buy or for each quantity, some of that goes there or to other charities, to help others in the background, that is, with a social focus [...]" (C20, buyer)

This actor also relates this strategy to the social implication of responsibility for the product, as indicated by buyer 18:

"Yes, they wrap them with recycled paper, so it is very nice [...] when they wrap the cups, they come in a box and inside, sorry, on top of that box, comes a bag, and that bag you chop it up and you can bury it because they are biodegradable" (C18, shopper)

Workers perceive this CE strategy as a transformation into something beautiful, relating it to supplier evaluation with social impact and the decent work that Green Glass provides, as the worker in video 2 indicates:

"Here we transform that rubbish into something beautiful. It is a job that goes beyond, beyond money, beyond the work one does, it is like a mission. And perhaps a wonderful revolution" (V2, worker)

The founding partner perceives this strategy as a transformative process that is favourable for the country, identifying social implications: supplier evaluation with social impact, responsibility for the product, and decent work, as the founding partner indicates in video 8:

"Green Glass' mission is to make every glass in the world a bottle glass, [...] we tell the world this story, that we want to change the face of recycling in Chile, that we want bottles not to be thrown away, and we propose these products that are very entertaining" (V8, founding partner)

3.3. Recycling Strategy

The stakeholders who perceived the recycling strategy were the buyers and the founding partner of Green Glass. These stakeholders recognise that Green Glass recovers materials that are waste for them, reprocesses them, and ensures that they are kept in a reuse cycle. This, in turn, connects to social aspects such as: supplier evaluation with social impact, and responsibility for the product. For example, buyer 11 refers to her perception of CE, focusing on its impact on society.

"Here we know perfectly well what the production process is, and also that the production process is a process that helps with recycling, care for the environment, etc." (C11, buyer)

And buyer 13 perceives it as a social implication with regard to the responsibility for the product, as she affirms:

"The positive thing is that it is recycled material, that it is easy to recycle again, and apart from being recycled it is beautiful, you can even give it as a gift" (C13, buyer)

On the other hand, the founding partner indicates and confirms the recycling strategy, focused on the supplier evaluation with social impact:

"From all the bottles that are received at the recycling centre, if we don't use them for glass, they go back to the glassworks where they will grind all the glass and transform it into bottle again" (V3, founding partner)

The synthesis of these results in the following Figure 4.

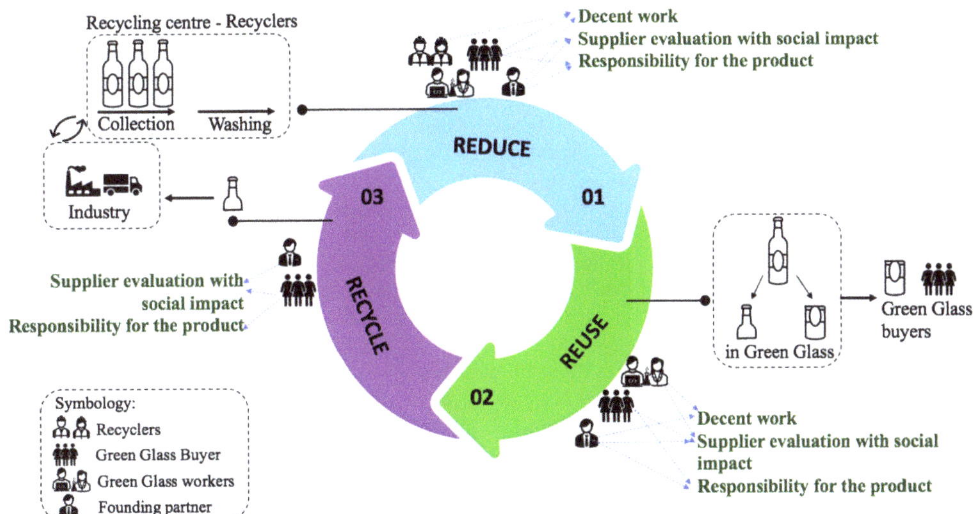

Figure 4. Results of stakeholder perception of the CE strategies: 3R and social aspects identified in the Green Glass company. Author's own elaboration.

4. Discussion

More and more companies are concerned about the environment and follow the philosophy of a CE model. For example, from the glass industry, for which the main environmental impacts from glass making are the emissions of combustion gases and the heat reaction of components, sustainability strategies can diminish their related impacts [74].

Given the persistent threat of global warming and social needs, it is important to reflect on the most common strategies of the model (3R) in a real company, through the perception of the key stakeholders involved, who see several social implications in these circular practices.

Using the case of the chilean company Green Glass, this study identified, on the one hand, that stakeholders: recyclers, buyers, workers, and the founding partner, perceive the presence of at least one of the CE strategies: 3R and, on the other hand, that most of them associate them with certain social implications.

First, beyond the findings of recent studies, which distinguish buyers and workers as stakeholders who observe and participate in CE practices, as in [26,75], we also identify key actors in the value chain of this model, such as waste pickers and the founding partner, who not only recognise the presence of the 3Rs, but also promote them from their roles in the company.

Second, we believe that the link that stakeholders have with the Green Glass company influences the perception and recognition of the CE strategy: 3R. However, we argue that the perception of the impact that these CE strategies have on society [31] can be explained by each stakeholder's experience with Green Glass. Thus, while reduction is perceived as the ability to reduce waste by recyclers, buyers, workers, and the founding partner, [8,32], the social implications of this strategy are mainly associated with decent work. CE business models have dignified work that has been discriminated against for decades, such as waste collection. The work of waste pickers is now recognised as an indispensable link that plays an important role in CE [76], considering them as a part of the community that provides guidance on the repair and replacement of the product [43]. In other words, the strategy of reducing enables recyclers' work to be recognised as decent work. This is similar to what authors call green employment in eco-innovative enterprises that promote the need to acquire and disseminate their environmental skills [77]. Therefore, as [78] points out, it is important to consider the informal recycling systems that already exist based on their practice, while working to improve efficiency and the working conditions of recyclers in developing countries. On the other hand, in CE, reducing is related to design and production, which involves eco-design, energy efficiency, and transparency in the production scale [43].

Another social implication that buyers and the founding partner associate with the strategy of reducing is supplier evaluation, with the social impact of supporting recyclers, and the social campaigns promoted by Green Glass. For example, through the Reforestemos foundation, where it is hoped that by planting trees, soil degradation, which has increased due to the accumulation of waste, will be remedied (even if only minimally). This social implication is in line with the efforts that civil society organisations are promoting to achieve the SDGs [1,79]. Likewise, buyers relate this strategy to the social implication of responsibility for the product, indicating that the Green Glass company is transparent in its processes, making visible on its website and social networks how they do things. Studied by other authors as the experience of buying circular products, where through the economy of experience, personal practices can be created that consumers value, generating a value proposition from the collection of waste in the CE [80].

The reuse strategy is perceived by workers, buyers and the founding partner as the transformation of the recovered glass bottles into a new product. This strategy allows for an extension of the life of the product [43], as well as helps to achieve closed loops of energy, as in the study in Sweden or Germany [43]. This transformation is linked to the social impact of the waste pickers and is related to decent work, which is also perceived by the workers involved in the production process at Green Glass. In fact, workers see their work as part of a mission that revolutionises what they are used to from traditional economic models that they simply discard. As indicated by [31], the CE has the potential to create new jobs, being the path to decent work; therefore, they should receive constant training and skills necessary to perform in this new economy.

Likewise, the reuse strategy is associated with the responsibility for the product, as it is perceived that there is a concern for caring for the environment with initiatives that guarantee sustainable waste management, for example, the use of paper packaging or biodegradable bags. This is in line with studies that highlight the importance of green packaging, which is the use of sustainable materials and designs for product packaging [81].

Regarding the recycling strategy, it is perceived only by buyers and the founding partner, who recognise the remanufacturing or reuse of waste by Green Glass. This strategy is linked to the supplier evaluation with social impact of caring for the environment by not generating waste or remanufacturing it [12]. The parts of the recycled bottles that are not used in the production process of a Green Glass tumbler are sent to a company that grinds the glass to make new bottles, which is relevant, as glass is 100% recyclable, unlike other tumblers that are made of glass material that require another process. Recycling is also linked to responsibility for the product by using packaging made from materials that are reused (e.g., cardboard and paper). Additionally, this strategy is linked by the energy recovery [35,43], even as strategies of the companies for which renewable energy is implemented in the CE process [43].

Finally, we believe that stakeholders' commitment and concern for environmental care guides a greater sensitivity to perceive the presence of CE strategies: 3R in the Green Glass company, e.g., consumers (e.g., [82]). The same can be said for the vision of the founding partner of Green Glass, who aims to lead the implementation of CE principles rather than acquire short-term financial gain [83], prioritising that his company's circular philosophy is recognised and shared by other stakeholders.

5. Conclusions

The aim of this paper was to provide a case study on CE based on the manufacturing of a product that contributes to 3R strategies, for which, the perceptions from recyclers, consumers, workers, and the founding partner were studied. Additionally, the practices by which these perceptions transcend into social aspects were analysed. This case study was carried out on the chilean company Green Glass. Based on a rigorous analysis using a qualitative methodology that allowed us to answer the research question of this study, it is concluded that stakeholders do perceive CE strategies by doing so through the actions of reducing, reusing, and recycling, with each stakeholder having a relevant role in the supply chain and process of the Green Glass company. In addition, the perceptions of CE detailed by each stakeholder allowed for the identification of their relationship to social aspects. These social aspects were grouped together in the supplier assessment for impacts on society, product responsibility, and decent work. However, their identification is not present in all strategies and is not perceived by all of the stakeholders considered.

Our work proposes a theoretical contribution to the framework of the CE strategies 3R: reduce, reuse and recycle, by establishing a relationship with social implications that are associated with the supplier evaluation with social impact, responsibility for the product, and decent work; all of which contribute to one of the pillars of the triple bottom line of sustainability: the social dimension. This is of great interest for the literature that indicates that, through CE, sustainability can be achieved. In doing so, this study contributes to the sustainability literature and to the SDGs proposed by the UN, mainly SDG 8: decent work and economic growth, and SDG 12: responsible production and consumption.

The managerial implications of the results of this study are aimed at confirming that, through CE, it is possible to capture the potential value of the three pillars of sustainability: economic, environmental, and social improvements promised by the 3R strategies of CE from the point of view of stakeholder perceptions and the social contribution offered by Green Glass. This is why we highlight the relevance and the example set by Green Glass to other companies and to emerging sustainable ventures and innovations that need to integrate waste recycling into their management system, which, if brought to formality and given the expected recognition, can provide a range of social benefits.

Our study, being based on a specific case study, has limitations and disadvantages in the generalisation of results and subjectivity. In this respect, the bias in the analysis of the data that could have occurred as a result of the selection of a qualitative methodology is mitigated by the use of diverse sources of information, by the coding from theory and by the triangulation carried out by the researchers. Another limitation is the geographical area analysed in an emerging country in Latin America and the specific study period.

On the in-depth understanding of a case, future research is suggested to look at similar success stories focused on CE in different economic sectors and in different contexts in Latin America, but also focusing on stories of failure. Additionally, to collect data for quantitative studies in order to analyse, for example, the intensity of social aspects in CE.

Author Contributions: All the authors designed the research. The idea of the research by K.M.-O. Conceptualization by K.M.-O. and F.J.-D. Methodology by N.G.-C. The data was collected by the authors and the research assistant. Analysis of data was performed by K.M.-O., F.J.-D. and N.G.-C. Writing, original draft preparation and editing by K.M.-O., F.J.-D. and N.G.-C. All authors have read and agreed to the published version of the manuscript.

Funding: This research received no external funding.

Institutional Review Board Statement: Not applicable.

Informed Consent Statement: Not applicable.

Data Availability Statement: Not applicable.

Acknowledgments: We are very grateful for the participation of Green Glass buyers in the interviews, and for the help of our research assistant: Paz Casaretto Palma, student at the School of Economic and Business of Universidad Finis Terrae, Chile.

Conflicts of Interest: The authors declare no conflict of interest. Green Glass had no role in the design of the study; in the collection, analyses, or interpretation of data; in the writing of the manuscript, and in the decision to publish the results. Also, the authors declare that they have no known competing financial interests or personal relationships that could have appeared to influence the work reported in this paper.

Appendix A

Table A1. Secundary Source: Analyzed Videos on Green Glass.

Video	Video Name	Year	Minutes	Video Link
V1	Historia, charla Icare	2019	27.20	https://bit.ly/3PJjVAI (accessed on 2 August 2022)
V2	Documental Netflix	2019	32.18	https://bit.ly/3oeCvoy (accessed on 2 August 2022)
V3	Prensa Mega, Mucho gusto	2019	22.35	https://bit.ly/3PqGHgI (accessed on 2 August 2022)
V4	Historia de Oscar Muñoz	2020	63.55	https://bit.ly/3uYDEnZ (accessed on 2 August 2022)
V5	Teletrece, cómo lo hizo	2018	7.11	https://bit.ly/3v4tbaq (accessed on 2 August 2022)
V6	Las cosas buenas	2018	2.13	https://bit.ly/3v1LoFy (accessed on 2 August 2022)
V7	Reforestemos	2018	1.28	https://bit.ly/3RNeaUg (accessed on 2 August 2022)
V8	TVN Buenos días a todos	2015	16.55	https://bit.ly/3ojrUIY (accessed on 2 August 2022)
V9	Green glass. Upcycled stories	2015	3.29	https://bit.ly/3aRJlxd (accessed on 2 August 2022)
V10	El éxito de Green Glass	2020	13.59	https://bit.ly/3OoP9f4 (accessed on 2 August 2022)
V11	Fundación quiltro	2019	1.40	https://bit.ly/3BlT1uP (accessed on 2 August 2022)
V12	Green glass on Tv	2012	2.44	https://bit.ly/3IWf4tO (accessed on 2 August 2022)
V13	Charla en Uruguay	2018	33.11	https://bit.ly/3OmZLeF (accessed on 2 August 2022)
V14	Inacap-video green glass	2017	1.33	https://bit.ly/3RR6rET (accessed on 2 August 2022)
V15	Café social-Canal13	2019	43.08	https://bit.ly/3PoYEfO (accessed on 2 August 2022)
V16	Entrevista a Oscar Muñoz	2016	4.59	https://bit.ly/3PrEUYU (accessed on 2 August 2022)
V17	Workcafé Santander	2020	71.16	https://bit.ly/3v3ioO1 (accessed on 2 August 2022)
V18	Green glass-la marca del año	2020	0.29	https://bit.ly/3v5wdLQ (accessed on 2 August 2022)
V19	Video FEN	2015	5.05	https://bit.ly/3cuSnRd (accessed on 2 August 2022)
V20	Emprendimiento AIEP	2020	86.05	https://bit.ly/3PrV3Oa (accessed on 2 August 2022)
V21	Charla inaugural CChC	2021	95.59	https://bit.ly/3oig2H9 (accessed on 2 August 2022)
V22	Green Glass in formula-E	2020	12.07	https://bit.ly/3oj06oc (accessed on 2 August 2022)
V23	Emprendimiento chileno	2019	9.27	https://bit.ly/3IWyA9h (accessed on 2 August 2022)
	Total		**544.66 min**	

Source: Own elaboration of the authors. Review carried out between August and September 2021.

Table A2. Secundary source: analyzed news on Green Glass.

News	News Name	Year	News Link
N1	Emprendimientos chilenos que crearon negocios siguiendo los conceptos de la EC	2020	https://bit.ly/3yXcVZX (accessed on 2 August 2022)
N2	Menos vidrio a los vertederos, la apuesta de Green Glass	2015	https://bit.ly/3zj5ii6 (accessed on 2 August 2022)
N3	Reflexiones sobre la EC en Chile	2018	https://bit.ly/3IUNjld (accessed on 2 August 2022)
N4	Dialogan sobre sustentabilidad y EC en la salmonicultura	2021	https://bit.ly/3v2KXv0 (accessed on 2 August 2022)
N5	Las iniciativas que buscan profundizar la EC en la industria	2021	https://bit.ly/3yNYpDK (accessed on 2 August 2022)
N6	Creador de green glass y haciendola.com participará en la cuarta jornada de amabilidad	2019	https://bit.ly/3v5dzUl (accessed on 2 August 2022)
N7	5 ideas para ser una empresa circular	2021	https://bit.ly/3RQHklm (accessed on 2 August 2022)
N8	Greenglass: el emprendimiento que ha salvado más de 60.000 botellas de convertirse en basura	2015	https://bit.ly/3okCoHS (accessed on 2 August 2022)
N9	En el día internacional del reciclaje realizan seminario de EC en Antofagasta	2019	https://bit.ly/3RZRf8u (accessed on 2 August 2022)
N10	Economía circular en la industria del vino	2021	https://bit.ly/3cyhDGl (accessed on 2 August 2022)
N11	Líderes en innovación social y comercio justo participarán en lanzamiento de programa birregional "Desafíos de sostenibilidad"	-	https://bit.ly/3yXe2sB (accessed on 2 August 2022)
N12	Agricultura circular: los desechos como materia prima	2021	https://bit.ly/3v0quXG (accessed on 2 August 2022)
N13	¿Qué es la EC?	-	https://bit.ly/3yVv7mT (accessed on 2 August 2022)
N14	El reciclaje es una moda y la ley de reciclaje es un somnífero	-	https://bit.ly/3zlWhom (accessed on 2 August 2022)
N15	Desafíos de sostenibilidad abre convocatoria para emprendedores y empresarios de Antofagasta y Tarapacá	2021	https://bit.ly/3ciOm20 (accessed on 2 August 2022)
N16	Fundador de Green Glass: quería tratar de cambiar el reciclaje	-	https://bit.ly/3PHWwQ1 (accessed on 2 August 2022)
N17	Green Glass: el emprendedor detrás del proyecto	2017	https://bit.ly/3PpISkS (accessed on 2 August 2022)
N18	Políticas y prácticas B	-	https://bit.ly/3yV3c6x (accessed on 2 August 2022)
N19	Green Glass: el emprendimiento chileno llega a Netflix	2018	https://bit.ly/3PpXqkl (accessed on 2 August 2022)
N20	Green Glass: distinción marca chilena emergente	-	https://bit.ly/3RRw208 (accessed on 2 August 2022)
N21	Green Glass: emprendedores con misión	2020	https://bit.ly/3v71yh1 (accessed on 2 August 2022)
N22	El joven embotellado	2017	https://bit.ly/2KZLn0l (accessed on 2 August 2022)
N23	Ejecutivos Sub 30, la nueva cepa de emprendedores	2017	https://bit.ly/3PmJI1O (accessed on 2 August 2022)
N24	Hernán Inssen, director de Hope	2019	https://bit.ly/3v2F5BS (accessed on 2 August 2022)

Source: Own elaboration of the authors. Review carried out between August and September 2021.

References

1. UNDP, United Nations Development Programme. What Are the Sustainable Development Goals? Available online: https://bit.ly/3cEdFf1 (accessed on 20 January 2022).
2. Morseletto, P. Targets for a circular economy. *Resour. Conserv. Recycl.* **2020**, *153*, 104553. [CrossRef]
3. Schroeder, P.; Anggraeni, K.; Weber, U. The relevance of circular economy practices to the sustainable development goals. *J. Ind. Ecol.* **2019**, *23*, 77–95. [CrossRef]
4. Hofmann, F. Circular business models: Business approach as driver or obstructer of sustainability transitions? *J. Clean. Prod.* **2019**, *224*, 361–374. [CrossRef]
5. Kjaer, L.L.; Pigosso, D.C.; Niero, M.; Bech, N.M.; McAloone, T.C. Product/service-systems for a circular economy: The route to decoupling economic growth from resource consumption? *J. Ind. Ecol.* **2019**, *23*, 22–35. [CrossRef]
6. Merli, R.; Preziosi, M.; Acampora, A. How do scholars approach the circular economy? A systematic literature review. *J. Clean. Prod.* **2018**, *178*, 703–722. [CrossRef]
7. Stahel, W.R. The circular economy. *Nature* **2016**, *531*, 435–438. [CrossRef]
8. Rhein, S.; Sträter, K.F. Corporate self-commitments to mitigate the global plastic crisis: Recycling rather than reduction and reuse. *J. Clean. Prod.* **2021**, *296*, 126571. [CrossRef]
9. European Commission. A European Strategy for Plastics in a Circular Economy. Available online: https://bit.ly/3b38chE (accessed on 30 November 2021).
10. Kirchherr, J.; Piscicelli, L.; Bour, R.; Kostense-Smit, E.; Muller, J.; HuibrechtseTruijens, A.; Hekkert, M. Barriers to the circular economy: Evidence from the European Union (EU). *Ecol. Econ.* **2018**, *150*, 264–272. [CrossRef]
11. Haupt, M.; Hellweg, S. Measuring the environmental sustainability of a circular economy. *Environ. Sustain. Indic.* **2019**, *1*, 100005. [CrossRef]
12. Kirchherr, J.; Reike, D.; Hekkert, M. Conceptualizing the circular economy: An analysis of 114 definitions. *Resour. Conserv. Recycl.* **2017**, *127*, 221–232. [CrossRef]
13. Prieto-Sandoval, V.; Jaca, C.; Ormazabal, M. Towards a consensus on the circular economy. *J. Clean. Prod.* **2018**, *179*, 605–615. [CrossRef]
14. Ellen MacArthur Foundation. Towards the Circular Economy. Economic and Business Rationale for an Accelerated Transition. Available online: https://bit.ly/3os12Xm (accessed on 30 November 2021).
15. MacArthur, E. Towards the circular economy. *J. Ind. Ecol.* **2013**, *2*, 23–44.
16. European Commission. Communication from the Commission to the European Parliament, the Council, the European Economic and Social Committee and the Committee of the Regions. Closing the Loop—An EU Action Plan for the Circular Economy. Available online: https://bit.ly/2xa2ztg (accessed on 30 November 2021).
17. Beaumont, N.J.; Aanesen, M.; Austen, M.C.; Börger, T.; Clark, J.R.; Cole, M.; Hooper, T.; Lindeque, P.K.; Pascoe, C.; Wyles, K.J. Global ecological, social and economic impacts of marine plastic. *Mar. Pollut. Bull.* **2019**, *142*, 189–195. [CrossRef]
18. Simon, B. What are the most significant aspects of supporting the circular economy in the plastic industry? *Resour. Conserv. Recycl.* **2019**, *141*, 299–300. [CrossRef]
19. Borrello, M.; Caracciolo, F.; Lombardi, A.; Pascucci, S.; Cembalo, L. Consumers' perspective on circular economy strategy for reducing food waste. *Sustainability* **2017**, *9*, 141. [CrossRef]
20. Bressanelli, G.; Saccani, N.; Pigosso, D.C.; Perona, M. Circular economy in the WEEE industry: A systematic review of the literature and a research agenda. *Sustain. Prod. Consum.* **2020**, *23*, 174–188. [CrossRef]
21. Puntillo, P.; Gulluscio, C.; Huisingh, D.; Veltri, S. Reevaluating waste as a resource under a circular economy approach from a system perspective: Findings from a case study. *Bus. Strategy Environ.* **2021**, *30*, 968–984. [CrossRef]
22. Elkington, J.; Rowlands, I.H. Cannibals with forks: The triple bottom line of 21st century business. *Altern. J.* **1999**, *25*, 42.
23. Korhonen, J.; Honkasalo, A.; Seppälä, J. Circular economy: The concept and its limitations. *Ecol. Econ.* **2018**, *143*, 37–46. [CrossRef]
24. Elia, V.; Gnoni, M.G.; Tornese, F. Measuring circular economy strategies through index methods: A critical analysis. *J. Clean. Prod.* **2017**, *142*, 2741–2751. [CrossRef]
25. Ghisellini, P.; Cialani, C.; Ulgiati, S. A review on circular economy: The expected transition to a balanced interplay of environmental and economic systems. *J. Clean. Prod.* **2016**, *114*, 11–32. [CrossRef]
26. Mies, A.; Gold, S. Mapping the social dimension of the circular economy. *J. Clean. Prod.* **2021**, *321*, 128960. [CrossRef]
27. Bocken, N.M.; De Pauw, I.; Bakker, C.; Van Der Grinten, B. Product design and business model strategies for a circular economy. *J. Ind. Prod. Eng.* **2016**, *33*, 308–320. [CrossRef]
28. Lewandowski, M. Designing the Business Models for Circular Economy-Towards the Conceptual Framework. *Sustainability* **2016**, *8*, 43. [CrossRef]
29. De Jesus, A.; Mendonça, S. Lost in transition? Drivers and barriers in the eco-innovation road to the circular economy. *Ecol. Econ.* **2018**, *145*, 75–89. [CrossRef]
30. Reike, D.; Vermeulen, W.J.; Witjes, S. The circular economy: New or refurbished as CE 3.0?—Exploring controversies in the conceptualization of the circular economy through a focus on history and resource value retention options. *Resour. Conserv. Recycl.* **2018**, *135*, 246–264. [CrossRef]

31. Padilla-Rivera, A.; Russo-Garrido, S.; Merveille, N. Addressing the social aspects of a circular economy: A systematic literature review. *Sustainability* **2020**, *12*, 7912. [CrossRef]

32. Van Langen, S.K.; Vassillo, C.; Ghisellini, P.; Restaino, D.; Passaro, R.; Ulgiati, S. Promoting circular economy transition: A study about perceptions and awareness by different stakeholders groups. *J. Clean. Prod.* **2021**, *316*, 128166. [CrossRef]

33. Demestichas, K.; Daskalakis, E. Information and communication technology solutions for the circular economy. *Sustainability* **2020**, *12*, 7272. [CrossRef]

34. Freeman, E.R. *Strategic Management: A Stakeholder Approach*; Pitman: Boston, MA, USA, 1984.

35. MMA, Ministerio del Medio Ambiente. Diagnóstico Producción, Importación y Distribución de Envases y Embalajes y el Manejo de los Residuos de Envases y Embalajes. Informe Final. 2010. Available online: https://bit.ly/3TiBmKy (accessed on 20 January 2022).

36. Statista. Ranking de Países de América Latina con Mayor Volumen de Vino Producido Entre 2017 y 2020. Available online: https://bit.ly/3glu7Tg (accessed on 20 September 2022).

37. MMA, Ministerio del Medio Ambiente. #ElijoReciclar. Available online: https://bit.ly/3RYIkny (accessed on 20 January 2022).

38. Mansilla-Obando, K.; Jeldes-Delgado, F.; Guiñez-Cabrera, N.; Ortiz-Henríquez, R. Modelo de negocio de economía circular: Caso tienda solidaria COANIQUEM. *Cuad. Adm.* **2021**, *37*, e2210822. [CrossRef]

39. MMA, Ministerio del Medio Ambiente. Ley de Fomento al Reciclaje. Available online: https://bit.ly/2lw5t66 (accessed on 20 January 2022).

40. País Circular. Ley REP Fija Meta de 60% de Reciclaje de los Envases y Embalajes en Chile al año 2030. Available online: https://bit.ly/2N2sR8G (accessed on 20 November 2021).

41. Potting, J.; Hekkert, M.P.; Worrell, E.; Hanemaaijer, A. *Circular Economy: Measuring Innovation in the Product Chain. Policy Report*; PBL Publishers: The Hague, The Netherlands, 2017.

42. Stewart, R.; Niero, M. Circular economy in corporate sustainability strategies: A review of corporate sustainability reports in the fast-moving consumer goods sector. *Bus. Strategy Environ.* **2018**, *27*, 1005–1022. [CrossRef]

43. Kalmykova, Y.; Sadagopan, M.; Rosado, L. Circular economy-from review of theories and practices to development of implementation tools. *Resour. Conserv. Recycl.* **2018**, *135*, 190–201. [CrossRef]

44. Sakai, S.I.; Yoshida, H.; Hirai, Y.; Asari, M.; Takigami, H.; Takahashi, S.; Tomoda, K.; Peeler, M.; Wejchert, J.; Schmid-Unterseh, T.; et al. International comparative study of 3R and waste management policy developments. *J. Mater. Cycles Waste Manag.* **2011**, *13*, 86–102. [CrossRef]

45. Hultman, J.; Corvellec, H. The European Waste Hierarchy: From the sociomateriality of waste to a politics of consumption. *Environ. Plan. A Econ. Space* **2012**, *44*, 2413–2427. [CrossRef]

46. Castellani, V.; Sala, S.; Mirabella, N. Beyond the throwaway society: A life cycle-based assessment of the environmental benefit of reuse. *Integr. Environ. Assess. Manag.* **2015**, *11*, 373–382. [CrossRef]

47. Reh, L. Process engineering in circular economy. *Particuology* **2013**, *11*, 119–133. [CrossRef]

48. European Commission. Reuse is the Key to the Circular Economy. Available online: https://bit.ly/3otxAjH (accessed on 20 January 2022).

49. Freeman, R.E.; Wicks, A.C.; Parmar, B. Stakeholder theory and the corporate objective revisited. *Organ. Sci.* **2004**, *15*, 364–369. [CrossRef]

50. Gupta, S.; Chen, H.; Hazen, B.T.; Kaur, S.; Gonzalez, E.D.S. Circular economy and big data analytics: A stakeholder perspective. *Technol. Forecast. Soc. Change* **2019**, *144*, 466–474. [CrossRef]

51. Jabbour, C.J.C.; Seuring, S.; de Sousa Jabbour, A.B.L.; Jugend, D.; Fiorini, P.D.C.; Latan, H.; Izeppi, W.C. Stakeholders, innovative business models for the circular economy and sustainable performance of firms in an emerging economy facing institutional voids. *J. Environ. Manag.* **2020**, *264*, 110416. [CrossRef]

52. Eisenhardt, K.M. Building theories from case study research. *Acad. Manag. Rev.* **1989**, *14*, 532–550. [CrossRef]

53. Eisenhardt, K.M.; Graebner, M.E. Theory building from cases: Opportunities and challenges. *Acad. Manag. J.* **2007**, *50*, 25–32. [CrossRef]

54. Gerring, J. What is a case study and what is it good for? *Am. Political Sci. Rev.* **2004**, *98*, 341–354. [CrossRef]

55. Yin, R.K. *Case Study Research: Design and Methods*, 3rd ed.; SAGE Publications: Thousand Oaks, CA, USA, 2003.

56. Yin, R.K. *Case Study Research: Design and Methods*, 5th ed.; SAGE Publications: Thousand Oaks, CA, USA, 2014.

57. Yin, R.K. *Case Study Research and Applications: Design and Methods*, 6th ed.; SAGE Publications: Thousand Oaks, CA, USA, 2018.

58. Green Glass. La Historia Real de Green Glass. Available online: https://bit.ly/3zuRwJd (accessed on 20 November 2021).

59. Grandes Marcas. Green Glass-Distinción Marca Chilena Emergente. Available online: https://bit.ly/3RRw208 (accessed on 20 November 2021).

60. Las Personas Más Comprometidas con el Medio Ambiente–Eco Actives–Llegarán al 43% de la Población Latina en 10 Años. Available online: https://bit.ly/3viYi2d (accessed on 20 January 2022).

61. Canal13C. Las Cosas Buenas-Green Glass. Available online: https://bit.ly/3v1LoFy (accessed on 20 November 2021).

62. Green Glass. Prensa Mega, Mucho Gusto. Available online: https://bit.ly/3PqGHgI (accessed on 20 November 2021).

63. Francamagazine. Hernán Inssen, Director de Hope. Available online: https://bit.ly/3v2F5BS (accessed on 20 November 2021).

64. Green Glass. Impacto. Available online: https://bit.ly/3b3qhMt (accessed on 20 November 2021).

65. Jick, T.D. Mixing qualitative and quantitative methods: Triangulation in action. *Adm. Sci. Q.* **1979**, *24*, 602–611. [CrossRef]

66. Flick, U. Triangulation in qualitative research. *Companion Qual. Res.* **2004**, *3*, 178–183.

67. Tarrow, S. Bridging the quantitative-qualitative divide in political science. *Am. Political Sci. Rev.* **1995**, *89*, 471–474. [CrossRef]
68. Glaser, B.G.; Strauss, A.L. *The Discovery of Grounded Theory: Strategies for Qualitative Research*; Routledge: New York, NY, USA, 1967; p. 282.
69. Strauss, A.; Corbin, J. *Basics of Qualitative Research*; Sage Publications: Thousand Oaks, CA, USA, 1990.
70. Rego, A.; Cunha, M.P.; Polónia, D. Corporate sustainability: A view from the top. *J. Bus. Ethics* **2017**, *143*, 133–157. [CrossRef]
71. Gioia, D.A.; Corley, K.G.; Hamilton, A.L. Seeking qualitative rigor in inductive research: Notes on the Gioia methodology. *Organ. Res. Methods* **2013**, *16*, 15–31. [CrossRef]
72. Olesen, V.; Droes, N.; Hatton, D.; Chico, N.; Schatzman, L. *Analysing Together. Analysing Qualitative Data*; Routledge: London, UK, 1994.
73. Batra, R.; Ahuvia, A.; Bagozzi, R.P. Brand love. *J. Mark.* **2012**, *76*, 1–16. [CrossRef]
74. Butler, J.; Hooper, P. *Glass Waste*; Elsevier: Amsterdam, The Netherlands, 2019; Chapter 15. [CrossRef]
75. Camacho-Otero, J.; Boks, C.; Pettersen, I.N. Consumption in the circular economy: A literature review. *Sustainability* **2018**, *10*, 2758. [CrossRef]
76. Tong, Y.D.; Huynh, T.D.X.; Khong, T.D. Understanding the role of informal sector for sustainable development of municipal solid waste management system: A case study in Vietnam. *Waste Manag.* **2021**, *124*, 118–127. [CrossRef]
77. Bassi, F.; Guidolin, M. Resource efficiency and Circular Economy in European SMEs: Investigating the role of green jobs and skills. *Sustainability* **2021**, *13*, 12136. [CrossRef]
78. Wilson, D.C.; Velis, C.; Cheeseman, C. Role of informal sector recycling in waste management in developing countries. *Habitat Int.* **2006**, *30*, 797–808. [CrossRef]
79. Mohan, M.; Rue, H.A.; Bajaj, S.; Galgamuwa, G.P.; Adrah, E.; Aghai, M.M.; Broadbent, E.; Khadamkar, O.; Sasmito, S.; Roise, J.; et al. Afforestation, reforestation and new challenges from COVID-19: Thirty-three recommendations to support Civil Society Organizations (CSOs). *J. Environ. Manag.* **2021**, *287*, 112277. [CrossRef]
80. Ta, A.H.; Aarikka-Stenroos, L.; Litovuo, L. Customer Experience in Circular Economy: Experiential Dimensions among Consumers of Reused and Recycled Clothes. *Sustainability* **2022**, *14*, 509. [CrossRef]
81. Wandosell, G.; Parra-Meroño, M.C.; Alcayde, A.; Baños, R. Green packaging from consumer and business perspectives. *Sustainability* **2021**, *13*, 1356. [CrossRef]
82. Testa, F.; Iovino, R.; Iraldo, F. The circular economy and consumer behaviour: The mediating role of information seeking in buying circular packaging. *Bus. Strategy Environ.* **2020**, *29*, 3435–3448. [CrossRef]
83. Sohal, A.; De Vass, T. Australian SME's experience in transitioning to circular economy. *J. Bus. Res.* **2022**, *142*, 594–604. [CrossRef]

Article

Social Impact Assessment of Circular Construction: Case of Living Lab Ghent

Nuri Cihan Kayaçetin *, Chiara Piccardo and Alexis Versele

Building Physics and Sustainable Design Unit, Department of Engineering Technology, Ghent Technology Campus, Katholieke Universiteit Leuven, 9000 Ghent, Belgium
* Correspondence: cihan.kayacetin@kuleuven.be

Abstract: The construction industry is considered to have a high potential in achieving the sustainable development goals. The circular economy is a promising framework that supports the shift from a linear-construction industry to an environmental-friendly and efficient sector. On the other hand, there is a lack of effort in measuring the impact of construction-related activities on users and society. The gap is greater when the context of social impacts is related to circular and bio-based construction. For this purpose, a social impact assessment framework was developed in the Interreg 2 seas CBCI project and tested on a residential prototype: Living Lab (LL) Ghent. Under 13 impact categories relevant to 4 stakeholder categories, circular and bio-based construction materials and methods were assessed for production and construction phases. Qualitative and quantitative data were collected through expert workshops and questionnaires. The results include identification of new indicators (urban mining, social economy, and post-intervention manuals) for several circular construction methods. The social impacts of the LL were discussed depending on each stakeholder category. It was seen that there are several positive impacts related to workers and the local community. Certain recommendations were also provided specifically on a construction-sector basis which may be integrated into existing social impact assessment guidelines.

Keywords: social life cycle assessment; circular economy; bio-based construction; living lab

Citation: Kayaçetin, N.C.; Piccardo, C.; Versele, A. Social Impact Assessment of Circular Construction: Case of Living Lab Ghent. *Sustainability* **2023**, *15*, 721. https://doi.org/10.3390/su15010721

Academic Editor: Ana Ramos

Received: 19 November 2022
Revised: 19 December 2022
Accepted: 28 December 2022
Published: 31 December 2022

Copyright: © 2022 by the authors. Licensee MDPI, Basel, Switzerland. This article is an open access article distributed under the terms and conditions of the Creative Commons Attribution (CC BY) license (https://creativecommons.org/licenses/by/4.0/).

1. Introduction

There is an increasing interest in measuring social impact from different sides, including public and private investors. The GECES [1] expert group on social economy and social enterprises gave the following definition to measuring social impact "The reflection of social outcomes as measurement, both long-term and short-term, adjusted for the effects achieved by others (alternative attribution), for effects that would have happened anyway (deadweight), for negative consequences (displacement) and for effects declining over time (drop off)".

The EU has developed standards for the assessment of the sustainability aspects of new and existing construction works (buildings and civil engineering works) [2]. They describe methodologies for the assessment of the sustainability of construction works covering the assessment of environmental, social, and economic performance (aspect and impacts). On the other hand, the previous research trends focused more on the technical (environmental and economic) aspects rather than the social performance. Some reasons can be referred to as the lack of data, difficulties in conducting social sciences, and lack of standard methods for quantifying social impacts.

The importance of quantifying social impacts is crucial for achieving stakeholder engagement for emerging concepts. The current trends in research and development focus on the transition from a linear construction into a circular and efficient sector [3,4]. The linear production model drives unmanageable economic growth which is unsustainable in the long run [5]. A circular economy may resolve this—and bring about a list of social benefits. These benefits may include new employment opportunities, increased cooperation and sense of community by participation in the sharing economy, and co-ownership of a physical product aiming for sharing functions and services by the user group [6].

In the scope of the circular economy, social assessment is not yet a well-developed or often-applied practice. Walker et al. [7] mention several challenges, and most are related to the difficulty of measuring social indicators. The most frequently observed reason for not including a social assessment was the lack of knowledge to execute one, followed by the complexity of the methodology, the lack of a standardised method, the available methods not being 'best practice' for social assessment, and lack of supply chain data. A low personnel number may further explain the lack of resources in SMEs to include social assessment activities.

In this context, this study adopts a social life cycle assessment (S-LCA) approach for achieving comprehensive and quantifiable outcomes. The aim of this study is to assess the social impact of a circular and bio-based housing prototype developed in KU Leuven Ghent Technology Campus in the scope of the Interreg 2 seas project CBCI [8]. For this purpose, existing social impact assessment studies were reviewed, and it was seen that there is a lack of research efforts on circular construction practices. Deriving from the existing literature, the study considers two hypotheses regarding the methodology:

- Circular construction methods have positive social impacts;
- There is a need for additional research for impact categories and indicators for emerging construction methods.

In order to explore these hypotheses, an assessment framework for circular and bio-based construction methods, including production and construction life cycle phases, was developed. For demonstrative reasons, the framework was applied to the case study and social impacts for several stakeholders were analyzed. In the end of the study, construction industry-specific suggestions were also provided to the existing guidelines.

2. Literature Review

In this section, existing documents regulating social impact assessment in the construction industry are reviewed together with the current literature.

2.1. S-LCA Standards and Guidelines

Social performance of a building is one of the pillars of the sustainability framework (together with environmental and economic performance). One of the prominent assessment methods for social impact is through a social life cycle assessment (S-LCA). The assessment of social performance differs from economic and environmental assessments as it requires, besides quantitative, also descriptive approaches. Where methods leading to a quantitative result are not available for assessment criteria and indicators, a checklist analysis approach is adopted to make the descriptive approach quantifiable.

The goal of an S-LCA is to quantify the social performance of the object of assessment by means of the compilation and application of information relevant to a description of the social quality of the object. The governing standard is EN 16309—sustainability of construction works—assessment of social performance of buildings [9] (hereafter referred to as 'standard'). The standard focuses on the use stage of a building by considering the following performance categories: accessibility, adaptability, health and comfort, impacts on the neighbourhood, maintenance, and safety and security. However, as the recommended life cycle module for S-LCA only covers the use stage, this approach may not be relevant for the assessment of a circular construction product.

Besides the standard, there have been efforts to develop the 'Guidelines for Social Life Cycle Assessment of Products and Organizations' [10] (hereafter referred to as 'guidelines'). The importance of the guidelines lies in the provision of necessary information on impact categories and indicators and data collection methods [11]. Currently, it does not include sector-specific recommendations or case studies from the construction sector [12].

2.2. S-LCA in the Construction Industry

In this section, an overview of S-LCA studies in the construction sector is provided. The keywords used to define the boundaries of this overview are 'life cycle' and 'social LCA' OR 'social assessment', and 'construction' OR 'buildings' OR 'built environment'. Only studies related to the building sector are considered. As shown in Table 1, the collection of studies is limited in number but geographically widespread. In addition, these studies commonly referred to the aforementioned S-LCA guidelines. The majority of the studies are conducted at the building level. Furthermore, there is a tendency to integrate S-LCA with the general sustainability frameworks.

Table 1. Social impact assessment studies in construction sector.

Title	Authors	Year	Keywords
Social life cycle assessment for material selection: a case study of building materials	Hosseinijou et al. [13]	2014	AHP, Material flow analysis, Hotspot analysis, S-LCA
A social life cycle assessment model for building construction in Hong Kong	Dong and Ng [14]	2015	Building construction, LCA, Precast concrete, sLCIA, Social LCA, SMoC
Comparative life cycle social assessment of buildings: Health and comfort criterion	Santos et al. [15]	2016	LCA, S-LCA, building performance, health and comfort criterion, comparative assessment
Assessment of health and comfort criteria in a life cycle social context: Application to buildings for higher education	Santos et al. [16]	2017	LCA, S-LCA, AHP, health and comfort criterion, comparative assessment, education buildings
Development of social sustainability assessment method and a comparative case study on assessing recycled construction materials	Hossain et al. [17]	2018	Construction materials, recycled materials, social life cycle assessment, SSG model
Evaluation of social life-cycle performance of buildings: Theoretical framework and impact assessment approach	Liu and Qian [18]	2019	Social sustainability assessment, Social LCA, Multi-stakeholder approach
Built environment design—social sustainability relation in urban renewal	Yıldız et al. [19]	2020	Urban renewal, social sustainability
Multi-criteria assessment approach for a residential building retrofit in Norway	Chen et al. [20]	2020	Building retrofit, cost optimal analysis, carbon emission, social assessment, MCA
Life cycle sustainability assessment analysis of different concrete construction techniques for residential building in Malaysia	Balasbaneh and Sher [21]	2021	Concrete buildings, construction techniques, life cycle sustainability, MCA
How to conduct consistent environmental, economic, and social assessment during the building design process	Soust-Verdaguer et al. [22]	2022	LCSA, LCA, LCI, LCC, BIM, triple bottom line sustainability assessment
BIM-based LCSA application in early design stages using IFC	LLatas et al. [23]	2022	LCSA, BIM, IFC, data structure, building design process
On the possibilities of multilevel analysis to cover data gaps in consequential S-LCA: Case of multistory residential building	Rizal Taufiq et al. [24]	2022	Consequential S-LCA, residential building, multilevel analysis
Developing a building performance score model for assessing the sustainability of buildings	Thanu et al. [25]	2022	Green building rating system, triple bottom line of sustainability, construction industry

A few studies have used S-LCA to compare different construction materials and techniques. Dong and Ng [14] developed a framework to support the local construction industry in the social impact assessment of construction works. The framework was tested on a case-study building in Hong Kong and compared several construction practices. Hosseinijou et al. [13] used a similar approach to assess the social impact of a case-study building assumed to have either a steel or concrete structural frame. The social impact assessment was integrated to a material flow analysis in order to identify relevant flows of materials in the life cycle, as well as social hot spots to analyze through interviews. Liu and Qian [18] developed a framework to compare steel and concrete structural frames, including not only the production and construction stages but also the operation and maintenance stages. Hossain [17] introduced a similar study on social sustainability for recycled construction materials. Balasbaneh and Sher [21] developed a multi-criteria assessment method in order to compare different types of concrete constructions, including environmental, economic, and social aspects, from the production to the operation stage.

Other studies have used multi-criteria assessment methods including social aspects in order to provide a full sustainability assessment of buildings. For example, Chen et al. [20] used criteria to assess the social impact of building renovations focusing on the operation stage of buildings. Thanu et al. [25] developed a similar approach to assess the social impacts of new buildings. Soust-Verdaguer et al. [22] and LLatas et al. [23] assessed environmental, economic, and social impacts of a case-study building in the early design stage and used BIM in order to facilitate the life cycle inventory. However, a limitation of these studies [22,23] is the use of a single parameter (i.e., working hours) to express social impacts.

Some studies explored indicators for the social impact assessment of buildings. For example, Santos et al. [15] and Yıldız et al. [19] used factor analysis to identify social aspects relevant for new buildings and building renovations, respectively. Finally, Rizal Taufiq et al. [24] assessed the social impacts of different materials commonly used in constructions at different levels (i.e., process, company, and country) in order to understand the implications of scaling up a social impact assessment. This is the only reviewed study including all the stages of the life cycle.

In these studies, the functional unit selection does not follow a common standard, in which some studies considered a whole building as their unit, whereas another adopted a common unit of 1 m². It was also observed that the set of indicators were individually unique for all studies. Depending on these insights, this study also explored the needs for conducting an S-LCA study on circular and bio-based building.

There are also several studies that focus on the social impacts of construction with other methodologies than S-LCA. Mesa et al. [26] provided a thorough overview of studies that focus on environmental and social life cycle impacts of construction demolition waste (CDW). Locurcio et al. [27] introduced a multi-criteria composite indicator to enable performance analysis of real estate investments, trying to respond whether the investors should renovate or reconstruct. Ibrahim et al. [28] and Wang et al. [29] developed frameworks to evaluate the socio-economic impacts of construction projects on society. Related to social impacts, there are several studies on the impact of circular economy methods for CDW [30–32].

It should also be noted that social impact assessment can still be considered as an emerging concept in the construction sector, and adapting it to circular construction would need further research and discussions. It would also be acceptable to claim that case study applications are one of the methods to close the gaps in the literature.

3. Materials and Methods

This study follows the suggestions of the available standard and guidelines, but also sought for innovation in S-LCA methodology. S-LCA is a comprehensive method which is also capable of providing quantifiable results. These characteristics match with the goal of this study: inclusion of several stakeholder groups and to display the social impacts in a tangible manner. As Figure 1 displays, the workflow of the study includes the definition of the case study called LL Ghent, two workshops for setting up the S-LCA, and the assessment of circular and bio-based construction practices via interviews with producers and constructors of LL Ghent.

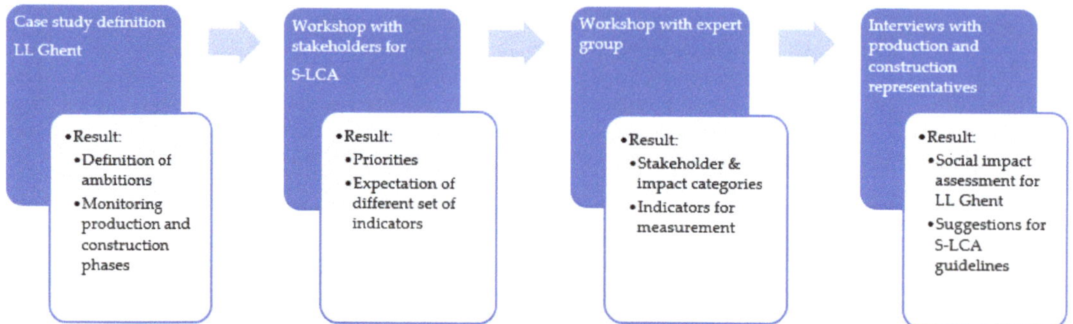

Figure 1. Workflow of the study.

In the literature, it was seen that existing categories or indicators may not be sufficient to consider some emerging processes. In order to ensure that this study assesses correct impacts categories and indicators, two workshops were conducted with different focus groups for determining the system boundaries of the study. The focus groups were categorized as (i) the expert group who draw the boundaries and (ii) the implementation group who validates the suggestions.

After determining adequate indicators, two sets of questionnaires were developed and an S-LCA was conducted on LL Ghent. The scope of the S-LCA considers the product level and assesses the impact of LL Ghent during production and construction phases.

3.1. Case Study Building—LL Ghent

In the aforementioned CBCI project, a case-study building called 'LL Ghent' was adopted to test the S-LCA methodology for circular and bio-based constructions.

LL Ghent was developed through a research through design (RtD) approach in order to explore technical, financial, and legal frameworks with the ultimate goal of creating an energy- and material-efficient building prototype. Additional preliminary studies were conducted, such as (i) an overview of evaluation tools to be utilized in the assessment of circularity [33] and (ii) an integrated decision support that combines environmental life cycle analysis and circularity assessment by using several impact categories that are monetized into a unit of $€/m^2$ [34]. Based on these preliminary studies, several ambitions for LL were identified:

- Bio-based $\rightarrow \rightarrow 75\%$ (in volume);
- Demountable $\rightarrow 0.7$ demountability index;
- Passive house $\rightarrow$ U value 0.15 $W/m^2 \cdot k$;
- Social economy $\rightarrow 10\%$ of investment.

A multi-criteria analysis was conducted to support the procurement of the LL Ghent. The criteria were based on output specifications rather than material specifications. A design-build (DB) tender was launched for the building structure and envelope and a design-build-operate-maintain (DBOM) tender for the technical devices. After both tenders, a consortium of partners were formed to co-create the LL Ghent as a circular and bio-based prototype in which several material and methods were tested. This study focuses only on the works that have been executed in the framework of the first tender.

The final product is a circular and bio-based living lab realized in KU Leuven Ghent Technology Campus with a hybrid structure that is composed of a prefabricated bio-based wall and floor panels with a steel frame structure. These panels are assembled to and around cross-laminated timber (CLT) staircase components. The geometry of LL was inspired by a typical 19th–20th century terraced house typology: two full storeys and one story under a sloping roof (see Figure 2). The front and rear façades have floor-to-ceiling windows. The roof is a classic pitched roof, with a skylight on one side. The side walls are typical blind waiting walls that enables for the final fitting of the prototype in an urban terraced house context. The footprint of the building is 33 m^2 and total gross floor area is 98.8 m^2. The total area of the front façades (west, east) is 56 m^2 and that of the side façades (north, south) is 129 m^2. The window area in all façades is 22.7 m^2. The roof has an area of 46 m^2.

(a)   (b)

Figure 2. Living Lab Ghent (**a**) during construction and (**b**) final stage in commissioning.

The production of the building components took place in off-site facilities and then assembly and finishing works were conducted on-site. Several sub-contractors were included together with a specific social worker company. In this context, as there were previous studies on the environmental impacts of LL, another aim of this study was to integrate a social assessment and achieve a full sustainability assessment.

3.2. Case Study Process

In order to initiate a framework for developing a life cycle inventory, two workshops were conducted. The first workshop was set to determine the goal and system boundaries of the S-LCA together with the stakeholders and project administration. The scope of the study was determined as the social impact assessment of production and construction phases due to the fact that circular methods in LL target a higher construction quality to increase reusability in the end of life. Use phase had to be omitted as the building was not commissioned during the period of this study.

A second workshop was held in order to identify the impact categories and indicators that are relevant for the construction methods utilized in LL Ghent. For this workshop, experts from the CBCI project environment and observing parties were invited. The participation analysis of the workshop can be seen in Table 2.

In this workshop, after an introduction to the S-LCA standards and guidelines, the participants from different stakeholder categories were requested to select impact categories (among the examples in the guidelines, or provide a new one) with respective importance rankings. The rankings are crucial to differentiate the specific priorities for the circular and bio-based approach. The participants were asked to rank the importance of impact categories on a scale of 1-to-10 (1 implies not important and 10 is means a most important category). Then, these rankings were utilized as weighting factors during the calculation. As a second step, they were also required to provide indicators for these impacts. There were several impact categories suggested that are not in the list of the guidelines. Those with similar characteristics were included in the existing categories (ease of assembly under health and safety, learning while working under occupational improvement). In addition, two new impact categories were considered of importance: affordability and use of local materials. Therefore, a list of 13 impact categories under 4 stakeholder categories were adopted as seen in Table 3.

Table 2. Overview of the expert group for the second workshop and interviews.

	Institution Role	Function	Number of Experts
	Research, design	Higher education	3
	Observing partner	Healthcare	1
Expert group	Dissemination	Network hub	2
	Research (bio-based)	Research center	2
	Prototyping	Research center	1
	Prototyping	Higher education	2
Interviews	Producer	Prefabrication	2
	Constructor	On-site assembly	2

Table 3. Impact categories and indicators for CBCI investments.

Stakeholder Category		Impact Category	Indicator	Weighting
Worker	1.	Health and safety	Incident/fatality rate	10
	2.	Working hours	>40 h = −1, <40 h = 1	7
	3.	Equal opportunity	Social economy use 10%	7
	4.	Occupational improvement	Asset for employment potential	7
Local community	5.	Safety and healthy living conditions	Construction waste, 70% recycling Healthy materials (bio-based)	10
	6.	Access to material resources	Improved IAQ	8
	7.	Community engagement	Complaints/compliments	8
	8.	Local materials	% of local materials	8
	9.	Affordability		8
	10.	Local employment	Improved employment rate, <75	6
Society	11.	End of life responsibility	Post-intervention manual	8
	12.	Commitment to sustainability	Participatory design Awareness rising	7
Consumer	13.	Transparency	Access to data and manuals	6

As the impact categories and indicators were determined, the experts appointed relevant impact categories to each construction method that was utilized in LL Ghent. Moreover, the construction methods were weighted for their importance. The impact categories that are related to construction methods and weighting factors can be seen in Table 4.

After the determination of impact categories and indicators, the means of data collection were prepared in the format of questionnaires. Two different sets of questionnaires were developed: the first one for production and second one for construction phase.

The first questionnaire for the production phase is based on the stakeholder and impact categories and related indicators that were previously determined (see Table 3). For each impact category, one or more questions were prepared. The questionnaire was conducted with a representative from the production contractor, workers, and CBCI experts. The second questionnaire for the construction phase is based on the construction methods and related indicators that were previously determined (see Table 4). Several impact categories were appointed to each construction method and questions were prepared for each impact category accordingly. The questionnaire was conducted with a representative from the construction contractor and workers.

Table 4. Indicators for the construction methods.

Construction Methods	Impact Category	Positive/Negative	Weighting
Bio-based construction	Access to material resources Safety and healthy living conditions Affordability Cultural heritage	Positive Positive Positive Negative	10
Demountable façade systems	Occupational improvement End of life responsibilities Transparency	Positive Positive Positive	8
Demountable technical services	End of life responsibility Occupational improvement	Positive Positive	8
Rearrangeable elements	Occupational improvement End of life responsibilities Community engagement	Positive Positive Positive	7
Pre-fabrication	Working hours Safety and healthy living conditions Community engagement Local employment	Positive Positive Positive Negative	9
Reused systems	Access to material resources Affordability Public commitment to sustainability	Positive Positive Positive	6
Social economy	Working hours Public commitment to sustainability Occupational improvement Equal opportunity Affordability	Positive Positive Positive Positive Positive	6

In Figure 3, the flow of the impact assessment framework can be seen as an overview. Based on this framework, the next section displays the results of the interviews and calculations.

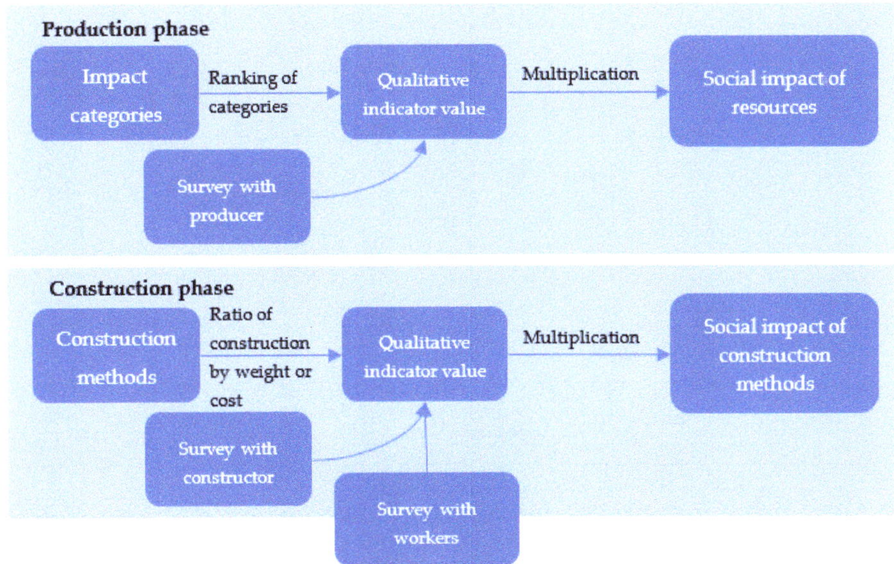

Figure 3. Impact assessment framework.

4. Results

In this section, the results of the questionnaires were first quantified for the production and construction phases of LL Ghent. They were interpreted according to responses for each impact category under four stakeholder categories. Then, the results are discussed from the perspective of each stakeholder category. This approach is expected to reflect stakeholder-centered thinking into the findings of the study.

4.1. Production Phase

The questionnaire responses were quantified based on a Likert scale (from 1 to 5). The scale implies that a significant positive social impact is denoted with 5, a score of 3 refers to a neutral state where there is no impact, and 1 refers to a significantly negative social impact. Then, these figures were normalized to a scale between −1 and 1. Each response on the Likert scale is transferred to a quartile between the scale of −1 and 1 (i.e., a Likert score of 1.5 is normalized to −0.75, and a Likert score of 4 is normalized to 0.5). In this way, the results would be comparable with other social LCA studies that follow the guidelines.

In the calculation, the weighting factors were utilized for prioritizing certain impact categories. These weighting factors reflect the current perspective of the expert group. It is foreseen that the weighting factors could vary depending on the purpose of the study or experts included. They may also change depending on the future developments of the construction sector.

The responses regarding the social impact of LL Ghent in the production phase are dominated by two keywords: (i) prefabrication and (ii) prototype. Prefabrication has an overall positive impact with regards to all stakeholder categories. Prototype has been associated with a few negative impacts such as a not fully automated production line, extra working hours due to unexpected reasons, etc. However, most of these negative impacts can be potentially mitigated or avoided in future follow-up productions similar to LL Ghent.

According to the general results in Table 5, it was seen that there was a positive impact towards the workers, local community, and society. Nevertheless, there was no significant impact on how consumers currently assess a product like LL Ghent.

Table 5. Impact assessment for production phase.

Stakeholder	Impact Category	Survey Results	Normalized Value	Weight Factor	Weighted Normal	Final Score
Worker	Health and Safety	2.6	−0.25	1.00	−0.25	
	Working hours	3.5	0.25	0.80	0.20	
	Equal opportunity	4.0	0.5	0.80	0.40	0.19
	Occupational improvement	4.0	0.5	0.80	0.40	
Local community	Safety and healthy living conditions	5.0	1	1.00	1.00	
	Access to material resources	4.0	0.5	0.90	0.45	
	Community engagement	3.5	0.25	0.90	0.23	
	Local materials	4.5	0.75	0.90	0.68	0.32
	Affordability	2.0	−0.5	0.90	−0.45	
	Local employment	3.0	0	0.90	0.00	
Society	End of life responsibility	4.0	0.5	0.90	0.45	
	Commitment to sustainability	4.5	0.75	0.80	0.60	0.53
Consumer	Transparency	3.0	0	0.70	0.00	0.00

4.2. Construction Phase

A similar calculation method was conducted for the construction phase, with a certain distinction. This time, several impact categories were appointed to a certain construction method and the impact of each construction method is calculated as a first step (see Table 6a). Then, the scores were weighted and normalized to a −1 and 1 scale. Finally, these scores were grouped under each stakeholder category by using the cost-wise ratio of each construction method in total cost of LL, so that it would represent the social perspectives (see Table 6b). This implies that one particular impact category (i.e., occupational improvement) may be affected by several construction methods (i.e., demountable façade and technical systems). In the end, a final score for each stakeholder category was achieved.

Table 6. (**a**) Impact assessment for construction phase. (**b**). Impact assessment for construction phase (grouping according to stakeholders).

(a)

Construction Methods	Impact Category	Survey Results	Normalized Value	Weight Factor	Weighted Normal	Cost % in LL
Bio-based construction	Access to material resources	5	1	1	1	30.00%
	Health and Safety	5	1		1	
	Safety and healthy living conditions	4	0.5		0.5	
	Affordability	1.5	−0.75		−0.75	
	Cultural heritage (negative)	4.5	0.75		0.75	
Demountable façade systems	Health and Safety	5	1	0.8	0.8	10.00%
	Occupational improvement	4	0.5		0.4	
	End of life responsibilities	3	0		0	
	Transparency	3	0		0	
Demountable technical systems	Health & Safety	3	0	0.8	0	5.00%
	Occupational improvement	3	0		0	
	End of life responsibilities	4	0.5		0.4	
Prefabrication	Health and Safety	4	0.5	1	0.5	45.00%
	Working hours	3	0		0	
	Safety and healthy living conditions	5	1		1	
	Community engagement	5	1		1	
	Local employment	4.3	0.5		0.5	
Reused systems	Health and Safety	3	0	0.7	0	10.00%
	Access to material resources	3.5	0.25		0.175	
	Affordability	1.5	−0.75		−0.525	
	Commitment to sustainability	2.5	−0.25		−0.175	

(b)

Stakeholder	Impact Category	Construction Method Results	Weight Factor	Weighted Normal	Final Score
Worker	Health and Safety	0.61	1.00	0.61	0.20
	Working hours	0.00	0.80	0.00	
	Equal opportunity	0.00	0.80	0.00	
	Occupational improvement	0.27	0.80	0.21	
Local community	Safety and healthy living conditions	0.80	1.00	0.80	0.51
	Access to material resources	0.79	0.90	0.71	
	Community engagement	1.00	0.90	0.90	
	Local materials	0.00	0.90	0.00	
	Affordability	0.24	0.90	0.22	
	Local employment	0.50	0.90	0.45	
Society	End of life responsibility	0.13	0.90	0.12	−0.01
	Commitment to sustainability	−0.18	0.80	−0.14	
Consumer	Transparency	0.00	0.70	0.00	0.00

The results include insights on five construction methods: bio-based materials, demountable façade and technical systems, prefabrication, and reused materials. The evaluation on the construction phase put forward a couple of highlights: (i) the adverse impact of circular materials on affordability and (ii) the overall positive impact of construction methods on the workers and local community. It must be noted that the negative impact of the affordability seems to be compensated by several other positive aspects related to the local community. However, uncertainties in the end of life yields an unexpected decrease on the social impact on society.

4.3. Stakeholder Analysis

In this sub-section, social impacts are discussed from the perspective of each stakeholder category. In the previous sub-sections, impacts of individual preferences on materials or methods during production and construction were discussed. Then, it is quite important to group the impact into each stakeholder category to discuss the impacts of the technical decision on social groups. Figure 4 provides a general overview of social impacts and shows that the LL Ghent has a positive impact for almost all stakeholder categories. In the following sub-sections, these impacts are analysed in detail.

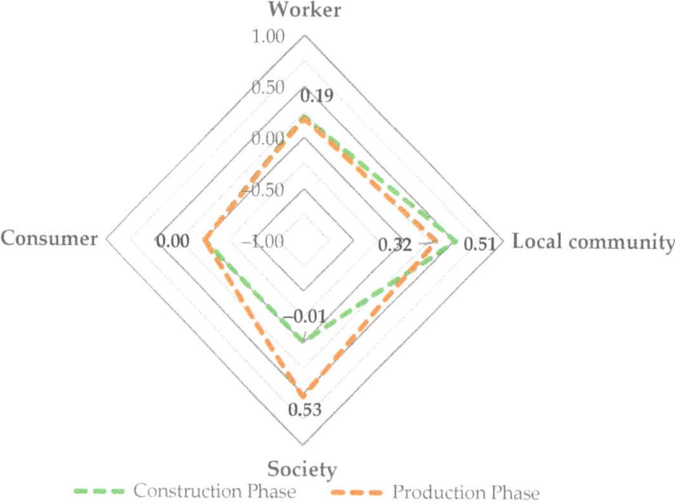

Figure 4. Social impact assessment for production and construction phases.

4.3.1. Workers

The impact of prefabrication was expected to be significant on the workers. This was also evident from the volume of discussion in this section during the interviews. However, the first two impact categories of health and safety and working hours were adversely affected by the fact that this production was a first-time for the producer. Hence, some of the safety aspects or optimization on the production line were not possible. For example, it was noted that the scores for these impact categories could be higher (by 1 point) for the next productions of the LL prototype. It was also acknowledged that the production process has provided a good learning opportunity for all workers.

The condition of workers on-site is greatly influenced by non-toxic bio-based materials usage with less precautions against skin irritations or mouth masks, and ease of assembly due to the modular and demountable prefabricated components. Nevertheless, the lack of certification and regulations on reused materials posed some risks on safety conditions. Moreover, the increase in the working hours due to unexpected delays in the first-time processes had a negative impact. The unique characteristic of the construction methods provided a great opportunity for occupational improvement for the workers.

4.3.2. Local Community

As the production of components took place in an off-site enclosed facility (significantly different than a conventional construction that takes place on-site), there are significant positive impacts on safety and healthy living conditions. It was seen that there was a considerable decrease in production waste. Due to the fact that all components of prefabricated wall panels are derived from local sources in Belgium, there is a positive impact on local materials and employment.

Nevertheless, it was acknowledged that the current market of bio-based materials is not favorable for affordable production. Although, it was also noted that possible follow-up production of similar buildings could be much more feasible for the producer with a fully automated production line.

The prefabricated bio-based components have a high positive impact on safety and healthy living conditions in the local community. As the construction duration on-site is significantly decreased when compared with conventional construction, there is a huge benefit from avoiding disturbances in the local area. Bio-based materials (especially the innovative blown-in insulation) have been proved to be high performing materials that provide a better indoor environment. On the other hand, reused materials have an uncertainty regarding thermal performance, so that caused a decrease for accessing better resources by the community.

It was acknowledged that there is a high recognition of benefits of prefabrication by the public. However, the low rate of affordability can be considered as the obstacle against having even higher positive impact on the local community.

4.3.3. Society

The society category considers the activities that ensure the end-of-life responsibilities and commitment to sustainability by the society. In the case of LL Ghent, the end-of-life responsibilities heavily rely on the special expertise of the producer. Without their involvement, it would not be feasible to repair or dismantle the components. Furthermore, it was foreseen that certain user manuals can be prepared in the future to enable self-builders to work on the LL components.

The end-of-life responsibilities are linked to the demountable façade and technical services. It was seen that due to the first-time experience in LL Ghent, some of the post-intervention documents for dismantling the components are not as elaborative as intended. A detailed end-of-life manual that is publicly shared would have a better impact. It was noted that the current rules and regulations do not require such documents during commissioning. This condition was considered as an unintentional obstacle against commitment to sustainability. This point is emphasized as a future improvement.

4.3.4. Consumer

The transparency of information on LL Ghent relevant for consumers was questioned in this category. Only the general information for advertisement and publicity is currently accessible by the users. Also, there is not a strategy foreseen by the producer for the dissemination of additional data to the public.

Due to the lack of public documents regarding the construction methods that are used in LL Ghent, the consumers may not have access to relevant data. This was considered as a non-impactful topic. However, as mentioned in the Society section, there are plans for improving the transparency of the information in the future.

5. Discussion

This study proposes an S-LCA framework for circular and bio-based constructions and provides an integrated view of experts, producers, contractors, and workers on the social impact of a circular building prototype in a quantified format. At the same time, this study raises several discussion points related to method development and limitations, as well as current gaps.

A few limitations of this study can be highlighted. First, the scope of the study does not include the use phase due to the fact that the LL Ghent was not commissioned at the time of the research. It is possible to conduct S-LCA on expected impacts (in the use phase) but there is no clear method on how to combine an actual impact with an expected impact. The social impact in the use phase of LL is a possibility for a future study.

Second, there is a lack of a validated set of indicators for buildings in the literature and in the global S-LCA method, as already highlighted by Alvarenga et al. [35]. For this reason, this study used workshops with focus groups to determine specific impact categories and indicators. This approach can cause a dependency on the expert groups who determine the impact categories and indicators. It is possible that a different set of impact categories and indicators could be selected by a different expert group. As the method also used weighting factors from the expert group, the mentioned impact may even be greater. Nevertheless, this approach provides a tangible link between the stakeholders and the object. Therefore, the method may be more successful when the aim is to assess the object from the perspective of specific stakeholders and not from a global perspective.

As this study evolves around the case study of LL Ghent, there is a significant impact of criteria definition on the content and results of this study; both impact categories and all construction methods were defined according to the LL Ghent. In this context, they are quite specific to circular and bio-based building components. Construction-related activities tend to differ significantly when compared to the other studies such as Dong and Ng [14]. This is related to the unique characteristics of each building typology, and the contribution of the results is specific to the building typology and construction method. On the other hand, this study could have benefited from common indicators or questionnaires.

Besides the limitations, the results of this study display similarities to previous case studies of Dong and Ng [14], Liu and Qian [18], and Balasbaneh and Sher [21] depending on the positive social impacts on worker and local community categories. The precast/prefabricated construction methods were the significant factor affecting the working conditions for workers and yield less disturbance for the local community. When compared with studies on circular economy/construction, the impact on the society through end-of-life responsibilities is similar with the study of Hossain et al. [17].

Regarding some general discussion in the literature, Hackenhaar et al. [36] discussed that existing LCA studies lack a criticality on raw material extraction. Even though this study does not possess a general raw material criticality, this aspect was considered through the assessment of reused systems. There is also a tendency to use hotspot analysis through quantitative databases on country-specific risks such as the social hotspot database (SHDB) [37]. On the other hand, such databases are useful to compare similar construction methods that take place in different regions. As this study required comparison of different construction methods that rely on local industries in a singular region, the level of detail in SHDB was not sufficient. A future study can compare the qualitative results of this study with a detailed input-output model specific for Belgium (Flanders).

6. Conclusions

This study shows that circular construction methods can have positive social impacts, such as (i) a healthy and safe working environment for workers and (ii) decreased on-site construction activities with positive effects on workers and local communities. Specifically, prefabrication methods have a significant impact on the well-being of the local community, as well as workers. Emerging concepts such as demountable construction methods provide learning opportunities for the workers. Overall, this study concluded that circular construction may have additional social benefits besides environmental benefits.

As the assessment in this study was conducted on a prototype, several points could be potentially improved. For example, low affordability and uncertainties on bio-based and reused materials were acknowledged by interviewees as the origins of the drawbacks below:

- Less accessible to public;
- Increased risk during production and construction.

Affordability could be certainly improved as the process is optimized and when mass production is initiated. This does not necessarily imply that such buildings would be accessible to the general public or low-income groups as long as positive social impacts are integrated to the current building design and commissioning processes. This is also a reason why social impact assessment should gradually be a legal requirement (just as environmental impact assessments are) for building permits. In that way, the value of social impacts can be acknowledged and this generated value may be fed back to the design and affordability of construction.

For facilitating such a reflection of the social impacts of construction to legal processes, there is a need for standardized indicators and databases. Together with the previous studies in the same domain, this study provided valuable impact categories and indicators that are specific for several construction methods. Researchers would greatly benefit from a collection of construction methods and materials in an industry-specific guideline.

In order to enhance the value of social impact on construction, there are several efforts to integrate S-LCA into multi-criteria assessments. For the case of LL Ghent, monetized environmental LCA results derived in a previous study [34] may be integrated with social LCA. For such a goal, there is a critical need for more data on the conversion of social risks and impacts into monetary values. Then, such a study could provide one way of integrating the environmental and social pillars at the building level. This is considered as a future study in the scope of life cycle sustainability assessment domain.

Author Contributions: Conceptualization, N.C.K. and A.V.; methodology, N.C.K.; validation, N.C.K., C.P. and A.V.; formal analysis, N.C.K.; investigation, N.C.K.; writing—original draft preparation, N.C.K.; writing—review and editing, C.P. and A.V.; visualization, N.C.K.; supervision, C.P. and A.V.; project administration, A.V.; funding acquisition, A.V. All authors have read and agreed to the published version of the manuscript.

Funding: The research is part of the Circular Bio-based Construction Industry (CBCI) project. CBCI is funded by the European Union Regional Development Fund Interreg 2 Seas Mers Zeeen (2S05-036).

Institutional Review Board Statement: Not applicable.

Informed Consent Statement: Informed consent was obtained from all subjects involved in the study.

Data Availability Statement: Not applicable.

Conflicts of Interest: The authors declare no conflict of interest.

References

1. Euclid Network GECES (Commission Expert Group on the Social Economy and Social Enterprises). Available online: https://euclidnetwork.eu/2020/09/geces/ (accessed on 15 November 2022).
2. *EN 15643-1:2010*; Sustainability of Construction Works—Sustainability Assessment of Buildings—Part 1: General Framework. CEN: Brussels, Belgium, 2010.
3. Ghisellini, P.; Ripa, M.; Ulgiati, S. Exploring environmental and economic costs and benefits of a circular economy approach to the construction and demolition sector. A literature review. *J. Clean. Prod.* **2018**, *178*, 618–643. [CrossRef]
4. Iacovidou, E.; Hahladakis, J.N.; Purnell, P. A systems thinking approach to understanding the challenges of achieving the circular economy. *Environ. Sci. Pollut. Res.* **2021**, *28*, 24785–24806. [CrossRef] [PubMed]
5. Ellen MacArthur Foundation. *Growth Within: A Circular Economy Vision for a Competitive Europe*; Ellen MacArthur Foundation: Cowes, UK, 2015.
6. Korhonen, J.; Honkasalo, A.; Seppälä, J. Circular Economy: The Concept and its Limitations. *Ecol. Econ.* **2018**, *143*, 37–46. [CrossRef]
7. Walker, A.M.; Opferkuch, K.; Roos Lindgreen, E.; Simboli, A.; Vermeulen, W.J.V.; Raggi, A. Assessing the social sustainability of circular economy practices: Industry perspectives from Italy and the Netherlands. *Sustain. Prod. Consum.* **2021**, *27*, 831–844. [CrossRef]
8. CBCI Interreg 2seas Circular Biobased Construction Industry. The European Union Regional Development Fund Interreg 2 Seas Mers Zeeen (2S05-036). Available online: https://www.interreg2seas.eu/nl/CBCI (accessed on 14 October 2022).
9. *EN 16309:2014*; Sustainability of Construction Works—Assessment of Social Performance of Buildings—Calculation Methodology. CEN: Brussels, Belgium, 2014.
10. UNEP. *Guidelines for Social Life Cycle Assessment of Products and Organizations*; Benoît Norris, C., Traverso, M., Neugebauer, S., Ekener, E., Schaubroeck, T., Russo Garrido, S., Berger, M., Valdivia, S., Lehmann, A., Finkbeiner, M., Arcese, G., Eds.; United Nations Environment Programme (UNEP): Paris, France, 2020.
11. Benoît-Norris, C.; Vickery-Niederman, G.; Valdivia, S.; Franze, J.; Traverso, M.; Ciroth, A.; Mazijn, B. Introducing the UNEP/SETAC methodological sheets for subcategories of social LCA. *Int. J. Life Cycle Assess.* **2011**, *16*, 682–690. [CrossRef]
12. Life Cycle Initiative. *Pilot Projects on Guidelines for Social Life Cycle Assessment of Products and Organizations*; Traverso, M., Mankaa, M.N., Valdivia, S., Roche, L., Luthin, A., Garrido, S.R., Neugebauer, S., Eds.; Life Cycle Initiative: Paris, France, 2022.

13. Hosseinijou, S.A.; Mansour, S.; Shirazi, M.A. Social life cycle assessment for material selection: A case study of building materials. *Int. J. Life Cycle Assess.* **2014**, *19*, 620–645. [CrossRef]
14. Dong, Y.H.; Ng, S.T. A social life cycle assessment model for building construction in Hong Kong. *Int. J. Life Cycle Assess.* **2015**, *20*, 1166–1180. [CrossRef]
15. Santos, P.; Gervásio, H.; Pereira, A.; Simões da Silva, L.; Bettencourt, A. Comparative life cycle social assessment of buildings: Health and comfort criterion. *Matériaux Tech.* **2016**, *104*, 601. [CrossRef]
16. Santos, P.; Carvalho Pereira, A.; Gervásio, H.; Bettencourt, A.; Mateus, D. Assessment of health and comfort criteria in a life cycle social context: Application to buildings for higher education. *Build. Environ.* **2017**, *123*, 625–648. [CrossRef]
17. Hossain, M.U.; Poon, C.S.; Dong, Y.H.; Lo, I.M.C.; Cheng, J.C.P. Development of social sustainability assessment method and a comparative case study on assessing recycled construction materials. *Int. J. Life Cycle Assess.* **2018**, *23*, 1654–1674. [CrossRef]
18. Liu, S.; Qian, S. Evaluation of social life-cycle performance of buildings: Theoretical framework and impact assessment approach. *J. Clean. Prod.* **2019**, *213*, 792–807. [CrossRef]
19. Yıldız, S.; Kıvrak, S.; Gültekin, A.B.; Arslan, G. Built environment design—Social sustainability relation in urban renewal. *Sustain. Cities Soc.* **2020**, *60*, 102173. [CrossRef]
20. Chen, X.; Qu, K.; Calautit, J.; Ekambaram, A.; Lu, W.; Fox, C.; Gan, G.; Riffat, S. Multi-criteria assessment approach for a residential building retrofit in Norway. *Energy Build.* **2020**, *215*, 109668. [CrossRef]
21. Balasbaneh, A.T.; Sher, W. Life cycle sustainability assessment analysis of different concrete construction techniques for residential building in Malaysia. *Int. J. Life Cycle Assess.* **2021**, *26*, 1301–1318. [CrossRef]
22. Soust-Verdaguer, B.; Bernardino Galeana, I.; Llatas, C.; Montes, M.V.; Hoxha, E.; Passer, A. How to conduct consistent environmental, economic, and social assessment during the building design process. A BIM-based Life Cycle Sustainability Assessment method. *J. Build. Eng.* **2022**, *45*, 103516. [CrossRef]
23. LLatas, C.; Soust-Verdaguer, B.; Hollberg, A.; Palumbo, E.; Quiñones, R. BIM-based LCSA application in early design stages using IFC. *Autom. Constr.* **2022**, *138*, 104259. [CrossRef]
24. Fauzi, R.T.; Lavoie, P.; Tanguy, A.; Amor, B. On the possibilities of multilevel analysis to cover data gaps in consequential S-LCA: Case of multistory residential building. *J. Clean. Prod.* **2022**, *355*, 131666. [CrossRef]
25. Thanu, H.; Rajasekaran, C.; Deepak, M. Developing a building performance score model for assessing the sustainability of buildings. *Smart Sustain. Built Environ.* **2022**, *11*, 143–161. [CrossRef]
26. Mesa, J.A.; Fúquene, C.E.; Maury-Ramírez, A. Life cycle assessment on construction and demolition waste: A systematic literature review. *Sustainability* **2021**, *13*, 7676. [CrossRef]
27. Locurcio, M.; Tajani, F.; Anelli, D.; Ranieri, R. A multi-criteria composite indicator to support sustainable investment choices in the built environment. *Valori E Valutazioni* **2022**, *30*, 85–100. [CrossRef]
28. Ibrahim, A.; El-Anwar, O.; Marzouk, M. Socioeconomic impact assessment of highly dense-urban construction projects. *Autom. Constr.* **2018**, *92*, 230–241. [CrossRef]
29. Wang, Y.; Han, Q.; de Vries, B.; Zuo, J. How the public reacts to social impacts in construction projects? A structural equation modeling study. *Int. J. Proj. Manag.* **2016**, *34*, 1433–1448. [CrossRef]
30. Nadazdi, A.; Naunovic, Z.; Ivanisevic, N. Circular Economy in Construction and Demolition Waste Management in the Western Balkans: A Sustainability Assessment Framework. *Sustainability* **2022**, *14*, 871. [CrossRef]
31. Tirado, R.; Aublet, A.; Laurenceau, S.; Habert, G. Challenges and Opportunities for Circular Economy Promotion in the Building Sector. *Sustainability* **2022**, *14*, 1569. [CrossRef]
32. Maury-Ramírez, A.; Illera-Perozo, D.; Mesa, J.A. Circular Economy in the Construction Sector: A Case Study of Santiago de Cali (Colombia). *Sustainability* **2022**, *14*, 1923. [CrossRef]
33. Kayaçetin, N.C.; Verdoodt, S.; Lefevre, L.; Versele, A. Evaluation of Circular Construction Works during Design Phase: An Overview of Valuation Tools. In *Sustainability in Energy and Buildings 2021*; Littlewood, J.R., Ed.; Smart Innovation, Systems and Technologies; Springer: Berlin, Germany, 2022; Volume 263, pp. 89–100.
34. Kayaçetin, N.C.; Verdoodt, S.; Lefevre, L.; Versele, A. Integrated decision support for embodied impact assessment of circular and bio-based building components. *J. Build. Eng.* **2023**, *63*, 105427. [CrossRef]
35. Alvarenga, R.A.F.; Cadena Martinez, E.; Zanchi, L.; Zamagni, A.; Sonderegger, T.; Ruiz, E.M. Critical Evaluation of Social Approaches. 2021. Available online: https://orienting.eu/publications/prova-2 (accessed on 16 May 2022).
36. Hackenhaar, I.; Alvarenga, R.A.F.; Bachmann, T.M.; Riva, F.; Horn, R.; Graf, R.; Dewulf, J. A critical review of criticality methods for a European Life Cycle Sustainability Assessment. *Procedia CIRP* **2022**, *105*, 428–433. [CrossRef]
37. Social Hotspot Database (SHDB). Available online: http://www.socialhotspot.org/ (accessed on 14 October 2022).

Disclaimer/Publisher's Note: The statements, opinions and data contained in all publications are solely those of the individual author(s) and contributor(s) and not of MDPI and/or the editor(s). MDPI and/or the editor(s) disclaim responsibility for any injury to people or property resulting from any ideas, methods, instructions or products referred to in the content.

Article

Research on the Restrictive Factors of Vigorous Promotion of Prefabricated Buildings in Yancheng under the Background of "Double Carbon"

Houchao Sun [1], Yuwei Fang [2], Minggan Yin [1] and Feiting Shi [1,*]

1 School of Civil Engineering, Yancheng Institute of Technology, Yancheng 224051, China
2 Quality Supervision Station, Yancheng Housing and Urban-Rural Development Bureau, Yancheng 224008, China
* Correspondence: shifeiting@ycit.cn

Abstract: In the field of construction, the promotion of prefabricated buildings has been strongly supported by the state due to its low-carbon, environmental protection and high-efficiency characteristics. The process of design, prefabrication, and installation, is restricted by factors such as unsound policy standards, insufficient technological innovation, lack of professional talents, and high costs, which have led to the slow development of prefabricated buildings in China. The main factors that restrict the development of prefabricated buildings in Yancheng are identified from the researcher's point of view by literature review and questionnaire survey method. The degree of centrality and cause of each constraint has been analyzed by the decision-making laboratory method (DEMATEL), and the interpretation structure method (ISM) was used to build a multi-level hierarchical structure model of constraints, the logical relationship, hierarchical relationship and relative importance of each constraint are clarified. It is concluded that industry policies, imperfect standards and insufficient government publicity are the fundamental reasons to hinder the development of prefabricated buildings in Yancheng. According to the order of the centrality, the main restrictive factors are determined, which benefits the establishment of the homologous counterplan for the vigorous promotion of prefabricated buildings in Yancheng.

Keywords: double carbon; prefabricated building; restrictive factors; DEMATEL; ISM

Citation: Sun, H.; Fang, Y.; Yin, M.; Shi, F. Research on the Restrictive Factors of Vigorous Promotion of Prefabricated Buildings in Yancheng under the Background of "Double Carbon". *Sustainability* **2023**, *15*, 1737. https://doi.org/10.3390/su15021737

Academic Editor: Ana Ramos

Received: 25 November 2022
Revised: 10 January 2023
Accepted: 11 January 2023
Published: 16 January 2023

Copyright: © 2023 by the authors. Licensee MDPI, Basel, Switzerland. This article is an open access article distributed under the terms and conditions of the Creative Commons Attribution (CC BY) license (https://creativecommons.org/licenses/by/4.0/).

1. Introduction

In order to achieve global sustainable development, the Chinese government has proposed to all countries in the world to strive to achieve "carbon peak" before 2030 and "carbon neutrality" before 2060, that is, "double carbon" goal. According to the "China Building Energy Consumption Research Report (2020)" issued by the Energy Conservation Committee of China Building Energy Conservation Association, the total carbon emissions of the whole process of China's construction in 2018 was 4.93 billion tons, accounting for 51.3% of the national carbon emissions. Therefore, whether the construction industry can achieve the "double carbon" goal is crucial to China's sustainable development.

In recent years, the prefabricated building industry are increasingly focused on "carbon peaking and carbon neutrality" in the construction field. Prefabricated building construction not only greatly saves water, energy, steel, wood and other resources, but also reduces construction waste and noise pollution. At the same time, prefabricated building construction can alleviate the shortage of labor force in the construction industry, and realize the transformation of the construction mode from the traditional extensive production and inefficient management to green, assembly, informatization and intelligentization. Therefore, it is particularly urgent to vigorously promote prefabricated buildings. At present, prefabricated buildings have achieved desired results in developed countries such as Europe, Japan, and Singapore, but promoting prefabricated buildings has been restricted to a certain extent in China [1].

Since 2016, the China State Council has issued a series of documents such as "Guiding Opinions on Vigorously Developing Prefabricated Buildings", which put forward the specific goal, which is to have prefabricated buildings account for 30% of new construction within a decade. The document requires that the application of prefabricated buildings should be enlarged with the three major urban agglomerations of Beijing-Tianjin-Hebei, Yangtze River Delta and Pearl River Delta as key areas. In 2017, the Ministry of Housing and Urban-Rural Development of China issued three major documents, namely, "The 13th Five-Year Plan for Prefabricated Buildings", "Administrative Measures for Prefabricated Building Demonstration Cities", and "Administrative Measures for Prefabricated Building Industry Bases", indicating that the development of prefabricated buildings has risen to the national strategic level [2]. In 2017, the Jiangsu Provincial Department of Housing and Urban-Rural Development issued the "Jiangsu Construction 2025 Action Outline", which promotes the development of engineering construction methods to the integration direction of the "four modernizations" including refining construction, information technology, green construction and industrialization, and vigorously put a policy into practice in four new construction methods, namely lean construction, digital construction, green construction and prefabricated construction.

The document, namely "Implementation Opinions on Accelerating the Development of Prefabricated Buildings", was released by the Yancheng government in 2017, and pointed out that the development path of government-guided, market-led, industry-driven and enterprise operation by the popularization and application of BIM technology to promote the development of prefabricated building design, production and construction. During the "Thirteenth Five-Year Plan" period, new prefabricated buildings of 3,795,900 square meters were built in Yancheng, and the ratio of prefabricated buildings area to new buildings area increased from 4% in 2018 to 30% in 2020. Six construction industry modernization demonstration bases have been settled in Yancheng, such as Funing Green Intelligent Building Industrial Base, and Jiangsu Shenggong Construction Co., Ltd. (240,000 cubic meters PC components). There are 32 manufacturers of prefabricated building components, such as Jiangsu Qianhe Prefabricated Building Technology Co., Ltd. and Jiangsu Jinmao Construction Group Co., Ltd., whose production covers concrete components, wood components and steel components. Yancheng Pioneer International Hotel project and Yancheng Hope Residential Community project were included in Jiangsu Province prefabricated building demonstration project in 2021. Yancheng government issued the document "Implementation Opinions on Vigorously Promoting the Development of Prefabricated Buildings", which proposed the development goal of prefabricated buildings. Prefabricated buildings should account for more than 35% of initiated construction buildings, and assembled decoration buildings should take a proportion of more than 10% of new housing in 2021. By the end of 2025, prefabricated buildings will reach 50% of initiated construction buildings, and assembled decoration buildings will exceed 30% of delivered housing.

In Yancheng, prefabricated buildings have been gradually promoted, and relevant documents and policies had been launched to stimulate the prefabricated construction market. but there are still constraints on the vigorous development of prefabricated buildings, and problems that need to be deeply studied and solved urgently, for example, the secondary deepening design is not matched with some standards and specifications, and prefabricated buildings are assembled only on "three plates"(prefabricated interior and exterior wall plate, prefabricated stair plate, prefabricated floor plate), The lack of qualified construction professionals and unadvanced operation tools lead to low work efficiency, and it is even difficult to guarantee the installation quality of prefabricated structures. Therefore, the cost of prefabricated construction is not advantageous. At present, prefabricated building is limited to concrete structure, prefabricated steel-concrete composite structure is less, the information level of prefabricated building is not systematized to fail to achieve information management, and so on. The author participated in 30-31# group project construction of Yancheng Institute of Technology dormitory. The designed and produced inner wall plate (Autoclaved Lightweight Concrete (ALC)) does not integrate water and electricity pipelines, which led to slotting ALC to arrange the pipelines. Simple installation equipment, unprofessional construction workers, and inefficient construction make it difficult to guarantee the construction quality of dormitory buildings.

2. Literature Review

Factors influencing the promotion of prefabricated buildings are classified into driving factors and restrictive factors. In terms of driving factors, Li et al. analyzed three aspects: direct factors, indirect factors and fundamental factors [3]. The results show that shortening the construction period, saving energy and reducing consumption, improving resource utilization and reducing on-site operations are the main driving factors. Wang et al. used a hybrid model to assess the environmental impact of prefabricated buildings and traditional cast-in-place buildings over the whole life cycle in Japanese construction cases [4]. It was found that the total energy consumption and carbon emissions of 40% assembly buildings downed 7%, and compared to those of traditional cast-in-place buildings. Prefabricated buildings cost less than traditional cast-in-place buildings, reducing the price per square meter by more than 10%. As the assembly rate increases, carbon emissions and costs drop, reaching the bottom when the assembly rate is 60%. Through literature reviews and industry interviews, zhang et al. have identified and discussed the factors reflecting major changes in the current Australian prefabricated construction [5]. These factors include prefabrication industry development, emerging benefits and challenges. The challenges identified from the interviews were grouped into eight fields, involving feasibility, design, manufacturing, transportation, site construction, standardization, skills and knowledge, funding and markets. With the increasing application of prefabricated structures in civil engineering, the seismic performance of prefabricated structures has attracted the wide attention of scholars all over the world. A novel seismic force-resisting system (SFRS) called Floor Isolated Re-entering Modular Construction System (FIRMOCS) has been proposed by Chen et al. for prefabricated modular mass timber (MT) construction [6]. The preliminary results showed that FIRMOC systems significantly reduced the seismic demand on the prefabricated modular MT construction using nonlinear dynamic time-history analysis according to the National Building Code of Canada, thus leading to an improved seismic response. Moreover, modular integrated construction(MiC) is an important process to integrate and assemble prefabricated prefinished modules. Wuni et al. pointed out the critical success factors (CSFs) for implementing MiC projects during the period 1993–2019 [7]. The study result showed that the US, UK, Malaysia, Australia, and Hong Kong are the largest contributors to the MiC CSFs. The further analysis generated 35 CSFs for implementing MiC projects. To improve the performance of the construction industry, design for manufacturing and assembly (DfMA) is introduced into the design of prefabricated buildings in the industrial building system (IBS) and prefabricated buildings, and combined with the parametric design of building information modeling (BIM). The main benefits, hindrances, and upsides of DfMA in the construction industry are discussed [8,9]. The implementation of the DfMA method in prefabricated construction projects is conducive to improving the level of green building activities. As can be seen from "World Green Building Trends 2021", the highest levels of current green building activity are in Australia or New Zealand, Canada and the US. And the highest levels of growth in those doing a majority of green projects are expected in Brazil, Colombia, Canada and Mexico.

Many scholars have different reasons for restricting the development of prefabricated buildings in China. Bian et al. used principal component analysis to identify key factors, including market environment, industrial organization, policies and regulations, talent factors, technical systems, etc. [10]. Based on the willingness of developers, Lu et al. found that high construction costs and low customer perceived value are the main factors restricting the development of prefabricated buildings [11]. Chen et al. pointed out from the perspective of the government that the problems in the implementation of prefabricated building policies included imperfect policy systems, insufficient environmental conditions for policy implementation, and low enthusiasm for policy executors [12]. Chen et al. proposed to increase incentives, improve incentive mechanisms, strengthen standardization, upgrade technical systems, and enhance publicity and education efforts to promote assembly development by building an evolutionary model of "government-developer" and "government-consumer" [13].

Zhai et al. concluded that low supply chain integration, the inability to freeze the design in the early stage, the high cost, and the lack of supportive policies and relevant standards are the main contradictions restricting the development of prefabricated buildings in China [14]. Zhang et al. used fuzzy analytic hierarchy process to analyze the main factors, and the conclusion was that the lack of professional talents and completive industrial chain, and the imperfection of relevant norms and standards restricted the development of China's prefabricated buildings [15]. Based on actual cases, Hong et al. analyzed and found that the main reason for the higher cost of prefabricated buildings in China than traditional buildings is the additional transportation and assembly costs of components [16]. Xue et al. found that little innovation is one major reason to obstruct the promotion of Chinese prefabricated buildings through social network analysis (SNA) and structural equation modeling [17]. Wu et al. studied the development of Chinese prefabricated buildings from the perspective of technology promotion and constructed a model of influencing factors from five aspects, including industry, company, technology, government, and market. The results showed that technology has more influence on prefabricated buildings than cost [18–22]. Lu et al. developed a lean-agile production system for prefabricated housing products based on market demand and production process to achieve a balance between market demand and production capacity [23].

The prefabricated building construction method is similar to automobile production, relying on pre-designed components that are factory-made, specialized, and standardized. The main work on construction sites is to install components, which can cut down the construction period, on-site wet work and formwork, water and electricity consumption, construction waste, and help to protect the environment. Since China's State Council promulgated "Guiding Opinions on Vigorously Developing Prefabricated Buildings" in 2016, and established 48 prefabricated building demonstration cities and 328 industrial demonstration bases [24,25]. the newly-started prefabricated building area in 2021 reached 740 million square meters, accounting for 24.5% of the new building area, and the newly-started prefabricated buildings in key construction areas accounted for 52.1% of the total in China [26].

However, the popularization of prefabricated buildings increases construction costs and lacks supply-side resources and market awareness. Meanwhile, due to government requirements, market development, and environmental protection, most developers are still in passive acceptance and are forced to build prefabricated buildings. Prefabricated building construction is representative of advanced productive forces that can best reflect the advantages of technological progress and industrialization. In China, a package of policies and guidance has been formulated in the stage of exploration and development. Many challenges remain in technology, market, operation, and management that must be addressed for prefabricated building construction to become industrialized.

In recent years, the research on prefabricated buildings is on the rise, and most studies on constraints to the development of prefabricated buildings in China are comprehensive [27,28]. Due to the vast territory of China, the basic conditions and development status of prefabricated buildings in different provinces and cities are different, so specific and quantitative studies are needed. There are few reports on the development status and influencing factors of prefabricated buildings in a certain city in China. The present research is in the stage of theoretical research and qualitative analysis, lacking quantitative analysis and objective data support. Research on the interrelationship and hierarchical relationship of the restrictive factors that affect the promotion of prefabricated buildings is difficult to reflect the deep connection, and very unfavorable to formulate corresponding measures to develop prefabricated buildings.

3. Methodology

This paper takes the prefabricated building construction in Yancheng City, Jiangsu Province as an example. Taking into account the characteristics of geography, humanity, environment and economy, the constraints on the promotion and implementation of prefabricated buildings are sorted out and categorized, a constraint model is created. Factor analysis was carried out by IBM SPSS 22.0 software, and the reliability of the model was verified. The DEMATEL-ISM method was used to analyze the relationship between the influencing factors qualitatively and quantitatively. The main restricting factors were determined and the corresponding countermeasures were put forward.

3.1. Ideas for Extracting and Identifying Restrictive Factors

Using comparative and analytical research methods, many literatures in "CNKI" and "Web of Science" databases on the development and promotion of prefabricated buildings are reviewed, five types of constraints are identified by PEST theory: market, economy, technology, society, and policy.

Based on the analysis of the initial list of restricting factors, the framework table of the restricting factor index system is classified, the amendments based on expert opinions are drawn up, and the revised index system of restricting factors for the promotion and implementation of prefabricated buildings is established.

According to the above index system, questionnaires and statistical regression analysis, the authors identify effective constraints and verify the rationality of the constraint index system.

3.2. DEMATEL-ISM Combined Model Analysis Method

3.2.1. DEMATEL Analysis Steps

(1)   Identifying influencing factors Sn.
(2)   The numbers 0–4 represent the association between the factors, where the number 0–4 indicates the level of association between the factors from none, low, average, relatively high and high, respectively.
(3)   Establishing the initial direct influence matrix.

The influence relationship between the factors can be shown from the matrix to create the initial direct influence matrix D. D_{ij} represents the degree of influence of S_i on S_j, $D_{ij} = 0,1,2,3,4$.

(4)   Normalizing the matrix.

The standardization matrix B is obtained by normalizing the direct influence matrix D, as shown in Equation (1).

$$B = \left[b_{ij}\right]_{n \times n} = \frac{1}{\max\limits_{1 \leq i \leq n} \sum\limits_{j=1}^{n} d_{ij}} D \tag{1}$$

(5)   Calculating normalized direct influence matrix to obtain the combined influence matrix T, as shown in Equation (2).

$$T = B + B^2 + B^3 + \ldots + B^n = B(I - B)^{-1} = t_{ij} \tag{2}$$

where element matrix I denotes the influence of the factor itself.

(6)   Calculating influence degree f_i, affected degree e_i, centrality degree z_i and cause degree y_i for each factor.

Influence degree f_i and affected degree e_i are equal to the sum of each value in row i and column i of combined influence matrix T, respectively. Centrality degree z_i which represents the degree of each factor's importance is the sum of influence degree f_i and affected degree e_i, influence degree f_i minus affected degree e_i is equal tocause degree y_i, The Equations (3)–(6) are as follows.

$$f_i = \sum_{j=1}^{n} t_{ij}(i = 1, 2, \ldots, n) \tag{3}$$

$$e_i = \sum_{j=1}^{n} t_{ji}(i = 1, 2, \ldots, n) \tag{4}$$

$$z_i = f_i + e_i(i = 1, 2, \ldots, n) \tag{5}$$

$$y_i = f_i - e_i(i = 1, 2, \ldots, n) \tag{6}$$

(7) Centrality degree z_i and cause degree y_i of each factor are respectively taken as horizontal and vertical coordinates in the factor-cause-result diagram, and the position of each constraint is marked in the coordinate axis for intuitive analysis of the importance of each constraint.

3.2.2. DEMATEL-ISM Method

After completing step (7) of DEMATEL model analysis, Dematel-ISM model analysis is carried out in the following steps.

(8) Calculating the overall influence matrix H. It is considered that the influence of each factor on itself is to be added, and the influence of the factors themselves can be reflected by the element matrix I, the overall influence matrix H can be obtained according to Equation (7).

$$H = I + T \tag{7}$$

(9) Calculating reachability matrix R according to specific threshold λ.

$$R = \left| r_{ij} \right|_{n \times n'} (i = 1, 2, \ldots, n; j = 1, 2, \ldots, n)$$
$$r_{ij} = \begin{cases} 1, h_{ij} \geq \lambda \\ 0, h_{ij} < \lambda \end{cases} \tag{8}$$

where λ is threshold value, in the process of selecting specific threshold, it is necessary to ensure the appropriate nodes and coordinate with the key nodes obtained by the DEMATEL method.

(10) Analysyzing hierarchical relation.

After solving reachable matrix R in step (9), reachable set $P(Si)$ and antecedent set $Q(Si)$ can be derived, $P(Si)$ is the column set containing 1 in row i of R, and $Q(Si)$ is rowset containing 1 in column i of R. When $P_{si} = P(S_i) \cap Q(S_i), i = 1, 2, \ldots, n$, 1st layer of factors can be obtained, and deleting this layer creates *R1*. In the same way, the *i*th layer of factors can be set up, and so on.

(11) Drawing the hierarchical diagram of constraints.

4. Results

4.1. Identifying the Constraints on the Vigorous Promotion of Prefabricated Buildings

The authors of this paper analyze and summarize the research results of many scholars on the development constraints of prefabricated buildings [3,10,25], 25 factors restricting the development of prefabricated buildings in China were identified, as shown in Table 1. Through a series of questionnaires,10 factors with negligible influence were eliminated from the 25 influence factors, and 15 factors with large influence on the vigorous promotion of prefabricated buildings in China were determined. As can be seen from Table 2, 15 factors are grouped into five categories: market, economy, technology, society, and policy according to the hierarchy and systemicity of constraints themselves and the connotation of the constraints by PEST theory.

Table 1. Yancheng prefabricated building promotion and implementation restriction index system framework.

Serial Number	Restrictive Factors
1	Insufficient motivation for enterprise transformation
2	The inefficiency of cooperation among participants
3	Imperfect supporting industrial chain
4	Traditional habit constraints
5	Insufficient market demand
6	Poor awareness of energy conservation
7	Weak willingness of construction enterprises to transform
8	High construction cost
9	High template cost and low utilization rate
10	Higher component production, installation costs and value-added tax
11	Large upfront investment
12	Lack of information technology application
13	Backward prefabricated building management model
14	Imperfect design standardization and modularization system
15	Lack of innovation
16	Insufficient investment in research and development
17	Insufficient technical support
18	The construction is difficult and the construction level is low
19	Talent shortage
20	low public awareness and purchase willingness
21	low degree of social acceptance
22	Insufficient government propaganda
23	Imperfect industry policy and standard system
24	Insufficient financial support
25	Imperfect industry standard system

Table 2. The revised index system of restrictive factors for the promotion and implementation of Yancheng prefabricated buildings.

Classification	No.	Restrictive Factors
	S1	Insufficient motivation for enterprise transformation
Marketaspect	S2	Imperfect supporting industry chain
	S3	Insufficient market demand
	S4	High construction cost, low cost performance
Economic management	S5	Large upfront investment
	S6	The prefabricated building management model is backward
	S7	Design standardization, modularization, and integrated systems are not perfect
Technical aspect	S8	Prefabricated building production, construction experience and technology are immature
	S9	Insufficient basic research and innovation of prefabricated buildings
	S10	Talent shortage
Social aspects	S11	Low public awareness and willingness to buy
	S12	low degree of social acceptance
	S13	Imperfect industry policy and standard system
Policyaspect	S14	Insufficient financial support
	S15	Insufficient government propaganda

To ascertain the reliability of the factors in Table 2, the authors designed questionnaires to survey senior professionals in the construction industry to collect feedback and form the original data. According to Clonbach reliability test and factor analysis by SPSS 22.0 software, the reliability of the constraint system had been tested.

(1) Questionnaire design and collection

The questionnaire is based on a cumulative scale table, in which the numbers 5, 4, 3, 2 and 1 represent high, relatively high, average, relatively low and low respectively according to the degree of importance from high to low. Please refer to Appendix A.

A total of 110 questionnaires were surveyed, and 102 completed questionnaires were returned. After eliminating 7 questionnaires with incomplete filling and obvious distortion of content, 95 valid questionnaires were actually obtained. Figures 1–3 show the basic statistics of 95 questionnaire.

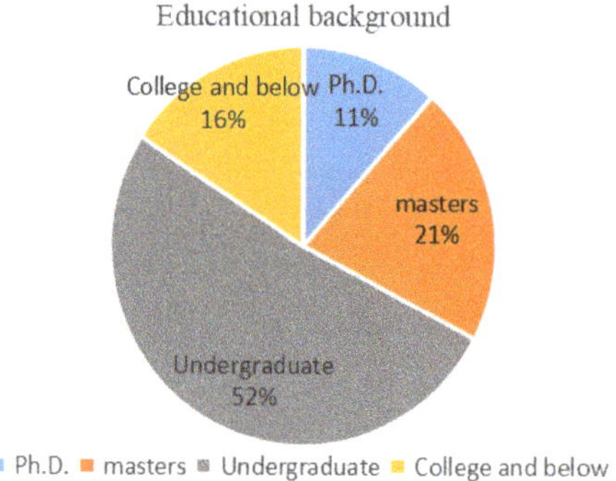

Figure 1. Educational background of interviewees.

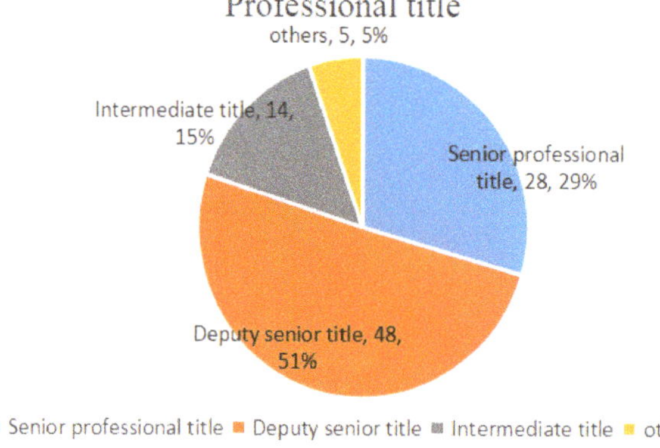

Figure 2. Professional title of interviewees.

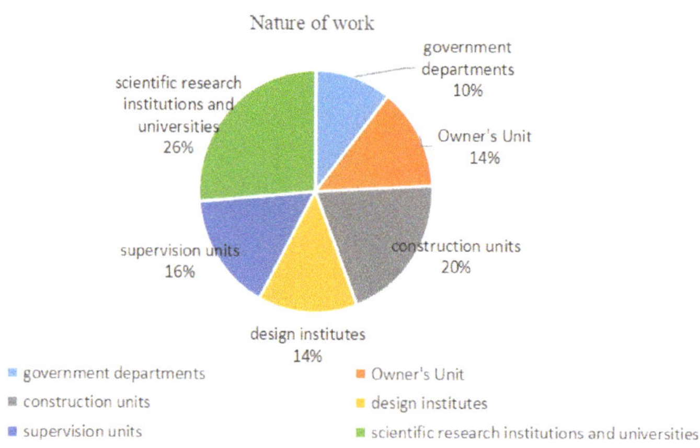

Figure 3. Work nature of interviewees.

(2) Reliability Analysis.

Reliability refers to the consistency and reliability of test results. In this reliability analysis, Cronbach's coefficient α was used to test the reliability of the questionnaire. The statistical significance of Cronbach's coefficient is shown in Table 3.

Table 3. Statistical significance of Cronbach's coefficient α.

Cronbach's Coefficient α	Statistical Significance
$\alpha < 0.6$	Insufficient reliability, unacceptable
$0.6 \leq \alpha < 0.7$	Acceptable reliability
$0.7 \leq \alpha < 0.8$	Certain reliability
$0.8 \leq \alpha < 0.9$	Relatively high reliability
$\alpha \geq 0.9$	High reliability

Using SPSS 22.0 software to analyze the reliability of the data, the following results are obtained.

Tables 4 and 5 show that all 95 questionnaires are valid, the Cronbach's coefficient α is 0.784, and the reliability of the survey data meets the requirements.

Table 4. Case Processing Summary.

		Number of Cases	%
Cases	Valid	95	100
	Exclusion[a]	0.0	0.0
	Total	95	100

Table 5. Reliability statistics.

Cronbach's Coefficient α	Cronbach's Coefficient α Based on Normalization Terms	Number of Items
0.782	0.769	15

(3) Component analysis.

Component analysis involves to group variables with common characteristics into a single factor by dimensionality reduction method, to reduce the number of variables, and to test hypotheses about relationships between variables. Before using SPSS 22.0 for component analysis, it is necessary to test the matching degree of the data by KMO and Bartlett sphericity test methods. The results are shown in Table 6.

Table 6. KMO and Bartlett test.

KMO Sampling Suitability Quantity		0.784
	Approximate chi-square	1107.234
Bartlett's test for sphericity	degrees of freedom	105
	Significance	0.000

It can be seen from Table 6 that the KMO value of the questionnaire is 0.784, which is suitable for Component analysis. The Bartlett sphericity test value is 1107.234, and $p = 0$, indicating that the results meet the significance standard to do factor analysis with common components in the matrix. Factor analysis was then performed to obtain the total variance in Table 7.

Table 7. Total Variance.

No.	Initial Eigenvalue			Extraction Sums of Squared Loadings			Rotation Sums of Squared Loadings		
	Total/%	Percentage Variance/%	Sum/%	Total/%	Percentage Variance/%	Sum/%	Total/%	Percentage Variance/%	Sum/%
1	5.999	39.996	39.996	5.999	39.996	39.996	4.930	32.869	5.999
2	2.784	18.560	58.556	2.784	18.560	58.556	2.347	15.646	2.784
3	1.365	9.100	67.655	1.365	9.100	67.655	2.199	14.661	1.365
4	1.184	7.895	75.551	1.184	7.895	75.551	1.599	10.663	1.184
5	0.883	5.887	81.438	0.883	5.887	81.438	1.140	7.599	0.883
6	0.774	5.161	86.600						
7	0.654	4.358	90.958						

According to Table 7, there are five components whose total initial eigenvalues are higher than 0.8, and their cumulative contribution rate is greater than 60%, which shows that 15 factors can be well converged into 5 common components. That is, the restrictive factors of Yancheng prefabricated buildings can be categorized into five aspects including market, economic management, technology, society, and policy.

4.2. DEMATEL-ISM Model Calculation

4.2.1. Establishing Direct Influence Matrix

According to model calculation requirements, 10 practitioners who have been engaged in the field of prefabricated construction for more than 15 years are invited to fill out the questionnaires on the relationship between the constraints. The statistical results of interviewed experts from different organizations are 2 from the construction owner unit, 2 from the design institute, 2 from the engineering consulting corporation, 2 from the construction enterprise and 2 from scientific research institutions. Based on profound experience over the years and the high professional quality of practitioners, they judged the mutual relations among the factors restricting the development of Yancheng prefabricated buildings in the questionnaires and scored a scale of (0,1,2,3,4) referring to Appendix B. To ensure the objectivity and accuracy of the scoring, the scoring of each item is averaged, then the direct influence matrix is constructed as shown in Table 8.

Table 8. Direct influence matrix.

Factor	S1	S2	S3	S4	S5	S6	S7	S8	S9	S10	S11	S12	S13	S14	S15
S1	0	3	3	2	0	1	2	2	2	0	1	4	0	0	0
S2	0	0	0	2	0	0	2	0	0	0	2	2	0	0	0
S3	4	3	0	0	0	0	0	0	0	2	3	3	1	0	0
S4	2	0	3	0	0	0	0	0	0	0	3	4	0	0	0
S5	2	0	0	3	0	2	0	3	3	0	1	1	1	0	0
S6	0	0	0	3	0	0	0	3	3	0	1	2	0	0	0
S7	2	2	3	3	0	1	0	0	0	0	3	3	0	0	0
S8	1	0	0	4	0	0	0	0	0	0	3	0	0	0	0
S9	1	1	2	2	2	4	2	3	0	0	0	0	3	0	0
S10	2	1	4	0	0	0	0	0	0	0	0	4	0	0	0
S11	3	3	0	0	0	0	0	0	0	0	0	0	0	0	0
S12	1	1	2	2	2	2	2	2	2	4	3	0	0	0	0
S13	3	3	1	1	1	2	2	2	2	4	1	3	0	0	0
S14	2	1	1	3	4	1	1	1	1	2	1	3	0	0	0
S15	2	3	1	3	0	4	3	4	4	0	1	2	0	0	0

4.2.2. Create Comprehensive Influence Matrix

(1)   Creating a normalized direct influence matrix.

According to the established direct influence matrix and Equation (1) calculation rules, the normalized direct influence matrix B is obtained as shown in Table 9.

Table 9. Normalized direct influence matrix.

Factor	S1	S2	S3	S4	S5	S6	S7	S8	S9	S10	S11	S12	S13	S14	S15
S1	0.000	0.111	0.111	0.074	0.000	0.037	0.074	0.074	0.074	0.000	0.037	0.148	0.000	0.000	0.000
S2	0.000	0.000	0.000	0.074	0.000	0.000	0.074	0.000	0.000	0.000	0.074	0.074	0.000	0.000	0.000
S3	0.148	0.111	0.000	0.000	0.000	0.000	0.000	0.000	0.000	0.074	0.111	0.111	0.037	0.000	0.000
S4	0.074	0.000	0.111	0.000	0.000	0.000	0.000	0.000	0.000	0.000	0.111	0.148	0.000	0.000	0.000
S5	0.074	0.000	0.000	0.111	0.000	0.074	0.000	0.111	0.111	0.000	0.037	0.037	0.037	0.000	0.000
S6	0.000	0.000	0.000	0.111	0.000	0.000	0.000	0.111	0.111	0.000	0.037	0.074	0.000	0.000	0.000
S7	0.074	0.074	0.111	0.111	0.000	0.037	0.000	0.000	0.000	0.000	0.111	0.111	0.000	0.000	0.000
S8	0.037	0.000	0.000	0.148	0.000	0.000	0.000	0.000	0.000	0.000	0.111	0.000	0.000	0.000	0.000
S9	0.037	0.037	0.074	0.074	0.074	0.148	0.074	0.111	0.000	0.000	0.000	0.000	0.111	0.000	0.000
S10	0.074	0.037	0.148	0.000	0.000	0.000	0.000	0.000	0.000	0.000	0.000	0.148	0.000	0.000	0.000
S11	0.111	0.111	0.000	0.000	0.000	0.000	0.000	0.000	0.000	0.000	0.000	0.000	0.000	0.000	0.000
S12	0.037	0.037	0.074	0.074	0.074	0.074	0.074	0.074	0.074	0.148	0.111	0.000	0.000	0.000	0.000
S13	0.111	0.111	0.037	0.037	0.037	0.074	0.074	0.074	0.074	0.148	0.037	0.111	0.000	0.000	0.000
S14	0.074	0.037	0.037	0.111	0.148	0.037	0.037	0.037	0.037	0.074	0.037	0.111	0.000	0.000	0.000
S15	0.074	0.111	0.037	0.111	0.000	0.148	0.111	0.148	0.148	0.000	0.037	0.074	0.000	0.000	0.000

(2)   Factor comprehensive influence matrix

According to Equation (2), calculate the matrix I-B first, then deduce the inverse of the matrix I-B, and finally obtain comprehensive influence matrix T, which is the matrix B multiplied by matrix I-B, as shown in Table 10.

Table 10. Comprehensive influence matrix.

Factor	S1	S2	S3	S4	S5	S6	S7	S8	S9	S10	S11	S12	S13	S14	S15
S1	0.094	0.187	0.190	0.168	0.028	0.084	0.124	0.126	0.114	0.055	0.155	0.253	0.021	0.000	0.000
S2	0.042	0.035	0.039	0.105	0.010	0.017	0.090	0.017	0.015	0.021	0.120	0.118	0.004	0.000	0.000
S3	0.215	0.186	0.074	0.068	0.020	0.036	0.051	0.045	0.041	0.117	0.186	0.207	0.045	0.000	0.000
S4	0.140	0.068	0.161	0.048	0.019	0.029	0.034	0.036	0.032	0.045	0.178	0.213	0.010	0.000	0.000
S5	0.141	0.057	0.069	0.193	0.022	0.118	0.039	0.167	0.151	0.032	0.120	0.123	0.057	0.000	0.000
S6	0.054	0.038	0.051	0.169	0.020	0.035	0.027	0.147	0.132	0.025	0.102	0.127	0.017	0.000	0.000
S7	0.156	0.149	0.178	0.173	0.019	0.068	0.042	0.042	0.037	0.046	0.203	0.209	0.011	0.000	0.000
S8	0.075	0.032	0.034	0.165	0.004	0.009	0.012	0.012	0.011	0.010	0.147	0.046	0.003	0.000	0.000
S9	0.131	0.111	0.147	0.181	0.092	0.191	0.115	0.177	0.060	0.048	0.111	0.122	0.127	0.000	0.000
S10	0.136	0.098	0.200	0.051	0.019	0.030	0.037	0.036	0.033	0.049	0.075	0.221	0.012	0.000	0.000
S11	0.126	0.136	0.025	0.030	0.004	0.011	0.024	0.016	0.014	0.008	0.031	0.041	0.003	0.000	0.000
S12	0.144	0.122	0.168	0.169	0.093	0.119	0.114	0.133	0.120	0.183	0.213	0.129	0.023	0.000	0.000
S13	0.216	0.205	0.153	0.157	0.067	0.133	0.136	0.148	0.133	0.201	0.163	0.258	0.023	0.000	0.000
S14	0.173	0.113	0.133	0.213	0.174	0.093	0.085	0.109	0.099	0.122	0.149	0.235	0.022	0.000	0.000
S15	0.180	0.201	0.144	0.252	0.032	0.212	0.173	0.230	0.207	0.047	0.181	0.215	0.029	0.000	0.000

(3)   Influence factor strength

According to Equations (3)–(6), the influence degree, affected degree, centrality degree and cause degree of each constraint factor are calculated, as shown in Table 11.

(4)   Centrality degree and cause degree distribution

According to the obtained centrality degree and cause degree, the scatter distribution diagram of each factor can be drawn, as shown in Figure 4. The attributes and importance of each factor can be analyzed more clearly and intuitively according to the position of each factor in Figure 4.

Table 11. Strength of each constraint factor.

Factor	Influence Degree f_i	Affected Degree e_i	Centrality Degree z_i	Cause Degree y_i
S1	1.598	2.022	3.620	−0.424
S2	0.633	1.738	2.371	−1.105
S3	1.292	1.766	3.057	−0.474
S4	1.015	2.143	3.158	−1.128
S5	1.290	0.624	1.914	0.666
S6	0.944	1.186	2.129	−0.242
S7	1.331	1.102	2.433	0.228
S8	0.558	1.441	1.999	−0.883
S9	1.612	1.197	2.809	0.415
S10	0.997	1.009	2.005	−0.012
S11	0.470	2.134	2.604	−1.664
S12	1.730	2.518	4.248	−0.788
S13	1.994	0.407	2.400	1.587
S14	1.719	0.000	1.719	1.719
S15	2.104	0.000	2.104	2.104

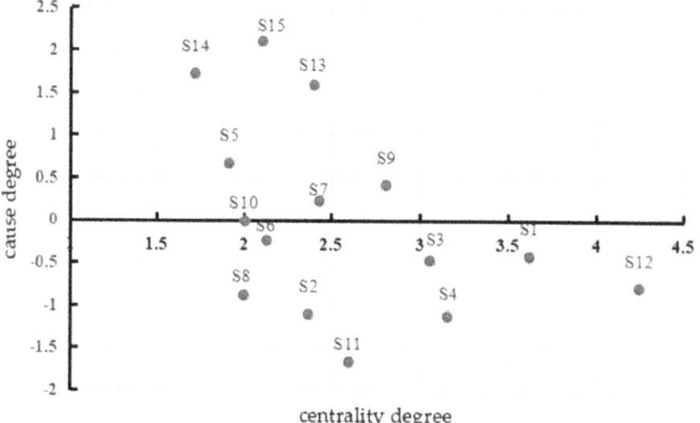

Figure 4. The distribution of centrality degree and cause degree of each factor.

4.2.3. Establishment of Reachability Matrix

The commonality between DEMATEL and ISM determines that the reachability matrix can be obtained from the comprehensive influence matrix. First, it is necessary to set up an appropriate threshold λ to simplify the structure of the system and parttition system layer. After several trial calculations, the final optimal value of λ is 0.16, which can optimize the system hierarchy. The reachability matrix R can be calculated by Equations (7) and (8), as shown in Table 12.

Table 12. Reachability matrix R.

Factor	S1	S2	S3	S4	S5	S6	S7	S8	S9	S10	S11	S12	S13	S14	S15
S1	1	1	1	1	0	0	0	0	0	0	0	1	0	0	0
S2	0	1	0	0	0	0	0	0	0	0	0	0	0	0	0
S3	1	1	1	0	0	0	0	0	0	0	1	1	0	0	0
S4	0	0	1	1	0	0	0	0	0	0	1	1	0	0	0
S5	0	0	0	1	1	0	0	1	0	0	0	0	0	0	0
S6	0	0	0	1	0	1	0	0	0	0	0	0	0	0	0
S7	0	0	1	1	0	0	1	0	0	0	1	1	0	0	0
S8	0	0	0	1	0	0	0	1	0	0	0	0	0	0	0
S9	0	0	0	1	0	1	0	1	1	0	0	0	0	0	0
S10	0	0	1	0	0	0	0	0	0	1	0	1	0	0	0
S11	0	0	0	0	0	0	0	0	0	0	1	0	0	0	0
S12	0	0	1	1	0	0	0	0	0	1	1	1	0	0	0
S13	1	1	0	0	0	0	0	0	0	1	1	1	1	0	0
S14	1	0	0	1	1	0	0	0	0	0	0	1	0	1	0
S15	1	1	0	1	0	1	1	1	1	0	1	1	0	0	1

4.2.4. Developing the Multi-Layer Hierarchical Model

After forming reachability matrix R, the components included in each level are obtained in turn according to step (10), as shown in Table 13. In this way, Table 14 can also be obtained. The hierarchical diagram of constraints at all levels is shown in Figure 5.

Table 13. Reachable set R and preceding set Q and their intersection A.

Factor	Reachable Set R	Preceding Set Q	Intersection set $A = R \cap Q$
S1	1, 2, 3, 4, 10, 11, 12	1, 3, 4, 5, 6, 7, 8, 9, 10, 12, 13, 14, 15	1, 3, 4, 10, 12
S2	2	1, 2, 3, 4, 5, 6, 7, 8, 9, 10, 12, 13, 14, 15	2
S3	1, 2, 3, 4, 10, 11, 12	1, 3, 4, 5, 6, 7, 8, 9, 10, 12, 13, 14, 15	1, 3, 4, 10, 12
S4	1, 2, 3, 4, 10, 11, 12	1, 3, 4, 5, 6, 7, 8, 9, 10, 12, 13, 14, 15	1, 3, 4, 10, 12
S5	1, 2, 3, 4, 5, 8, 10, 11, 12	5, 14	5
S6	1, 2, 3, 4, 6, 10, 11, 12	6, 9, 15	6
S7	1, 2, 3, 4, 7, 10, 11, 12	7, 15	7
S8	1, 2, 3, 4, 8, 10, 11, 12	5, 8, 9, 14, 15	8
S9	1, 2, 3, 4, 6, 8, 9, 10, 11, 12	9, 15	9
S10	1, 2, 3, 4, 10, 11, 12	1, 3, 4, 5, 6, 7, 8, 9, 10, 12, 13, 14, 15	1, 3, 4, 10, 12
S11	11	1, 3, 4, 5, 6, 7, 8, 9, 10, 11, 12, 13, 14, 15	11
S12	1, 2, 3, 4, 10, 11, 12	1, 3, 4, 5, 6, 7, 8, 9, 10, 12, 13, 14, 15	1, 3, 4, 10, 12
S13	1, 2, 3, 4, 10, 11, 12, 13	13	13
S14	1, 2, 3, 4, 5, 8, 10, 11, 12, 14	14	14
S15	1, 2, 3, 4, 6, 7, 8, 9, 10, 11, 12, 15	15	15

Note: Numbers represent a certain element, for example, 2 represents the second factor.

Table 14. Hierarchical Decomposition.

Level	Factor
Layer 1 (Top)	S2-Imperfect supporting industry chain, S11-Low public awareness and willingness to buy
Layer 2	S1-Insufficient motivation for enterprise transformation, S3-insufficient market demand, S4-High construction cost and low cost performance, S10-lack of talent-rich professionals, S12-low degree of social acceptance
Layer 3	S6- Backward prefabricated building management mode, S7-The low system of design standardizationmodularization, and integration, S8-Immature construction experience, S13-Imperfect industry policy and standard system
Layer4	S5-Large upfront investment, S9-Insufficient basic research and innovation of prefabricated buildings
Layer5 (bottom)	S14-Insufficient financial support, S15-Insufficient government propaganda

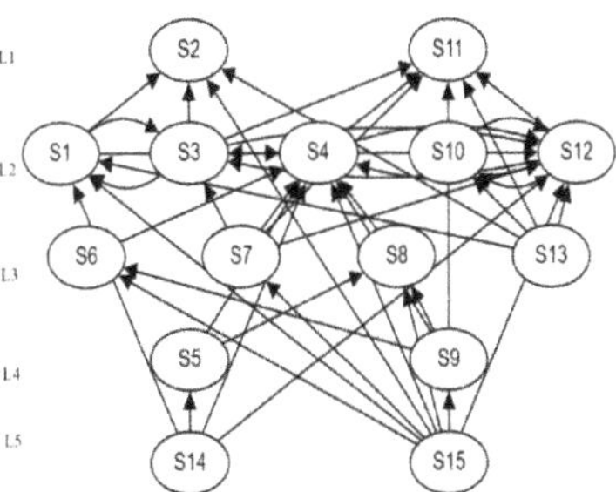

Figure 5. Multi-layer hierarchical diagram of restrictive factors for the promotion and implementation of Yancheng prefabricated buildings.

5. Discussion

5.1. DEMATEL Analysis Results

(1) Influence degree Analysis

The larger the influence degree value of a factor is, the more the factor influences other factors. Compared to the value of influence degree, the top five factors are S15, S13, S12, S14, and S9, as shown in Table 15.

Table 15. Ranking listof influence degree.

Sort	Factor	Influence Degree
1	S15	2.104
2	S13	1.994
3	S12	1.730
4	S14	1.719
5	S9	1.612
6	S1	1.598
7	S7	1.331
8	S3	1.292
9	S5	1.290
10	S4	1.015
11	S10	0.997
12	S6	0.944
13	S2	0.633
14	S8	0.558
15	S11	0.470

(2)   Affected degree analysis

The higher the affected degree value of a factor is, the greater the factor is constrained by other factors. As can be seen from Table 16, the top five most affected factors are S12 (insufficient market demand), S4 (insufficient enthusiasm for enterprise transformation), S11 (low public awareness and willingness to buy), and S1 (high construction cost and low-cost performance), S3 (low social acceptance).

Table 16. Ranking list of affected degree.

Sort	Factor	Affected Degree
1	S12	2.518
2	S4	2.143
3	S11	2.134
4	S1	2.022
5	S3	1.766
6	S2	1.738
7	S8	1.441
8	S9	1.197
9	S6	1.186
10	S7	1.102
11	S10	1.009
12	S5	0.624
13	S13	0.407
14	S14	0.000
15	S15	0.000

(3)   Centrality degree Analysis

It can be concluded from Table 17, the larger the centrality degree is the closer the relationship between this factor and other factors in the system. In Table 17, the top five factors with centrality values from large to small are S12, S1, S4, S3 and S9. S12 with the largest value of centrality degree is the most important factor, which closely relates to other constraints in the entire system.

Table 17. Ranking list and weight of the centrality degree.

Sort	Factor	Centrality Degree	Weight
1	S12	4.248	0.110
2	S1	3.620	0.094
3	S4	3.158	0.082
4	S3	3.057	0.079
5	S9	2.809	0.073
6	S11	2.604	0.068
7	S7	2.433	0.063
8	S13	2.400	0.062
9	S2	2.371	0.061
10	S6	2.129	0.055
11	S15	2.104	0.054
12	S10	2.005	0.052
13	S8	1.999	0.052
14	S5	1.914	0.050
15	S14	1.719	0.045

(4)   Cause degree Analysis

The larger the value of the degree of cause, the greater the influence this element has on other elements in the system. The reason degree sorting table is shown in Table 18. It can be seen from the table that the top five causes are: S15, S14, S13, S5, and S9. Among them, S15 (Insufficient publicity by the government) has the highest degree of reason, indicating that the insufficient publicity of policies and regulations by the government in terms of prefabricated buildings has the greatest impact on other factors in the entire restrictive factor system.

Table 18. Ranking list of cause degree.

Sort	Factor	Cause Degree
1	S15	2.104
2	S14	1.719
3	S13	1.587
4	S5	0.666
5	S9	0.415
6	S7	0.228
7	S10	−0.012
8	S6	−0.242
9	S1	−0.424
10	S3	−0.474
11	S12	−0.788
12	S8	−0.883
13	S2	−1.105
14	S4	−1.128
15	S11	−1.664

The factors with high centrality are more closely related to other factors in the system and have greater influence, which are used as the basis to screen these important factors. According to the value of cause degree to rank 15 factors in Table 18, the top five factors are A15, S14, S13, S5 and S9, which should be paid attention to develop Yancheng prefabricated buildings.

The factor with a high affected degree is defined as the resulting factor, which is easily affected by other factors in the system during the development of prefabricated buildings. In the process of formulating measures, it is concluded from Figure 4 that it is necessary to take targeted measures for these factors.

Yancheng is a relatively underdeveloped city in Jiangsu Province, the Yancheng government has limited financial resources for the development of prefabricated buildings every year. Although the policies and documents have been formulated and promulgated to vigorously promote prefabricated buildings, the publicity of the policies only remains within the construction industry. Due to the lack of effective research and innovation system, the public generally believes that the structure of prefabricated buildings is unsafe, which restricts the promotion of Yancheng. From the perspective of long-term development, increasing financial investment, improving research & innovation, and strengthening public awareness and acceptance, all of which are the basis for vigorously promoting Yancheng prefabricated buildings.

5.2. ISM Analysis Results

Figure 5 shows the multi-layer hierarchical relation obtained by the ISM method, then an important conclusion can be drawn that the constraint factor system of Yancheng prefabricated buildings is a five-layer hierarchical structure, and the hierarchy of the constraint factors is from simple to complex and from top to bottom.

The first layer of factors is the superficial restrictive factor, also known as the cause of proximity, including imperfect supporting industrial chain, low public awareness and low purchase willingness. In terms of the imperfection of the industrial chain, it is mainly reflected in the insufficient design capabilities, as well as the insufficient number and production capacity of component manufacturers. It is mainly reflected that the public with low awareness and willingness hardly initiate buying prefabricated houses and believing that the structure is safe. These factors are the direct reasons for the vigorous promotion of Yancheng prefabricated buildings.

The second to fourth layers belong to the middle-level constraints, also known as transition causes, including insufficient enthusiasm for enterprise transformation, insufficient market demand, high construction costs, low-cost performance, lack of talents, low social recognition, backward manufacturing management mode, unsound design standardization, modularization, and integrated systems, insufficient construction experience, imperfect industry policy and standard system, the relatively large initial investment, and the insufficient basic research and innovation. While these factors which belong to the middle part of the internal influence path of the system directly or indirectly affect surface factors, and are also affected by bottom constraint factors. In Yancheng, the cost and construction period of prefabricated buildings are higher than those of cast-in-place buildings, and the integrated design of various specialties has not been truly achieved. There is a lack of professional technicians and skilled installation workers. The on-site management mode still relies on traditional methods and lacks an intelligence management platform, etc. The factors that cause these problems are involved in market, economy, technology, society, and policy, with a wide distribution and a large number.

The fifth layer of factors belongs to the bottom constraints, also known as essential causes, including insufficient financial support and insufficient government publicity. The bottom-level constraints are the most complicated factors affecting the promotion of prefabricated buildings in Yancheng, and they have a relatively large impact on other factors. The scope of action is mainly the middle-level constraints, and the mechanism of action is direct or indirect. The two constraints included in the bottom-level constraints are factors at the policy level, so it is crucial to promote prefabricated buildings in Yancheng. Therefore, the Yancheng government needs to increase financial support, especially in terms of basic innovation and market cultivation, and strengthen publicity efforts. so that good policies can take root and improve public awareness and willingness to buy prefabricated houses. If the attention of the government, the inclination of policies and the investment of various resources, the underlying constraints will have been improved, it has a positive impact on the middle and surface constraints, and thus the healthy development of prefabricated buildings will be greatly gone on.

From the above, the following points are obtained:

(1) In the DEMATEL method, the reasons for the high centrality include insufficient basic research and innovation of prefabricated buildings, imperfect industry policies and standard systems, insufficient government publicity, lack of talents, and relatively low initial investment in scientific research. Large and insufficient financial support. These factors are likely to affect other factors in the system and play a greater role in the system. However, low social recognition, high construction cost and low-cost performance, low public awareness and purchase willingness are the primary result factors with high centrality, which are easily affected by other factors in the system.

(2) In the ISM method, Imperfect supporting industry chain, low public awareness and low willingness to purchase are superficial factors that directly affect the promotion and implementation of prefabricated buildings in Yancheng. Insufficient financial support and insufficient government publicity, as the underlying factors of the hierarchical structure, are the fundamental factors affecting the development of prefabricated buildings in Yancheng. They restrict the transitional factors and surface influencing factors within the system directly or indirectly. The problems must be solved with the promotion of prefabricated buildings in Yancheng.

6. Conclusions and Suggestions

Under the background of the "double carbon target" proposed by China, the development of the construction industry is towards green, low carbon, environmental protection and energy saving. The vigorous development of prefabricated buildings is in line with china's sustainable development strategy. Yancheng in China, as an area for the promotion of prefabricated construction, has made some achievements in the development of the prefabricated construction industry, but there are still many unfavorable factors for its development. Through a series of studies in this paper, the index system of restricting factors affecting the development of the prefabricated construction industry is proposed and modified. the authors analyze the main restrictive factors of the prefabricated construction industry in Yancheng through the DEMATEL-ISM method and find out the key restrictive factors of vigorously promoting the prefabricated construction industry. It provides certain precision-oriented policies for promoting the development of prefabricated buildings in Yancheng. The main conclusions are as follows:

(1) Through literature review, it is concluded that prefabricated buildings are being gradually promoted around the world due to their low carbon, high efficiency, energy saving, green, and environmental protection. Among them, the United States, Canada, Australia and other countries have relatively high development levels by new methods of DfMA. The relatively rapid and high-quality development of the industrial chain in design, production and installation has been an important reference for the promotion of prefabricated buildings in China.

(2) Establishing the index system of restrictive factors of prefabricated buildings in Yancheng. Firstly, 25 restrictive factors of prefabricated buildings are preliminarily summarized and sorted out through literature analysis, and the general framework of the restrictive factor index system for the promotion and implementation of prefabricated buildings in Yancheng is initially formed. Then, through the form of asking for expert opinions and drawing on the experience of experts, the restrictive factors are revised to form an index system of restrictive factors for the promotion and imple-

mentation of prefabricated buildings in Yancheng. To ensure that the index system is reasonable and reliable, a large number of questionnaires are designed and issued to collect relevant data, and the reliability and validity analysis and factor analysis of the questionnaires are carried out to verify the scientificity of the index system of restrictive factors for the promotion and implementation of prefabricated buildings in Yancheng. Finally, the index system of restrictive factors for the promotion and implementation of prefabricated buildings in Yancheng is determined to be composed of 15 major restrictive factors in five aspects: market, economy, technology, society and policy.

(3) Analyzing the main constraints of prefabricated buildings in Yancheng. Based on the application of the DEMATEL method, we analyze the logical relationship and influence degree among the restrictive factors, and find out the cause factors and result factors in the constraint index system.; On the basis of the DEMATEL comprehensive matrix, ISM is used to establish an reachability matrix, and a hierarchical model including surface-level influencing factors, middle-level transition factors and bottom-level influencing factors is constructed, and the comprehensive level of each restrictive factor in the system is obtained. The key restrictive factors for the vigorous promotion of prefabricated buildings in Yancheng have been identified.

Based on the current situation of the development of prefabricated buildings in Yancheng, according to the research results: In terms of policies, regulations and standard systems, the government needs to continue to supplement and improve the system of policies and regulations related to prefabricated buildings, improve the implementability of various policies and standards, and strengthen the implementation of policies and standards.; In terms of government financial support, the Yancheng government needs to strengthen financial support and guarantee measures, and further refine and formulate specific incentive measures; In terms of government publicity, it is necessary to strengthen the exposure and promotion of major media, increase the enthusiasm of enterprise transformation and improve public awareness; In terms of technology and innovation, enterprises are encouraged to set up research and development institutions, focusing on the research and development of standardized design and construction technologies for prefabricated buildings.It is necessary to gradually promote the DfMA design method; In terms of personnel training, the government should encourage the cultivation of professional talents in colleges and universities in Yancheng, and strengthen the training of talents combining production, education and research; In terms of market cultivation, the government should focus on developing prefabricated construction industrial bases and parks, cultivate market production entities, and improve the industrial chain.

Author Contributions: Conceptualization, F.S. and H.S.; methodology, H.S.; software, H.S.; validation, H.S., Y.F. and M.Y.; formal analysis, H.S.; investigation, Y.F.; resources, M.Y.; data curation, F.S.; writing—original draft preparation, H.S.; writing—review and editing, F.S.; visualization, H.S.; supervision, F.S.; project administration, F.S.; funding acquisition, H.S. All authors have read and agreed to the published version of the manuscript.

Funding: This research was funded by the Yancheng Natural Science Soft Project Foundation of China, grant number yckxrkt2022-29.

Institutional Review Board Statement: Not applicable.

Informed Consent Statement: Not applicable.

Data Availability Statement: The data used to support the findings of this study are available from the corresponding author upon request.

Acknowledgments: The authors gratefully acknowledge the financial support of YCIT, and the testing services from Analysis and Test center of YCIT.

Conflicts of Interest: The authors declare no conflict of interest.

Appendix A

Questionnaire on the Importance of Restricting Factors for the Promotion and Implementation of Prefabricated Buildings in Yancheng

Dear experts:

I am from the research group of Yancheng Institute of Technology "Promotion and Application of Prefabricated Buildings in Yancheng under the Background of 'Double Carbon'". I am doing a questionnaire on "Restricting Factors in the Promotion and Implementation of Prefabricated Buildings in Yancheng". We need to conduct a survey on these factors Importance evaluation (scored in the form of numbers 1–5, numbers 5, 4, 3, 2, and 1 represent high, relatively high, general, relatively low, and low according to the degree of importance from high to low). thank you for your support!

Educational background: ☐Ph.D. ☐masters ☐Undergraduate ☐College and below

Professional title: ☐Senior professional title ☐Deputy senior title ☐Intermediate title ☐others

Nature of work: ☐government departments ☐Owner's Unit☐construction units ☐design institutes ☐supervision units ☐scientific research institutions and universities

Table A1. Survey Questionnaire on the Importance of Restricting Factors for the Promotion and Implementation of Prefabricated Buildings in Yancheng.

Restrictive Factors		Degree of Importance				
		High	Relatively High	Average	Relatively Low	Low
		5	4	3	2	1
marketaspect	insufficient motivation for enterprise transformation					
	imperfect supporting industry chain					
	insufficient market demand					
economic management	high construction cost, low cost performance					
	large upfront investment					
	the prefabricated building management model is backward					
technical aspect	design standardization, modularization, and integrated systems are not perfect					
	Prefabricated building production, construction experience and technology are immature					
	Insufficient basic research and innovation of prefabricated buildings					
social aspects	talent shortage					
	low public awareness and willingness to buy					
	low degree of social acceptance					
Policyaspect	imperfect industry policy and standard system					
	insufficient financial support					
	insufficient government propaganda					

Appendix B

Questionnaire on the influence relationship of restrictive factors in the vigorous promotion and implementation of prefabricated buildings in Yancheng

Dear Expert:

I am from the research group of Yancheng Institute of Technology "Promotion and Application of Prefabricated Buildings in Yancheng under the Background of 'Double Carbon'". Evaluation of the influence relationship among them (the influence of row elements on column elements is scored in the form of numbers 0–4, and the numbers 0, 1, 2, 3, and 4 represent no influence, low influence, average impact, relatively high impact, high impact, respectively, ignore the impact on itself). thank you for your support.

Table A2. Questionnaire for the Influencing Relationship of Restricting Factors in Vigorously Promoting and Implementing Prefabricated Buildings in Yancheng.

	S1	S2	S3	S4	S5	S6	S7	S8	S9	S10	S11	S12	S13	S14	S15
S1															
S2															
S3															
S4															
S5															
S6															
S7															
S8															
S9															
S10															
S11															
S12															
S13															
S14															
S15															

Note: S1-insufficient motivation for enterprise transformation; S2-imperfect supporting industry chain; S3-insufficient market demand; S4-high construction cost, low cost performance; S5-large upfront investment; S6-the prefabricated building management model is backward; S7-design standardization, modularization, and integrated systems are not perfect; S8-prefabricated building production, construction experience and technology are immature; S9-insufficient basic research and innovation of prefabricated buildings; S10-talent shortage; S11-low public awareness and willingness to buy; S12-low degree of social acceptance; S13-imperfect industry policy and standard system; S14-insufficient financial support; S15-insufficient government propaganda.

References

1. Wang, J.; Zhao, J.; Hu, Z. Review and thinking on development of building industrialization in China. *J. China Civ. Eng. J.* **2016**, *49*, 18. (In Chinese)
2. Zhang, B.; Wei, C.; Xu, Q. Industrial Status and Development Path of Regional Prefabricated Buildings in China. *J. Nantong Vocat. Univ. J.* **2020**, *34*, 97–100. (In Chinese)
3. Li, D.; Li, X. Research on the Driving Factors of Prefabricated Buildings Based on Interpretative Structural Model (ISM). *Constr. Econ. J.* **2019**, *40*, 87–91. (In Chinese)
4. Wang, H.; Zhang, Y.; Gao, W.; Kuroki, S. Life Cycle Environmental and Cost Performance of Prefabricated Buildings. *Sustainability* **2020**, *12*, 2609. [CrossRef]
5. Zhang, Z.; Tan, Y.; Shi, L.; Hou, L.; Zhang, G. Current State of Using Prefabricated Construction in Australia. *Buildings* **2022**, *12*, 1355. [CrossRef]
6. Chen, Z.; Popovski, M.; Chun, N. A novel floor-isolated re-centering system for prefabricated modular mass timber construction-Concept development and preliminary evaluation. *Eng. Struct.* **2020**, *222*, 111168. [CrossRef]
7. Wuni, I.Y.; Shen, G.Q. Critical success factors for modular integrated construction projects: A review. *Build. Res. Inf.* **2020**, *48*, 763–784. [CrossRef]

8. Yuan, Z.; Sun, C.; Wang, Y. Design for Manufacture and Assembly-oriented parametric design of prefabricated buildings. *Autom. Constr.* **2018**, *88*, 13–22. [CrossRef]

9. Razak, M.I.A.; Khoiry, M.A.; Badaruzzaman, W.H.W.; Hussain, A.H. DfMA for a Better Industrialised Building System. *Buildings* **2022**, *12*, 794. [CrossRef]

10. Bian, J.; Wang, Z.; Liu, X. Research on Restrictive Factors of Assembly Building Development based on Principal Components Analysis Method. *Constr. Econ. J.* **2021**, *42*, 76–80. (In Chinese)

11. Lu, R.; Qing, K.; Tan, Y.; Xu, S. The main problems and countermeasures in the development of prefabricated buildings from the perspective of developers. *Build. Struct. J.* **2021**, *51*, 1134–1138. (In Chinese)

12. Chen, W.; Wu, Z. Practical Dilemma and Improvement Path of Prefabricated Building Policy Implementation in China: From the Perspective of Perform System Model of Meter-Horn Policy. *J. Constr. Econ.* **2021**, *42*, 77–80. (In Chinese)

13. Chen, Y.; Lin, S.; Shi, Y. Evolutionary Game on Incentive Policy for Prefabrication. *J. Civ. Eng. Manag.* **2018**, *35*, 155–160. (In Chinese)

14. Zhai, X.; Reed, R.; Mills, A. Factors impeding the offsite production of housing construction in China: An investigation of current practice. *Constr. Manag. Econ.* **2014**, *32*, 40–52. [CrossRef]

15. Zhang, X.; Skitmore, M.; Yi, P. Exploring the challenges to industrialized residential building in China. *Habitat Int.* **2014**, *41*, 176–184. [CrossRef]

16. Hong, J.; Shen, G.; Li, Z. Barriers to promoting prefabricated construction in China: A cost–benefit analysis. *J. Clean. Prod.* **2018**, *172*, 649–660. [CrossRef]

17. Xue, X.; Zhang, X.; Wang, L. Analyzing collaborative relationships among industrialized construction technology innovation organizations: A combined SNA and SEM approach. *J. Clean. Prod.* **2018**, *173*, 265–277. [CrossRef]

18. Wu, G.; Yang, R.; Li, L. Factors influencing the application of prefabricated construction in China: From perspectives of technology promotion and cleaner production. *J. Clean. Prod.* **2019**, *219*, 753–762. [CrossRef]

19. Zhang, S.; Li, Z.; Ma, S.; Li, L.; Yuan, M. Critical Factors Influencing Interface Management of Prefabricated Building Projects: Evidence from China. *Sustainability* **2022**, *14*, 5418. [CrossRef]

20. Shang, Z.; Wang, F.; Yang, X. The Efficiency of the Chinese Prefabricated Building Industry and Its Influencing Factors: An Empirical Study. *J. Sustain.* **2022**, *14*, 10695. [CrossRef]

21. Zhao, W.; Zhang, B.; Yang, Y. Empirical study of comprehensive benefits for prefabricated buildings: A case study of Hefei city. *Int. J. Electr. Eng. Educ.* **2020**, 1–17.

22. Li, Q.; Chen, R.; Ma, M. Research on the Constraints of the Development of Prefabricated Buildings Based on DEMATEL-ISM. *J. Eng. Manag.* **2020**, *34*, 30–43. (In Chinese)

23. Lu, W.; Olofsson, T.; Stehn, L. A lean-agile model of homebuilders' production systems. *Constr. Manag. Econ.* **2011**, *29*, 25–35. [CrossRef]

24. Department of Housing and Urban Rural Development of Guangdong Province of the People's Republic of China. Letter of the General Office of the Ministry of Housing and Urban-Rural Development on the Identification of the First Batch of Prefabricated Building Demonstration Cities and Industrial Bases (Jian Ban Ke Han [2017] No. 771). [EB/OL]. (6 December 2017). Available online: http://zfcxjst.gd.gov.cn/xxgk/wjtz/content/post_1394948.html (accessed on 30 October 2022).

25. Ministry of Housing and Urban-Rural Development of the People's Republic of China. Notice of the General Office of the Ministry of Housing and Urban-Rural Development onthe Identification of the Second Batch of Model Cities and Industrial Bases of Prefabricated Buildings (Jian Ban Biao Han [2020] No. 470). [EB/OL]. (15 September 2020). Available online: https://www.mohurd.gov.cn/gongkai/fdzdgknr/tzgg/202009/20200915_247204.html (accessed on 28 October 2022).

26. Meng, N. The Development of Prefabricated Buildings is Accelerating. *J. Constr. Archit.* **2022**, *29*, 14–17. (In Chinese)

27. Gu, Q. Study on Obstacle Factors Analysis and Countermeasures of Prefabricated Building Promotion in China. Master's Thesis, Guangzhou University, Guangzhou, China, 2020. (In Chinese)

28. Wang, M. Research on the Restrictive Factors and Countermeasures of Popularization of Prefabricated Buildings in Anhui Province. Master's Thesis, An Hui Jian Zhu University, Hefei, China, 2021. (In Chinese)

Disclaimer/Publisher's Note: The statements, opinions and data contained in all publications are solely those of the individual author(s) and contributor(s) and not of MDPI and/or the editor(s). MDPI and/or the editor(s) disclaim responsibility for any injury to people or property resulting from any ideas, methods, instructions or products referred to in the content.

sustainability

Article

Do Central Inspections of Environmental Protection Affect the Efficiency of the Green Economy? Evidence from China's Yangtze River Delta

Haisheng Chen [1,2,3,4] and Manhong Shen [1,2,3,*]

1   Institute of Ecological Civilization, Zhejiang A&F University, Hangzhou 311300, China
2   Institute of Carbon Neutrality, Zhejiang A&F University, Hangzhou 311300, China
3   College of Economics and Management, Zhejiang A&F University, Hangzhou 311300, China
4   Zhejiang Provincial Credit Center, Hangzhou 310007, China
*   Correspondence: smh@zafu.edu.cn

Abstract: As an important part of China's ecological civilization, the impact of the Central Inspections of Environmental Protection (CIEP) on the development of a green economy has been widely recognized. This article uses the first round of the Central Inspections of Environmental Protection (CIEP) and the "look-back" in cities above the prefecture level in China's Yangtze River Delta as a quasi-natural experiment to construct more scientific green economic efficiency indicators based on OH (2010), and employs a multi-period spatial DID (difference-in-differences) model to empirically investigate the impact of the CIEP on the urban green economic efficiency. This study confirms that: (1) The Central Inspections of Environmental Protection have a significant contribution to the green economic efficiency of cities, and the "look-back" is of great significance to the long-term green development of cities. (2) The Central Inspections of Environmental Protection have had a positive impact on the building of a pro-clear government–business relationship in coastal and riverine areas, promoting the application of green technology research and development, and, thus, improving the green economic efficiency of cities. (3) Under the constraints of the central environmental protection inspection system, the southern Jiangsu region has been effective in promoting the green transformation of enterprises to enhance the efficiency of the city's green economy due to its location endowment and historical tradition of opening ports and trading in the late Qing Dynasty. (4) Under the pressure of environmental regulation, some enterprises chose to relocate their production to non-inspected areas, which had a negative spillover effect on the green economic efficiency of the cities they moved into. Policy Implications: The impact of central environmental inspections on the efficiency of urban green economies varies from time to time and place to place, and it is important to regulate the use of administrative resources and strengthen inter-provincial coordination to promote synergy and cooperation across provincial environmental inspection systems. This paper provides ideas for understanding the logical starting point for the implementation of the central environmental inspection system, and for better promoting the green transformation and high-quality development of regional economies based on national characteristics.

Keywords: Central Inspections of Environmental Protection; green economic efficiency; "look-back"; regional heterogeneity; spatial spillover; distortion effect

Citation: Chen, H.; Shen, M. Do Central Inspections of Environmental Protection Affect the Efficiency of the Green Economy? Evidence from China's Yangtze River Delta. *Sustainability* **2023**, *15*, 747. https://doi.org/10.3390/su15010747

Academic Editor: Antonio Boggia

Received: 22 November 2022
Revised: 28 December 2022
Accepted: 28 December 2022
Published: 31 December 2022

Copyright: © 2022 by the authors. Licensee MDPI, Basel, Switzerland. This article is an open access article distributed under the terms and conditions of the Creative Commons Attribution (CC BY) license (https://creativecommons.org/licenses/by/4.0/).

1. Introduction

The concept of "green development" was first introduced by the United Nations Development Program in 2002 and is widely regarded by society as the ideal path to achieve the organic integration of the economy and environment (Adam, 2009) [1]. Since 2006, China has been the world's largest emitter of carbon emissions, and with the positive progress of the domestic economic transformation, both total carbon emissions and carbon emissions per unit of GDP have slowed down to a certain extent. To this end, China proposed, in

September 2020, the "double carbon" target of "peak carbon" by 2030 and "carbon neutral" by 2060, which is both a solemn commitment to the world and a timetable and roadmap for China's environmental governance and green development. The report of the 20th Party Congress emphasizes that "Chinese modernization is a modernization in which people and nature live in harmony" and that "we must firmly establish and practice the concept that green water and green mountains are golden mountains and plan development in the context of the harmonious coexistence of people and nature". Putting in hard work to solve environmental problems is both a highly sensitive political proposition and an important area of institutional innovation for Chinese-style governance modernization [2].

In fact, the Central Environmental Protection Inspectorate is an important institutional design made by the Central Government to promote ecological protection and green development. In July 2015, the Central Leading Group for Comprehensively Deepening Reform considered and adopted the Environmental Protection Inspectors Program (for Trial Implementation), an innovative proposal to build an environmental protection inspection mechanism. Compared to previous environmental governance initiatives, such as the "Ten Articles of the Atmosphere" [3], "Trading of Pollution Rights" [4] and "Environmental Protection Talks" [5], the Central Inspections of Environmental Protection have authoritative and mandatory features, with a more stringent accountability mechanism, and an innovative governance practice in China's green system from "supervising enterprises" to the party and government being equally responsible. In December 2015, the Central Inspections of Environmental Protection launched a pilot scheme in Hebei Province, and, in August 2017, completed the first round of environmental protection inspections, covering 31 provinces and municipalities. In 2019, the central government studied and introduced China's first party regulation in the field of ecology and environment, the Central Ecological and Environmental Protection Inspectors' Regulations, which stipulate that within each term of the party's Central Committee, the party committees and governments of all provinces and municipalities, relevant departments of the State Council and relevant central enterprises shall be routinely inspected and, where appropriate, rectified via "look back" and special inspections. This provides institutional safeguards for the normalization of the Central Inspections of Environmental Protection.

In terms of the practice of central environmental inspections, the focus includes three stages, including the stationing of inspections, the placing of inspectors in local municipalities and the sorting and analysis of the results. The central government selects the head of the inspection team (the deputy head of the team is usually the current deputy head of the state environmental protection department) and leads the team into the provinces and municipalities to carry out environmental inspection work for one month. During their stay, the inspection teams combine reports from the public, information review and on-site spot checks to inspect outstanding environmental problems in the development process of localities. Phenomena such as inaction on the part of party committees and governments in environmental protection are also the focus of the Central Inspections of Environmental Protection. After the completion of the inspection, the inspection team is given a deadline to complete the inspection report and provide feedback to the provincial party committees and governments, which will submit the rectification and reform plan to the State Council for review within 30 working days and make the environmental rectification plan and its implementation available to the public.

In the context of the accelerated construction of a modern Chinese green institutional system, examining the impact of the Central Inspections of Environmental Protection on the green economic efficiency at the city level is of great theoretical and practical value in grasping the practical effects of the green premium (meaning the absolute difference between how much people should pay for fossil fuel-based energy and renewable energy) [6] and green economic development under the constraints of the environmental protection inspection system, as well as in formulating and improving Chinese environmental public policies in the next phase. The purpose of this paper is to answer the following questions: firstly, what is the level of efficiency of China's green economy, represented by the

Yangtze River Delta region, under the central environmental protection inspection system? Secondly, does the central environmental inspection system have an impact on the green economic efficiency of the Yangtze River Delta cities? Finally, if the central environmental inspection system has an impact on the green economic efficiency, does this impact vary over time and across regions, i.e., is the impact heterogeneous over time and space?

The innovations include the following three parts. Firstly, the focus on the Central Inspections of Environmental Protection, an important institutional arrangement for promoting ecological protection in China and its economic impact, has inspired ideas for further understanding the logic of ecological governance in China. Second, it focuses on the Yangtze River Delta, one of the most dynamic, open and innovative regions in China, and selects a more appropriate research case for a better analysis of the policy effects of the Central Inspections of Environmental Protection. Third, a diverse and heterogeneous discussion of the role of the Central Inspections of Environmental Protection on the urban green economic efficiency, both at the temporal and spatial levels, provides an empirical basis for better improving central environmental protection inspection policies.

The research structure of this paper includes the following three main aspects. It begins with a literature review of existing studies, then constructs core indicators and sets up a multi-period spatial DID model, and, finally, conducts a focused analysis on the temporal variability, locational variability, geographical variability and spillover variability of the impact of the Central Inspections of Environmental Protection on the green economic efficiency, respectively, in order to examine the possible impact of the Central Inspections of Environmental Protection on the urban green economic efficiency.

2. Literature Review

From the literature, environmental governance policies are divided into two main categories: administrative and market-based. From theoretical analysis, market-based environmental governance policies, such as carbon emission rights, emissions trading and pollution taxes, can help reduce pollution emissions and promote technological innovation [7] and are ideal for combating environmental pollution, but empirical evidence shows that the improvement of environmental problems in the US and Canada is mainly attributed to strictly enforced environmental regulations and accompanying provisions [8]. The limited role of market transactions in reducing pollutant emissions in China [9] means there is a need for greater environmental enforcement to achieve effective governance. Although studies have confirmed that administrative environmental policies have been a key factor in the sustained improvement in environmental quality since China's reform and opening up [10], the sustainability of administrative policies has been questioned, and the reduction in pollutant emissions may only be short-term [11], with the possibility of a significant rebound in pollution after the relaxation of environmental regulations [12]. In addition, under the "pollution sanctuary effect" hypothesis, the analysis of policy games between local governments may also lead to difficulties in the implementation of national-level environmental regulation policies [13]. Based on the above considerations, the selection of appropriate and effective environmental policy instruments is a challenging task. In fact, the combination of the high cost of implementing environmental regulations and the significant variation in pollution emissions across sectors and regions has led to the uneven implementation of environmental policies. For example, Gibson evaluated the effectiveness of the Clean Air Act in the US and found that companies changed the form of their pollution emissions under the system, impacting on land and water resources [14]. The Clean Water Act is not as effective as the Clean Air Act [15].

The main reason for the emergence of the central environmental inspection system is the information asymmetry between the central government and local authorities, and the resulting incentive incompatibility of the principal–agent relationship between the two levels. On the one hand, the central government has "delegated responsibility but not power" to local governments [16] and local governments are not only the implementers of central government policies, but also the providers of local economic and social development and public services [17]. This poses a serious challenge to local governments in terms of the quality and quantity of policy objectives, and the further down the hierarchy you go, the more difficult it becomes. On the other hand, economic development goals are still an important criterion when examining officials [18], which has led to a tendency for local governments to "conspire" with local enterprises after environmental protection mandates have been issued [19], using the advantage of local information, and different levels of local governments may form "offensive and defensive alliances" between themselves [20]. These coping strategies may weaken the implementation of central environmental policies at the local level, and to a certain extent affect the spatial layout of local productivity and the efficiency of the green economy. From an empirical study, W Wang et al. [21] (2021) used data from 290 prefecture-level cities in China to evaluate the air improvement effects of the central environmental inspection mechanism, and found that under conditions of information symmetry, the central government can design incentive contracts to enable local governments to reach the Pareto optimal level of effort. Within a very short period of time, both the first round of environmental supervision and the return visit of environmental supervision significantly improved the air quality and significantly reduced major single pollutants such as $PM_{2.5}$ and PM_{10}. Compared with the first round of environmental protection inspections and environmental protection inspection returns, the latter had higher levels of air pollution reduction and better results. Ruxin Wu et al. [22] (2019) used the effect of central environmental protection inspections on air pollution management as an example and used regression discontinuity to find that central environmental protection inspections have a positive effect on the air quality index (AQI), but this effect is only short-term and unsustainable. In addition, there are inter-provincial differences. Because of the performance and promotion of local officials, and for accountability reasons, specific environmental assessments by the central government through local governments are more effective than central environmental inspections. Ruoqi Li et al. [23] (2020) found that the gradual improvement in the scientific level of the accountability system of the central environmental inspection team, with more precise accountability targets, improved the environmental quality and had a. Zhigao Luo et al. [24] (2019) extracted environmental intention words from Chinese central and provincial government work reports through data mining from the perspective of environmental federalism, and used an instrumental variables approach to conduct an empirical experiment on the battle between the centralization and decentralization of environmental governance. The results show that there is a negative correlation between the central government's environmental governance intentions and the provincial environmental quality, while there is a positive correlation between the provincial government's environmental governance intentions and the provincial environmental quality. Environmental centralization, together with its political, economic and cultural factors, has transformed provincial governments into proponents of environmental pollution, while the central government's ongoing campaign to check environmental protection has forced provincial governments to play a role to some extent.

From the above analysis of the literature, it can be seen that the Central Inspections of Environmental Protection are a large-scale governance activity in China to deal with environmental governance, and an integrated innovation of environmental governance tools to promote the modernization of the harmonious coexistence between human beings and nature. Under the "party-run" national vertical management structure [25], the central government is committed to holding local governments accountable for the environmental quality of their regions through the environmental inspection system, adopting a multi-principal governance approach that is "party-led, business-led and socially engaged". However, studies have mainly focused on the impact of the Central Inspections of Environmental Protection on environmental quality improvement but lacked the impact on the quality development of the green economy, especially the efficiency of the green economy. In terms of the practical effects of the inspections, did the first round of the Central Inspections of Environmental Protection and the 'look-back' really promote high-quality regional development and enhance the green economic efficiency? Taking the Yangtze

River Delta region of China as an example, this paper examines the heterogeneous effects of the first round of the Central Inspections of Environmental Protection and the 'look-back' as a quasi-natural experiment, using a multi-period spatial DID model to test the effects of the Central Inspections of Environmental Protection on the urban green economic efficiency.

3. Models, Methods and Data

3.1. Green Economy Efficiency Measurements

The exclusion of resource factors does not fully reflect the characteristics of economic development. Therefore, some scholars have incorporated resource and environmental factors into productivity measurement models and used different methods to measure green economic efficiency [26–32]. For example, in the work of Chung et al. (1997) [33] measuring total factor productivity (TFP) in Swedish pulp mills, the first SBM (Slacks-Based Measure) model with pollution emissions as a non-desired output and a directional distance function was developed. Tone (2003) [34] innovated a hybrid distance function EBM (Epsilon-Based Measure) model that incorporates slack variables into the function and considers both a CCR (A. Charnes & W.W. Cooper & E. Rhodes) model with radial factors and a non-radial SBM model with slack variables. The non-radial SBM model weakens the measurement error that may arise from a single distance function. Following OH (2010) [35], the results of the GML (Global Malmquist–Luenberger) index of the SBM model were used to characterize the green economic efficiency of the city. In this regard, the mathematical expression of the SBM model is

$$\rho^* = \min \frac{\frac{1}{m}\sum\limits_{i=1}^{m}\frac{\overline{x}_i}{x_{i0}}}{\frac{1}{S_1+S_2}\left(\sum\limits_{r=1}^{S_1}\frac{\overline{y}_r^g}{y_{r0}^g}+\sum\limits_{r=1}^{S_2}\frac{\overline{y}_r^b}{y_{r0}^b}\right)}, s.t. \left\{ \begin{array}{l} \overline{x} \geq \sum\limits_{j=1,\neq k}^{n}\theta_j x_j \\ \overline{y}^g \leq \sum\limits_{j=1,\neq k}^{n}\theta_j y_j^g \\ \overline{y}^b \geq \sum\limits_{j=1,\neq k}^{n}\theta_j y_j^b \\ \overline{x} \geq x_0, \overline{y}^g \leq y_0^g, \overline{y}^b \geq y_0^b, \overline{y}^g \geq 0, \theta \geq 0 \end{array} \right. \tag{1}$$

$$x \in R^m, y^g \in R^{S_1}, yb \in R^{S_2}$$

$$X = [x_1, x_2, \ldots, x_n] \in R^{m \times n}, Y^g = [y_1^g, y_2^g, \ldots, y_n^g] \in R^{S_1 \times n}, Y^b = [y_1^b, y_2^b, \ldots, y_n^b] \in R^{S_2 \times n}$$

The SBM model is premised on the assumption of constant size, represents input, desired and undesired output slack, and the p objective function value characterizes the decision unit efficiency value.

The mathematical expression for the GML index is

$$GML^{t,t+1} = \frac{1 + \overrightarrow{D}_o^G(x^t, y^t, b^t; y^t, b^t)}{1 + \overrightarrow{D}_o^G(x^{t+1}, y^{t+1}, b^{t+1}; y^{t+1}b^{t+1})}$$

$$= \frac{1 + \overrightarrow{D}_o^G(x^t, y^t, b^t; y^t, b^t)}{1 + \overrightarrow{D}_o^G(x^{t+1}, y^{t+1}, b^{t+1}; y^{t+1}b^{t+1})} \times \frac{1 + \overrightarrow{D}_o^G(x^t, y^t, b^t; y^t, b^t)/1 + \overrightarrow{D}_o^G(x^t, y^t, b^t; y^t, b^t)}{1 + \overrightarrow{D}_o^G(x^{t+1}, y^{t+1}, b^{t+1}; y^{t+1}b^{t+1})/1 + \overrightarrow{D}_o^G(x^{t+1}, y^{t+1}, b^{t+1}; y^{t+1}b^{t+1})} \tag{2}$$

$$= \frac{TE^t(x^t, y^t, b^t; y^t, b^t)}{TE^{t+1}(x^{t+1}, y^{t+1}, b^{t+1}; y^{t+1}b^{t+1})} \times \frac{BPG^{G,t}(x^t, y^t, b^t; y^t, b^t)}{BPG^{G,t+1}(x^{t+1}, y^{t+1}, b^{t+1}; y^{t+1}b^{t+1})}$$

$$= GMLEC^{t,t+1} \times GMLTC^{t,t+1}$$

The GML index can be decomposed into technical efficiency (EC) and technical progress (TC). $\overrightarrow{D}_o^G(x^t, y^t, b^t; y^t, b^t)$, $\overrightarrow{D}_o^G(x^{t+1}, y^{t+1}, b^{t+1}; y^{t+1}, b^{t+1})$ denote the efficiency values of the decision unit in period t and period $t + 1$ in period t, respectively.

In the measurement of the urban green economy efficiency, the selection of input indicators includes: (1) Capital input. The physical capital stock of the city is selected for measurement. Due to the lack of corresponding statistical indicators, this paper uses the method of Shan, Haojie (2008) [36], using the data of the fixed asset investment flow of each city and deflating it, with 2010 as the base period, in which the depreciation rate is set at 10.96%. (2) Labor input. The total number of employees in the secondary and tertiary industries was selected as the indicator of the labor force in a particular city. (3) Resource and energy inputs. Total water supply and social electricity consumption were chosen as the indicators for measuring the resource and energy inputs in the economic development of the city, respectively. (4) Desired output. Considering the level of economic development and the quality of life of urban residents as the main indicators of the desired output, the real gross regional product and the greening coverage of built-up areas were selected as the proxy variables for the above two indicators, respectively. (5) Non-desired output. Focusing on the selection of industrial smoke emissions, industrial wastewater emissions, industrial SO_2 emissions and $PM_{2.5}$ concentrations as indicators to measure the pollution situation in the process of urban economic development. From the kernel density plot (Figure 1), although the measured green economic efficiency of the Yangtze River Delta cities shows a right-skewed distribution overall, the difference with the normal distribution (Normal distribution) is not obvious, and the measurement results generally meet the statistical requirements.

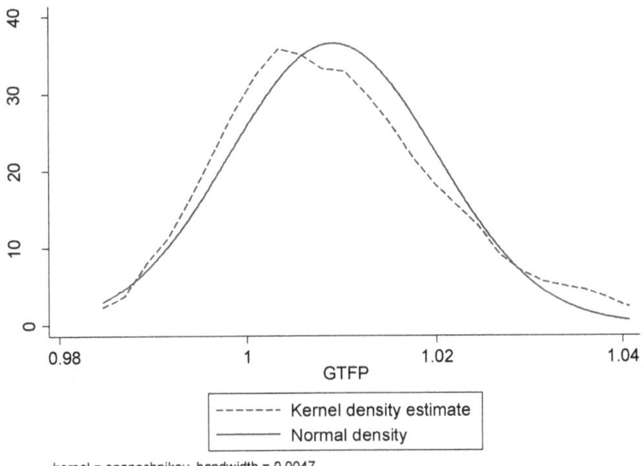

Figure 1. Kernel density diagram of urban green economic efficiency measurement results.

3.2. Central Inspections of Environmental Protection

The first round of the Central Inspections of Environmental Protection consisted of four batches, of which the first batch started in July–August 2016 and involved eight provinces and municipalities, including Inner Mongolia, Heilongjiang, Jiangsu, Jiangxi, Henan, Guangxi, Yunnan and Ningxia. April–May 2017 involved seven provinces and municipalities, including Tianjin, Shanxi, Liaoning, Anhui, Fujian, Hunan and Guizhou, and the fourth batch started in August–September 2017, involving eight provinces and municipalities, including Jilin, Zhejiang, Shandong, Hainan, Sichuan, Tibet, Qinghai and Xinjiang. After the completion of the first round of inspections in 2018, 20 more provinces and urban areas were "looked back", including the fourth round of the Central Inspections of Environmental Protection, from 5 June to 5 July 2018, which carried out a "look back" at the rectification and reform of the first round of the Central Inspections of Environmental Protection in Jiangsu Province. This paper will, therefore, look back at the 2016 reports of the Yangtze River Delta region. Therefore, in this paper, the 13 municipalities in Jiangsu and Shanghai, which were inspected by the Central Inspections of Environmental Protection in 2016 in the Yangtze River Delta region, are used as the treatment group of the multi-period double difference model, while the other 27 municipalities are used as the control group of the study. An analysis of the characteristics of the cities in the Yangtze River Delta involved in the first round of Central Environmental Protection Inspectors is shown in Table 1.

Table 1. Analysis of the characteristics of the cities in the Yangtze River Delta involved in the first round of the Central Inspections of Environmental Protection (batches 1 and 2).

City	Experimental Group	Control Group	Are Coastal and Riverine	Geographical Area	City	Experimental Group	Control Group	Are Coastal and Riverine	Geographical Area
Shanghai	Yes	No	Yes	–	Quzhou	No	Yes	No	–
Nanjing	Yes	No	Yes	Sunan	Zhoushan	No	Yes	Yes	–
Wuxi	Yes	No	Yes	Sunan	Taizhou	No	Yes	Yes	–
Xuzhou	Yes	No	No	Northern Sudan	Lishui	No	Yes	No	–
Changzhou	Yes	No	Yes	Sunan	Hefei	No	Yes	No	–
Suzhou	Yes	No	Yes	Sunan	Huaibei	No	Yes	No	–
Nantong	Yes	No	Yes	Suzhou-China	Bozhou	No	Yes	No	–
Lianyungang	Yes	No	Yes	Northern Sudan	Cebu	No	Yes	No	–
Huai'an	Yes	No	No	Northern Sudan	Bengbu	No	Yes	No	–
Yancheng	Yes	No	Yes	Northern Sudan	Fuyang	No	Yes	No	–
Yangzhou	Yes	No	Yes	Suzhou-China	Huainan	No	Yes	No	–
Zhenjiang	Yes	No	Yes	Sunan	Chuzhou	No	Yes	No	–
Taizhou	Yes	No	No	Suzhou-China	Lu'an	No	Yes	No	–
Suqian	Yes	No	No	Northern Sudan	Ma On Shan	No	Yes	Yes	–
Hangzhou	No	Yes	Yes	–	Wuhu	No	Yes	Yes	–
Ningbo	No	Yes	Yes	–	Xuancheng	No	Yes	No	–
Wenzhou	No	Yes	Yes	–	Tongling	No	Yes	Yes	–
Huzhou	No	Yes	No	–	Chizhou	No	Yes	Yes	–
Jiaxing	No	Yes	Yes	–	Anqing	No	Yes	Yes	–
Introduction	No	Yes	Yes	–	Huangshan	No	Yes	No	–
Jinhua	No	Yes	No	–	–	–	–	–	–

3.3. Control Variables

Following Taskin and Zaim (2001) [37] and other studies, control variables such as firm size (X1), per capita savings (X2), government influence on the economy (X3), strength of foreign ties (X4), per capita investment in education (X5) and unemployment insurance coverage (X6) were selected to be included in the model for econometric estimation. The higher the savings, the lower the demand for low-quality goods and services and the lower the tolerance for counterfeit products, forcing producers and operators to consciously keep their promises and improve the quality of goods and services. In the case of increasing efforts to build an honest government, the higher the degree of government influence on the economy, the stronger the demonstration drive on the market society and the greater the role in enhancing the city's business credit environment. The higher the intensity of external linkages, the greater the focus on benchmarking international rules to improve the city's business environment. In addition, investment in education is closely related to residents' awareness of integrity, and the improvement and expansion of unemployment insurance is also conducive to stabilizing the market development expectations of micro- and small enterprises and individual entrepreneurs and other micro-operators, and optimizing the city's business credit environment.

The scale of enterprises (X1) is measured by the ratio of industrial assets above the scale to the number of industrial enterprises above the scale; the government's influence on the economy (X3) and the strength of external ties (X4) are measured by the general budget expenditure of the local finance and the proportion of the total import and export of goods to the regional GDP, respectively; the per capita savings of residents (X2), the per capita investment in education (X5) and the unemployment insurance coverage rate (X6) are measured by the ratio of the year-end balance of urban and rural residents' savings, education expenditure and unemployment insurance coverage to the total population. In summary, the inclusion of institutions, factors and the environment as control variables is used to mitigate the problem of missing explanatory variables.

3.4. Description of Data

The research object of this paper is 41 cities above prefecture level in the Yangtze River Delta region, and the research interval is set from 2011 to 2019, considering the research needs and data availability. The PM2.5 concentration data were obtained from the satellite remote sensing data published by NASA, and the 1:4 million Chinese basic geographic information data provided by the National Center for Basic Geographic Information were cropped to obtain the average PM2.5 concentration values of the cities in the past years. For missing data, the interpolation method was used to process the data.

3.5. Measurement Models

In examining the impact of the Central Inspections of Environmental Protection on the efficiency of the green economy in cities, the traditional multi-period DID model has certain advantages for the analysis of the effect of policy implementation, with cities belonging to the treatment group when they are affected by policy implementation and to the control group when they are not affected by policy implementation; at the same time, a time dummy variable for policy implementation is introduced, with the year before policy implementation taking the value of 0 and the year after policy implementation taking the value of 1. The basic model formulation is as follows.

$$GTFP_{it} = \beta_0 + \sum_{k=1}^{K} X_{it,k}\beta_k + DID_{it}\beta_{k+1} + \varepsilon_{it} \tag{3}$$

where *GTFP* represents the level of the green economy efficiency and β represents the regression coefficient of each variable. t represents the time of implementation of the CIEP and takes values between [1, T], T = 10. $X_{it,\,k}$ represents the k control variables in the model and takes values between [1, K], K = 8. DID_{it} represents the dummy variable interaction term, which is the policy effect parameter to be estimated, and is obtained by multiplying the values of the dummy variables for the group attributes and the time of the CIEP and by centralization. The dummy variables taken for the group attributes and the time of the CIEP are multiplied together and centralized. ε_{it} represents the random error term.

However, the magnitude of the effect of the Central Inspections of Environmental Protection on the green economic efficiency of cities, in the context of the construction of a large national unified market, gradually weakens the influence of provincial administrative fragmentation on the development of the market economy, and the green development of cities is increasingly influenced by neighboring cities, especially in the context of the integrated development of the Yangtze River Delta. Rising as a national strategy, the spatial interaction of the Central Inspections of Environmental Protection in the implementation process will also be more obvious. Spatial correlations of the core variables confirm these conjectures, with global univariate Moran indices of 0.6672 and −0.1511 for the interaction term and green economic efficiency, respectively, and a global bivariate Moran index of 0.0457 for the interaction term and green economic efficiency in 2019 (Table 2), confirming that there is a significant spatial correlation between the urban green economic efficiency and the effect of the central environmental inspections on the implementation of the green economic efficiency. Both have a more significant spatial correlation. In this case, the above traditional multi-period DID model may become inapplicable to the analysis of policy effects, and spatial factors must be incorporated into the model in order to more scientifically measure and analyze the possible impact of the Central Environmental Protection Inspectorate on the efficiency of the green economy.

Table 2. Spatial correlation tests for policy implementation effects: the Moran index.

Measurement Indicators	2015	2016	2019
Green economic efficiency	−0.0627	−0.1061	−0.1511
Interaction term (DID) × Green economic efficiency	−0.2004	−0.0346	0.0457

In view of this, a more scientific multi-period spatial DID model is constructed on the basis of the traditional multi-period DID model, taking into account the requirement of spatial multi-collinearity avoidance, to conduct a targeted analysis of the impact of the Central Inspections of Environmental Protection on the efficiency of the urban green economy. In particular, the spatial lagged model (SLM) is formulated as

$$GTFP_{it} = \sum_{it=1}^{NT} \rho_{SLM} (\xi \otimes \pi)_{it,it} GTFP_{it} + \beta_0 + \sum_{k=1}^{K} X_{it,k} \beta_k + DID_{it} \beta_{k+1} + \varepsilon_{it} \quad (4)$$

The spatial lag model (SEM) is formulated as

$$GTFP_{it} = \beta_0 + \sum_{k=1}^{K} X_{it,k} \beta_k + DID_{it} \beta_{k+1} + \mu_{it} \quad (5)$$

$$\mu_{it} = \sum_{it=1}^{NT} \rho_{SEM} (\xi \otimes \pi)_{it,it} \mu_{it} + \varepsilon_{it} \quad (6)$$

where ξ and π denote the temporal and spatial weight matrices after row normalization, respectively. $\xi \otimes \pi$ represents the endogenous spatio-temporal weight matrix, which is measured according to the city–location relationship, and the matrix element is 1 if the two cities are adjacent in a spatial location and 0 otherwise. In addition, $i = 1, 2, \ldots , N$, $N = 41$, denotes the 41 cities above prefecture level in the Yangtze River Delta region; ρ represents the spatial correlation coefficient in the spatial econometric model, and represents the normally distributed random error vector.

4. Estimation of Measurement Results

4.1. Central Inspections of Environmental Protection and Green Economy Efficiency: Time Impact Variability

Based on the spatial correlation test of the core variables, the spatial DID method was used to test the possible impact of the Central Inspections of Environmental Protection on the green economic efficiency of the city, and the estimation results are reported in Table 3, where models (1)–(3) and (6)–(9) represent the regression results before and after the Central Inspections of Environmental Protection, respectively, and models (4)–(6) represent the results of the benchmark analysis, reflecting the impact of the Central Inspections of Environmental Protection on the city's green economic efficiency in general, and OLS, SLM and SEM denote the ordinary least squares model, spatial lag model and spatial error model, respectively. Models (3), (5) and (9) were selected for further analysis by combining the magnitude of the Log-likelihood, AIC and SC values, respectively.

Table 3. Central environmental inspections and green economy efficiency: variability in time impact.

	Before_2015			Benchmarking_2016			After_2019		
	(1)	(2)	(3)	(4)	(5)	(6)	(7)	(8)	(9)
Model type	OLS	SLM	SEM	OLS	SLM	SEM	OLS	SLM	SEM
Interaction items (DID)	−0.014 (−1.085)	−0.011 (−0.911)	−0.013 * (−1.39)	0.017 (1.138)	0.025 ** (1.878)	0.012 (1.092)	−0.006 (−0.441)	−0.004 (−0.28)	−0.004 * (−0.468)
Control variables	Control	Control	Control	Control	Control	Control	Control	Control	Control
R-squared	0.126	0.132	0.253	0.193	0.259	0.264	0.175	0.178	0.378
Log-likelihood	90.422	90.576	92.185	89.106	90.822	90.2	90.883	90.948	94.463
AIC	−160.845	−159.152	−164.371	−158.212	−159.645	−160.4	−161.766	−159.897	−168.925
SC	−143.709	−140.302	−147.235	−141.076	−140.796	−143.264	−144.631	−141.048	−153.789

Note: **, * denote 5%, 10% significance levels, respectively; t/z values in parentheses. Regression results for control variables are omitted for space constraints and are available from the authors upon request.

The results show that the treatment effect coefficients of the central environmental protection inspection on the green economic efficiency of the city are -0.013, 0.025 and -0.004, respectively, which indicates that the central environmental protection inspection shows a more significant positive relationship with the green economic efficiency of the city. This is consistent with the findings of Li Feng et al. (2022) [38] that the 'Central Inspections of Environmental Protection improve environmental quality and reduce wastewater discharge'. However, the green institutional arrangement has a negative impact on the green economic efficiency of the city both before and after the central environmental protection inspection, and the time of the central environmental protection inspection. The closer the time to the Central Inspections of Environmental Protection, the greater the negative impact on the green economic efficiency. The Central Inspections of Environmental Protection focus on monitoring environmental protection under the local government–enterprise interaction, and, by collecting clues from society, they identify and deal with outstanding environmental problems in a timely manner, which creates some institutional incentives for environmentally friendly enterprises and has a supportive effect on the urban green economic efficiency in general. As the Central Inspections of Environmental Protection cover matters such as the party and government's responsibility for ecological protection and the promotion of the implementation of double responsibility for ecological protection, in response to the environmental protection inspections, local governments set out to rectify environmental pollution problems before the inspections, and in order to quickly achieve the desired results, a "one-size-fits-all" approach is inevitable at the implementation level, which has an impact on the normal market economic order and even reduces the green economic efficiency. This can have an impact on the normal market economy and even reduce the efficiency of the green economy. In addition, the Central Inspections of Environmental Protection have urged local governments to continue to address environmental problems, which, on the one hand, reflects the importance of 'ecological priority', but, on the other hand, can lead to market uncertainty about 'green development'. On the one hand, this reflects the importance of 'ecological priority', but, on the other hand, it can lead to a market that is hesitant about 'green development', potentially reducing the efficiency of cities' green economies.

4.2. Central Inspections of Environmental Protection and Green Economy Efficiency: Locational Impact Variability

Considering the spatial distribution and historical tradition of industrial development in China since the reform and opening up of the country, cities in different geographical locations will, to some extent, influence the effect of the Central Inspections of Environmental Protection on the efficiency of the urban green economy. For example, in the late Qing Dynasty and early Republican period, Chongqing, Wuhan, Nanjing, Shanghai and Anqing along the Yangtze River were known as the 'Five Tigers of the Yangtze' and played an important role in the modern history of China. In fact, coastal riverside regions have a more export-oriented economy than non-coastal riverside regions, and, with capital accumulation and international benchmarking of the market development environment, the economic structure has been gradually transformed and optimized and may appear more positive and optimistic in their response to the central environmental protection inspection system. Therefore, the interaction term between whether coastal and riverine areas and the Central Inspections of Environmental Protection was introduced in the benchmark model to discuss the mechanism of the role of the Central Inspections of Environmental Protection on the urban green economic efficiency, and the estimated results are collated in Table 4.

Combining the Log-likelihood, AIC and SC value magnitudes, models (2) and (6) were selected for the next step of the analysis. The estimation results show that the interaction term (DID) regression coefficient, i.e., the treatment effect coefficient, is 0.027 for coastal riverine areas, which passes the 5% significance level test, while the treatment effect is insignificant for non-coastal riverine areas, indicating significant locational variability in the impact of the Central Inspections of Environmental Protection on the urban green economic efficiency. The economic explanation is as follows: for coastal cities such as Shanghai and Nantong and riverine cities such as Nanjing, Suzhou and Wuxi, the Central Inspections of Environmental Protection have an important role in raising the level of the green economic efficiency in cities; this is in line with Hong Tao et al.'s [39] study that "centralized environmental governance and strict legal regulations may provide strong constraints on local government deregulation of the environment". On the one hand, the environmental protection inspectors' rectification of polluting enterprises that have been strongly reflected by the public is conducive to eliminating "black" industries and reshaping the green industrial system to improve the efficiency of the city's green economy; on the other hand, the accountability, notification and treatment of possible corruption are conducive to optimizing the political ecology and building a new type of pro-clear government–business relationship. This will have a positive impact on the city's ability to move up the value chain, develop technology-intensive industries and enhance development efficiency.

Table 4. Central environmental inspections and green economy efficiency: variability in locational impacts.

	Coastal and Riverine Areas			Non-Coastal Riverine Areas		
	(1)	**(2)**	**(3)**	**(4)**	**(5)**	**(6)**
Model type	OLS	SLM	SEM	OLS	SLM	SEM
Interaction Item (DID) × Zone	0.021 (1.4)	0.027 ** (2.086)	0.013 (1.118)	−0.006 (−0.291)	−0.004 (−0.227)	0.006 (0.342)
Control variables	Control	Control	Control	Control	Control	Control
R-squared	0.21	0.273	0.252	0.162	0.194	0.264
Log-likelihood	89.525	91.213	90.115	88.323	89.128	89.714
AIC	−159.05	−160.426	−160.23	−156.647	−156.256	−159.428
SC	−141.914	−141.577	−143.094	−139.511	−137.406	−142.293

Note: ** denote 5% significance levels; t/z values in parentheses. Regression results for control variables are omitted for space constraints and are available from the authors upon request.

4.3. Central Inspections of Environmental Protection and Green Economy Efficiency: Geographical Impact Variability

In order to more comprehensively reflect the impact of the Central Inspections of Environmental Protection on the efficiency of urban green economies, it is important to take full account not only of the variability in the location of particular cities, but also to discuss the heterogeneity of particular regions with different development histories. In Jiangsu province, for example, the southern region of Jiangsu, with Nanjing, Suzhou, Wuxi, Changzhou and Zhenjiang as representative cities, the central region of Jiangsu, with Yangzhou, Taizhou and Nantong as representative cities, and the northern region of Jiangsu, with Xuzhou, Lianyungang, Suqian, Huaian and Yancheng as representative cities, show a decreasing spatial distribution pattern in terms of the degree of economic development, the proportion of green industries in the economic structure and the sensitivity to environmental regulation. There are also differences in the share of green industries in the economic structure and their sensitivity to environmental regulation. Therefore, the interaction term × geographical characteristics was introduced into the baseline model, and the results are reported in Table 5. Models (2), (6), (9) and (12) were selected for the next stage of analysis by combining the magnitudes of the Log-likelihood, AIC and SC values, respectively.

Table 5. Conduction path verification results.

Type	Southern Sudan			Su-Central Region			Northern Jiangsu Province			Shanghai Area		
	(1)	(2)	(3)	(4)	(5)	(6)	(7)	(8)	(9)	(10)	(11)	(12)
	OLS	SLM	SEM	OLS	SLM	SEM	OLS	SLM	SEM	OLS	SLM	SEM
Interaction item (DID) × Geographical area	0.02 (0.856)	0.029 ** (1.401)	0.017 (0.899)	0.008 (0.38)	0.012 (0.656)	−0.002 (−0.108)	0.008 (0.396)	0.01 (0.583)	0.016 (1.124)	0.002 (0.345)	−0.002 (−0.038)	−0.058 ** (−1.503)
Control variables	Control	Control	Control	Control	Control	Control	Control	Control	Control	Control	Control	Control
R^2	0.179	0.231	0.263	0.163	0.202	0.26	0.164	0.2	0.291	0.16	0.194	0.323
Log-Likelihood	88.746	90.065	90.04	88.363	89.317	89.667	88.371	89.271	90.242	88.268	89.103	90.468
AIC	−157.492	−158.13	−160.081	−156.725	−156.635	−159.332	−156.741	−156.542	−160.485	−156.536	−156.206	−160.937
SC	−140.356	−139.28	−142.945	−139.59	−137.785	−142.196	−139.606	−137.693	−143.349	−139.401	−137.357	−143.801

Note: ** denote 5% significance levels, respectively; t/z values in parentheses. Regression results for control variables are omitted for space constraints and are available from the authors upon request.

The results show that the regression coefficients of the interaction term × geographical characteristics, i.e., the treatment effect coefficients, pass the significance test (coefficients of 0.029 and −0.058, respectively) for the southern and Shanghai regions; while the central and northern regions of Suzhou do not show significant results, the findings are consistent with those of Y Tan et al. [40] regarding the impact of the Central Inspections of Environmental Protection on the air quality in China. In fact, compared to the central and northern regions, the southern region of Suzhou is at a higher level of economic and social development due to its good location and historical tradition of opening ports and trading in the late Qing and early Ming dynasties, and its economic structure is dominated by collective enterprises and foreign-funded enterprises, which tend to be stronger and have the ability and willingness to install pollution control equipment through research and development or the purchase of green innovative technologies under the central environmental protection inspection system arrangement, and to respond to the government's This micro-level green decision making drives the efficiency of the city's green economy. For Shanghai, the economic development leader in the Yangtze River Delta region, despite positive results in the development of the new economy, the impact of the Central Inspections of Environmental Protection on traditional industries, of which COSCO Shipping, China Baowu Steel and China Electric Equipment are the backbone, remains high, with increased environmental protection expenditure crowding out productive capital expenditure to some extent and inhibiting corporate R&D, constraining the city's green economic efficiency.

Table 5 shows the central environmental inspections and green economy efficiency: variability in geographical impact.

4.4. Central Inspections of Environmental Protection and Green Economy Efficiency: Spillover Impact Variability

Given the spatial spillover, it is necessary to take into account the impact of the Central Inspections of Environmental Protection on the green economic efficiency of non-inspected areas when discussing the effect of the Central Inspections of Environmental Protection on the green economic efficiency of cities. In fact, the integrated development of the Yangtze River Delta was proposed in 2010, and, in 2018, the central government supported the integrated development of the Yangtze River Delta region and elevated it to a national strategy, promoting a large number of major infrastructure developments and a series of institutional policy synergies through the release of a phased development plan outline, with an increasingly rational interaction between the three provinces and one city's industrial division of labor. Therefore, the interaction term × other regional characteristics was introduced into the benchmark model and the results are reported in Table 6. Combining the magnitude of the Log-likelihood, AIC and SC values, models (2), (5) and (9) were selected for the next stage of analysis, respectively.

Table 6. Central environmental inspections and green economy efficiency: spillover impact variability.

	Non-Inspector Areas			Zhejiang Province			Anhui Province		
	(1)	(2)	(3)	(4)	(5)	(6)	(7)	(8)	(9)
	OLS	SLM	SEM	OLS	SLM	SEM	OLS	SLM	SEM
Interaction item (DID) × Other area	−0.017 (−1.138)	−0.025 ** (−1.878)	−0.012 (−1.092)	−0.01 (−0.784)	−0.014 * (−1.246)	−0.005 (−0.524)	−0.012 (−0.422)	−0.016 (−0.657)	−0.018 * (−0.865)
Control variables	Control	Control	Control	Control	Control	Control	Control	Control	Control
R-squared	0.193	0.259	0.264	0.176	0.224	0.254	0.164	0.202	0.279
Log-likelihood	89.106	90.822	90.2	88.67	89.875	89.78	88.385	89.317	90.013
AIC	−158.212	−159.645	−160.4	−157.34	−157.75	−159.559	−156.77	−156.633	−160.026
SC	−141.076	−140.796	−143.264	−140.204	−138.901	−142.423	−139.634	−137.784	−142.891

Note: **, * denote 5%, 10% significance levels, respectively; t/z values in parentheses. Regression results for control variables are omitted for space constraints and are available from the authors upon request.

The results show that the regression coefficient of the Central Inspections of Environmental Protection interaction term × non-inspected region characteristics, i.e., the treatment effect coefficient, is -0.025, and the regression coefficients of the Central Inspections of Environmental Protection interaction term on Zhejiang and Anhui, i.e., the treatment effect coefficients, are -0.014 and -0.018, respectively, indicating that the Central Inspections of Environmental Protection had a negative impact on the green economic efficiency of the non-inspected regions, and the degree of impact on Anhui is higher than that on Zhejiang. In fact, the Central Inspections of Environmental Protection will have an impact on the green development status of the inspected areas for a certain period of time, and some enterprises choose to relocate their production across locations to other regions in order to avoid environmental regulations. H Wang et al. [41] even measured that the number of polluting enterprises shrank by 48% after the implementation of the Central Inspections of Environmental Protection. This is the reason why the green economy efficiency of cities in neighboring provinces such as Zhejiang and Anhui decreased after Shanghai and Jiangsu were inspected by the Central Environmental Protection Inspectorate, and it also confirms the existence of the "pollution refuge" phenomenon in the Yangtze River Delta region [42]. The transfer of industries from Shanghai and Jiangsu to Zhejiang follows a specific stepped path, and the costs of both explicit land use and implicit green regulation are clearly higher than those in Anhui, making the transfer of a small proportion of industries to Zhejiang and most to Anhui the next best option. As the transfer of industries across provinces (cities) tends to be light green or even black enterprises with low technology intensity, the greater the number of incoming industries, the more significant the negative impact on the efficiency of the local urban green economy.

Using the Yangtze River Delta cities as a case study, this paper analyses the heterogeneous impact of the Central Inspections of Environmental Protection on the efficiency of the green economy, which has certain applications in the formulation and improvement of environmental policies in China, such as paying attention to the expected impact on the economy before and after the release of policies. In the process of policy implementation, it is necessary to take full account of the industrial structure factors of cities in different zones and adopt differentiated initiatives, as well as to consider the dynamic distribution of economic activities within and outside the region and adopt cross-regional policy synergies. The limitation of this paper is mainly the sample size issue. The article's research targets 41 cities in the Yangtze River Delta region and is concentrated in eastern China, which may not allow for a comprehensive analysis of the impact of the Central Inspections of Environmental Protection on the efficiency of China's green economy. To overcome the limitations of the study, a more comprehensive and detailed assessment and analysis for cities in eastern, central and western China could be considered in the next step.

5. Conclusions and Insights

The existing literature on the effects of the Central Inspections of Environmental Protection focuses on the provincial level and lacks variation at the prefecture level. This article uses the first round of the Central Inspections of Environmental Protection and "look-back" in prefecture-level and above cities in the Yangtze River Delta region as a quasi-natural experiment to establish more reasonable green economic efficiency indicators based on OH (2010), taking prefecture-level and above cities in the Yangtze River Delta as research samples and combining the spatial heterogeneity characteristics in the process of policy implementation [43], and empirically investigates the impact of the Central Inspections of Environmental Protection on the urban green economic efficiency using a multi-period spatial DID approach, and further discusses the heterogeneity of the impact in the time before and after the inspections, in different zones and regions of the inspected areas, and in the non-inspected areas.

This study found that: (1) The Central Inspections of Environmental Protection collect opinions from society and weigh them against each other, both to deter black enterprises to a certain extent and to create institutional incentives for green enterprises, which in turn support the efficiency of urban green economies. In response to the Central Inspections of Environmental Protection's "look-back", local governments often adopt "one-size-fits-all" measures such as simple shutdowns and rectification in the process of environmental remediation in order to quickly achieve the desired results, which may affect the normal economic order and reduce the green economic efficiency. (2) In terms of regional differences, the Central Inspections of Environmental Protection have a significant positive impact on coastal and riverine areas, but not on non-coastal and riverine areas. Compared with non-coastal riverine areas, coastal riverine areas have a higher proportion of export-oriented industries, a more complete green industry system and stronger industrial agglomeration capacity [44], and are more optimistic and confident in responding to the Central Inspections of Environmental Protection, which have a positive effect on localities' efforts to build a new type of pro-clear government–business relationship, promote enterprises' commitment to green technology research and application and improve the urban green economic efficiency. The central environmental inspections have had a positive effect on the development of a new type of pro-business relationship, the promotion of enterprises' commitment to green technology research and development and the improvement of the city's green economic efficiency. (3) In terms of geographical variability, the Central Inspections of Environmental Protection had a positive and negative impact on the green economic efficiency in southern Jiangsu and Shanghai, respectively, but not in central and northern Jiangsu. Due to its good location and historical tradition of opening ports and trading in the late Qing and early Ming dynasties, the southern Jiangsu region is dominated by collective and foreign-invested economies, and under the constraint of the Central Inspections of Environmental Protection, it has developed and purchased green innovative technologies to install pollution control equipment through technological innovation [45], which has promoted the green transformation of enterprises and improved the efficiency of the city's green economy at the same time. (4) In terms of spatial spillover differences, the Central Inspections of Environmental Protection had a negative impact on the green economic efficiency of non-inspected regions, with a higher degree of impact in Anhui than in Zhejiang. In order to reduce the cost of pollution control under strict environmental regulations, some enterprises choose to relocate their production across locations to non-inspected areas. This cross-regional allocation of productivity [46] has a negative impact on the efficiency of the green economy in the cities they move to. Compared to Zhejiang, Anhui, a member of the Yangtze River Delta, has a cost advantage in absorbing low green technology value-added enterprises, but the green economic efficiency is also more negatively affected by the spillover effects of the Central Inspections of Environmental Protection.

This study reveals that, firstly, the Central Inspections of Environmental Protection can not only promote the government, market and society to reach a consensus on "ecological priority and green development", but also become an important institutional arrangement affecting the high-quality development of the urban economy by enhancing the efficiency of the green economy, which provides a new way of understanding "promoting the harmonious coexistence of human beings and nature". This provides a better understanding of the factors influencing the efficiency of the green economy [47] and to address the current challenges of multiple governance under environmental regulation in China. Secondly, improving the design of environmental regulations accompanying the Central Environmental Protection Inspectorate, properly handling the relationship between the inspections and development for different time cut-off points, reasonably regulating the use of administrative resources, and reducing inappropriate administrative intervention in the market allocation of resource factors is not only a basic requirement for integrated regional development, but also an important prerequisite for building a large national unified market and improving the efficiency of China's green economy. Once again, considering the existence of regional development path dependencies and the obvious differences in industrial institutions between regions at different stages of development [48], and the different degrees of influence of the Central Inspections of Environmental Protection on the green economic efficiency [49], it is important to study the implementation of differentiated environmental protection inspection policies based on regional heterogeneity characteristics to better achieve regional economic development goals. Finally, central environmental inspections should establish mechanisms for collecting information and clues on environmental issues across regions, urging rectification mechanisms and initiating accountability mechanisms, and strengthening inter-provincial coordination to achieve the parallel implementation of regional green policies as far as possible, which will not only help to reduce the distortion of the effectiveness of central environmental inspections by administrative fragmentation, but will also be beneficial in improving the overall green economic efficiency of the region and, thus, achieving a profound transformation of the Chinese economy.

Author Contributions: Writing—original draft, H.C.; Writing—review & editing, M.S. All authors have read and agreed to the published version of the manuscript.

Funding: This research was funded by the Major Special Project for Philosophical and Social Science Research of the Ministry of Education of the People's Republic of China (No. 2022JZDZ009); the Major Commissioned Project of Zhejiang Provincial Social Science Foundation of China "Let green be the most moving color of Zhejiang's development".

Institutional Review Board Statement: Not applicable.

Informed Consent Statement: Not applicable.

Data Availability Statement: All data generated or analyzed during this study are included in this published article.

Conflicts of Interest: The authors declare no conflict of interest.

References

1. Adam, Z. *Green Development-Environment and Sustainability in a Developing World*; Routledge: London, UK, 2009.
2. Qi, J.; Yu, O. On the "Central Environmental Protection Inspector System" as a Movement-based Governance Mechanism: A Discussion with Professor Chen Haisong. *Theor. Discuss.* **2018**, *2*, 157–164.
3. Luo, Z.; Li, H. The impact of the implementation of the "Ten Atmospheric Articles" policy on air quality. *China Ind. Econ.* **2018**, *9*, 136–154.
4. Fu, J.; Si, X.; Cao, X. The impact of emission right trading mechanism on green development. *China Popul. Resour. Environ.* **2018**, *28*, 12–21.
5. Shi, Q.; Chen, S.; Guo, F. Interviews by the Ministry of Environmental Protection and environmental governance: An example of air pollution. *Stat. Res.* **2017**, *34*, 88–97.
6. Bill, G. *How to Avoid a Climate Disaster: The Solutions We Have and the Breakthroughs We Need*; Knopf, A.A., Ed.; Penguin Books: New York, NY, USA, 2021.
7. Shi, G.; Zhou, L.; Zheng, S.; Zhang, Y.G. Environmental subsidies and pollution control-an empirical study based on the power industry. *Economics* **2016**, *15*, 1439–1462.
8. Deschenes, O.; Greenstone, M.; Shapiro, J. Defensive Investments and the Demand for Air Quality: Evidence from the NOx Budget Program. *Am. Econ. Rev.* **2017**, *107*, 2958–2989. [CrossRef]
9. Tu, Z.; Chen, R. Can the emissions trading mechanism achieve the Porter effect in China? *Econ. Res.* **2015**, *50*, 160–173.
10. Song, H.; Sun, Y.; Chen, D. Assessment of government air pollution control effects: An empirical study from the construction of "low-carbon cities" in China. *Manag. World* **2019**, *35*, 95–108, 195.
11. Cao, J.; Wang, X.; Zhong, X. Has the traffic restriction policy improved the air quality in Beijing? *Economics* **2014**, *13*, 1091–1126.
12. Shi, Q.; Guo, F.; Chen, S. The "political blue sky" in haze control: Evidence from the local "two sessions" in China. *China Ind. Econ.* **2016**, *5*, 40–56.
13. King, K.; Shen, K. Neighbor as beggar or neighbor as partner?—Environmental regulation enforcement interaction and urban productivity growth. *Manag. World* **2018**, *34*, 43–55.
14. Matthew, G. Regulation-Induced Pollution Substitution. *Rev. Econ. Stat.* **2019**, *101*, 827–840.
15. Keiser, D.A.; Shapiro, J. Consequences of the Clean Water Act and the Demand for Water Quality. *Q. J. Econ.* **2019**, *134*, 349–396. [CrossRef]
16. Zhang, M. The logic of environmental governance of grassroots governments in the context of environmental protection inspectors. *J. Huazhong Agric. Univ. (Soc. Sci. Ed.)* **2020**, *4*, 20–28.
17. He, Y.; Wang, G.; Chen, S. An analysis of the politics of urban government spending in China. *China Soc. Sci.* **2014**, *7*, 87–106.
18. Zhou, L.; Liu, C.; Li, X.; Weng, X. "Cascading" and officials' incentives. *World Econ. J.* **2015**, *1*, 1–15.
19. Zhou, L. A study on the promotion tournament model of local officials in China. *Econ. Res.* **2007**, *7*, 36–50.

20. Li, Q.; Zhang, K. Environmental governance model innovation under the new normal—How effective is the policy of Central Inspections of Environmental Protection? *Nankai J. (Philos. Soc. Sci. Ed.)* **2022**, *5*, 50–62.
21. He, D.; Kong, F. The Chinese experience of public policy implementation. *China Soc. Sci.* **2011**, *5*, 61–79.
22. Wang, W.W.; Sun, X.R.; Zhang, M. Does the central environmental inspection effectively improve air pollution?—An empirical study of 290 prefecture-level cities in China. *J. Environ. Manag.* **2021**, *286*, 112274.
23. Wu, R.X.; Hu, P. Does the "miracle drug" of environmental governance really improve air quality? Evidence from China's system of central environmental protection inspections. *Int. J. Environ. Res. Public Health* **2019**, *16*, 850. [PubMed]
24. Li, R.Q.; Zhou, Y.C.; Bi, J.; Liu, M.M.; Li, S.S. Does the central environmental inspection actually work? *J. Environ. Manag.* **2020**, *253*, 109602. [CrossRef] [PubMed]
25. Luo, Z.G.; Hu, X.Y.; Li, M.M.; Yang, J.R.; Wen, C.H. Centralization or decentralization of environmental governance-Evidence from China. *Sustainability* **2019**, *11*, 6938. [CrossRef]
26. Yu, J.; Liu, Y. The inspection system in vertical intergovernmental relations:A study of the Central Inspections of Environmental Protection. *Acad. Mon.* **2020**, *52*, 69–80.
27. Oskam, A. Productivity Measurement, Incorporating Environmental Effects of Agricultural Production. *Dev. Agric. Econ.* **1991**, *7*, 186–204.
28. Reinhard, S.; Lovell, C.; Thijssen, G. Econometric Estimation of Technical and Environmental Efficiency: An Application to Dutch Dairy Farms. *Am. J. Agric. Econ.* **1999**, *81*, 44–60. [CrossRef]
29. Hur, T.; Kim, I.; Yamamoto, R. Measurement of Green Productivity and Its Improvement. *J. Clean. Prod.* **2004**, *12*, 673–683. [CrossRef]
30. Li, R.; Li, N.; Yan, X. Mechanisms of electricity market integration on the efficiency of regional green economy. *Resour. Sci.* **2022**, *44*, 523–535.
31. Zhang, H.-Y.; Ding, Y. Global value chain embedding and green total factor energy efficiency—Evidence from Chinese manufacturing industry. *Zhejiang Soc. Sci.* **2022**, *2*, 4–13.
32. Li, J.; Chen, L.; Liu, M. The impact of Internet development on regional green economic efficiency in China. *China Popul. Resour. Environ.* **2021**, *31*, 149–157.
33. Chung, Y.H.; Fare, R.; Grosskopf, S. Productivity and Undesirable Output: A Directional Distance Function Approach. *J. Environ. Manag.* **1997**, *51*, 229–240. [CrossRef]
34. Tone, K. Dealing with Undesirable Outputs in DEA: A Slacks Based Measure (SBM) Approach. *Energy Policy* **2003**, *35*, 6323–6331.
35. Oh, D. A global Malmquist-Luenberger productivity index. *J. Product. Anal.* **2010**, *34*, 183–197. [CrossRef]
36. Shan, H. Re-estimation of capital stock K in China: 1952–2006. *Quant. Econ. Tech. Econ. Res.* **2008**, *25*, 17–31.
37. Taskin, F.; Zaim, O. The Role of International Trade on Environmental Efficiency: A DEA Approach. *Econ. Model.* **2001**, *18*, 1–17. [CrossRef]
38. Feng, L.; Chen, Z.; Chen, H. Does the Central Environmental Protection Inspectorate Accountability System Improve Environmental Quality? *Sustainability* **2022**, *14*, 6575. [CrossRef]
39. Hong, T.; Yu, N.N.; Mao, Z.G. Does environment centralization prevent local governments from racing to the bottom?—Evidence from China. *J. Clean. Prod.* **2019**, *231*, 649–659. [CrossRef]
40. Tan, Y.T.; Mao, X.Q. Assessment of the policy effectiveness of Central Inspections of Environmental Protection on improving air quality in China. *J. Clean. Prod.* **2020**, *288*, 125100. [CrossRef]
41. Wang, H.G.; Fan, C.H.; Chen, S.C. The impact of campaign-style enforcement on corporate environmental action: Evidence from china's central environmental protection inspection. *J. Clean. Prod.* **2021**, *290*, 125881. [CrossRef]
42. Chen, H.; Ni, D.; Zhu, S.; Ying, Y.; Shen, M. Does the National Credit Demonstration Policy Affect Urban Green Economy Efficiency? Evidence from the Yangtze River Delta Region of China. *Int. J. Environ. Res. Public Health* **2022**, *19*, 9926. [CrossRef]
43. Lin, Q.; Ling, H. Study on Green Utilization Efficiency of Urban Land in Yangtze River Delta. *Sustainability* **2021**, *13*, 11907. [CrossRef]
44. Ge, K.; Zou, S.; Ke, S.; Chen, D. Does Urban Agglomeration Promote Urban Land Green Use Efficiency? Take the Yangtze River Economic Zone of China as an Example. *Sustainability* **2021**, *13*, 10527. [CrossRef]
45. Hu, S.; Zeng, G.; Cao, X.; Yuan, H.; Chen, B. Does Technological Innovation Promote Green Development? A Case Study of the Yangtze River Economic Belt in China. *Int. J. Environ. Res. Public Health* **2021**, *18*, 6111. [CrossRef]
46. Wang, H.; Yang, G.; Qin, J. City Centrality, Migrants and Green Inovation Efficiency: Evidence from 106 Cities in the Yangtze River Economic Belt of China. *Int. J. Environ. Res. Public Health* **2020**, *17*, 652. [CrossRef]
47. Chen, H.; Zhu, S.; Sun, J.; Zhong, K.; Shen, M.; Wang, X. A Study of the Spatial Structure and Regional Interaction of Agricultural Green Total Factor Productivity in China Based on SNA and VAR Methods. *Sustainability* **2022**, *14*, 7508. [CrossRef]
48. Xiang, Y.; Wang, S.; Zhang, Y.; Dai, Z. Green Development Efficiency Measurement and Influencing Factors of the Paper Industry in the Yangtze River Economic Belt. *Water* **2021**, *13*, 1286. [CrossRef]
49. Liu, K.; Qiao, Y.; Zhou, Q. Analysis of China's Industrial Green Development Efficiency and Driving Factors: Research Based on MGWR. *Int. J. Environ. Res. Public Health* **2021**, *18*, 3960. [CrossRef]

Disclaimer/Publisher's Note: The statements, opinions and data contained in all publications are solely those of the individual author(s) and contributor(s) and not of MDPI and/or the editor(s). MDPI and/or the editor(s) disclaim responsibility for any injury to people or property resulting from any ideas, methods, instructions or products referred to in the content.

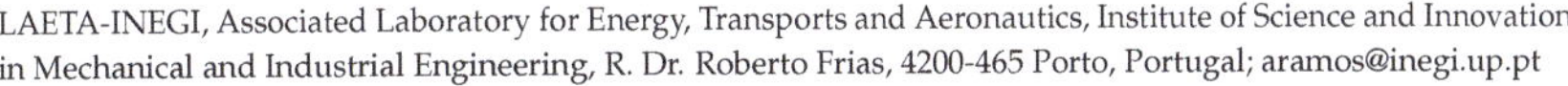

Editorial

Afterword for the Special Issue "Circular Economy Strategies for Sustainable Development: Applications and Impacts"

Ana Ramos 

LAETA-INEGI, Associated Laboratory for Energy, Transports and Aeronautics, Institute of Science and Innovation in Mechanical and Industrial Engineering, R. Dr. Roberto Frias, 4200-465 Porto, Portugal; aramos@inegi.up.pt

Citation: Ramos, A. Afterword for the Special Issue "Circular Economy Strategies for Sustainable Development: Applications and Impacts". *Sustainability* **2024**, *16*, 311. https://doi.org/10.3390/su16010311

Received: 4 December 2023
Revised: 15 December 2023
Accepted: 19 December 2023
Published: 29 December 2023

Copyright: © 2023 by the author. Licensee MDPI, Basel, Switzerland. This article is an open access article distributed under the terms and conditions of the Creative Commons Attribution (CC BY) license (https://creativecommons.org/licenses/by/4.0/).

1. Background

Circular economy (CE) is a holistic approach to sustainable development that aims to minimize waste and make the most of resources. It involves designing products, processes, and systems with an emphasis on all strategies leading to a recirculation of materials into new production cycles or flows so that it is possible to take advantage of their highest value and the maximum number of cycles [1,2]. Potting et al. [3] suggest the 9R methodology, which, within this SI, perfectly serves as a compilation of strategies to pursue the proposed thematic of circularity through innovative applications and impact measurement. This is an environmentally preferred hierarchical approach for closing material loops. The tighter the loop (lower R), the lesser external inputs are needed to close it, and the more circular the strategy. The longer the loop (higher R), the less circular it is, and the less we it should be prefered. Indeed, the approach is based on a hierarchical framework from the most circular to the most linear systems, under three greater groups of strategies: smarter product use and manufacture, extend the lifespan of products and its parts, and useful application of materials. The first set comprises strategies such as refusing, rethinking and reducing, while the second group spans from reusing, repairing, refurbishing, remanufacturing and repurposing; and the third one deals with recycling and recovering options.

Measuring the impacts of the newly proposed solutions is obviously crucial, as this enables an effective assessment of the benefits achieved, also accounting for resource efficiency, waste reduction and the regeneration of natural systems. Environmental indicators together with economic and social benefits associated with these circular strategies can be evaluated in a holistic manner, promoting more informed decisions and policy-making under a life cycle thinking perspective [4]. Long-term planning as well as benchmarking and comparisons are also anchored in impact measurement, which is a key aspect of sustainability. In summary, measuring the impacts of circular economy strategies is integral to accountability, improvement, and the overall success of sustainable practices, supporting transparency, and contributing to the ongoing evolution of circular economy principles.

These topics collectively highlight the multidimensional nature of CE strategies, incorporating environmental, social, and economic considerations. Research in these areas contributes to the ongoing efforts to create a more sustainable and circular approach to resource use and economic development.

2. Overview of the Special Issue

This Special Issue belongs to the Section "Sustainable Management" of the journal *Sustainability* and was open between January 2021 and January 2023. Its Guest Editor is Ana Ramos (Institute of Science and Innovation in Mechanical and Industrial Engineering), senior researcher in the area of environmental engineering who specialized in waste-to-energy techniques, sustainability, life cycle thinking and circular economy. The keywords for this Special Issue include circular economy, sustainability, environmental assessment, climate change, sustainable development goals, socio-economic impacts, sustainability indicators, energetic efficiency, industrial engineering, and innovation. This Special Issue focuses on the development and impact measurement of circular strategies to foster sustainable development and to promote the regeneration of resources, while enabling pollution decrease and restoring biodiversity.

After the call to contribute a paper to this Special Issue (Contribution 1), several authors have accepted the invitation, and a compilation of eleven works was achieved, representing a wide set of geographies (Poland, Ukraine, Germany, Mexico, Belgium, Chile, China, Portugal, The Netherlands, Italy, and Switzerland).

From these, a review on seaweed functionality as a sustainable bio-based material was proposed by Pranav Nakhate and Yvonne van der Meer (Contribution 2), highlighting a new feedstock possibility to maximize value creation in bio-based materials. Indeed, developmental process, by-product promotion, financial assistance, and social acceptance approaches were also put forward in order to promote seaweed as a circular material accounting for several sustainability dimensions. Matheus Oliveira, Ana Ramos, Eliseu Monteiro and Abel Rouboa (Contribution 3) reported an improved purification process for crude glycerol, as a by-product in the production of biodiesel, towards higher circularity in biofuel production. A 99.77% purification degree enabled the final product to meet the specifications for pharmaceutical uses, simultaneously allowing the development of new products with greater added-value and contributing to the zero-waste principles.

Under the technical point of view, Baharam Roy, Peter Kleine-Möllhoff and Antoine Dalibard (Contribution 4) developed an innovative biorefinery concept that includes superheated steam drying and the torrefaction of biomass residues, recovering the volatile fraction as well as several valuable platform chemicals (e.g., 5-(hydroxymethyl)furfural and acetic acid). The authors also conducted the economic evaluation of the proposed solution (Contribution 5), comparing two systems: one that supplies various platform chemicals and torrefied biomass and the other providing thermal energy for external consumers in addition to platform chemicals. Both options have shown to be profitable if the focus of the platform chemicals produced is on high quality and higher-priced segments. With this, a viable business model could be developed.

Product-wise, Virginia Lama, Serena Righi, Brit Maike Quandt, Roland Hischier and Harald Desing (Contribution 6) suggested the recently developed resource pressure method to assess the environmental sustainability of various carpet design alternatives. The product system was evaluated in relation to its consumption of primary resources and the final generation of waste, with a good correlation for most of the considered impact categories being observed. Thus, the method has proved to be simple and effective in predicting environmental impacts, accounting for the environmental pillar of sustainability.

From a more tactical-related perspective, Aleksandra Kuzior, Olena Arefieva, Zarina Poberezhna and Oleksiy Ihumentsev (Contribution 7) described the mechanism of formation of the strategic potential of an enterprise in circular economy, providing the potential for compliance with the strategic goals of the company, and allowing decisions to be made on the implementation of measures to better meet these goals. The conditions for the transition to a circular economy at the macro level were stated after a review of the circular economy concept and its impact on business and resource conservation, as well as in environmental protection. Furthermore, Carlos Scheel and Bernardo Bello (Contribution 8) developed a roadmap to transform linear value chains into an industrial ecology cluster of zero-waste chains, enabling institutions to exploit a circular value extended system. Their systemic approach enables the possibility to create a socially inclusive, environmentally resilient, and economically viable system of capital, with guidelines to transform linear value chains into a cluster of circular economy systems being provided for a dedicated case study in the mining industry in British Guyana.

Enduring the social sphere of sustainability, Katherine Mansilla-Obando, Fabiola Jeldes-Delgado and Nataly Guiñez-Cabrera (Contribution 9) analyzed stakeholders' perception of circular economy strategies using the 3Rs framework and stakeholder theory, conducting a case study for a glass company through a qualitative methodology. The main takeaway was that stakeholders perceive the contribution of the 3Rs approach to have social implications, such as supplier evaluation with social impact, responsibility for the product, and decent work conditions. In a different area, Nuri Cihan Kayaçetin, Chiara Piccardo and Alexis Versele (Contribution 10) proposed a social impact assessment framework tested on a living lab for the construction sector. A multitude of impact categories relevant to different stakeholder groups were qualitatively and quantitatively assessed for production and construction phases. New indicators for circular construction methods were identified, with the social impacts being discussed for each stakeholder category. Also, recommendations were provided from a construction-based view, with the capability to be integrated into existing social impact assessment guidelines. Also, in the field of construction, Houchao Sun, Yuwei Fang, Minggan Yin and Feiting Shi (Contribution 11) assessed the promotion of prefabricated buildings from a low-carbon, environmental protection and high-efficiency perspective under the "Double Carbon" framework. A decision-making laboratory method together with an interpretation structure method were used to build a multi-level hierarchical structure model of constraints, the logical relationship, hierarchical relationship and relative importance of each constraint being clarified.

Finally, Haisheng Chen and Manhong Shen (Contribution 12) analyzed the effect of the central inspections of environmental protection on the development of a green economy, verifying that it varies from time to time and place to place. The authors state that it is important to regulate the use of administrative resources and strengthen inter-provincial coordination to promote synergy and cooperation across environmental inspection systems. This provided insights for understanding the logics behind the implementation of such environmental inspection mechanisms, and for promoting the green transformation of regional and national economy in China.

3. Concluding Remarks

Indeed, this series of works confirmed that circular economy is a topic to be embraced at different levels, by distinct stakeholders and within diverse applications. Correspondingly, the impact assessment of the circular strategies was evaluated, a multitude of examples spanning from the environmental concerns to the economic viability, the technical feasibility, and the social impacts being described.

The primary overarching message conveyed in this work is that linear economy should be replaced by circular systems, and waste cannot simply be discarded because it persists on our planet, impacting our lives in various ways. It is unavoidable to recognize the existing situation as an opportunity for evolution towards a more environmentally friendly direction. Therefore, individual behavioral changes can play a pivotal role in reducing both the volume of generated waste and the energy demands linked to a comfortable and affluent lifestyle, as well as the replacement of fossil-based products by more sustainable alternatives. Plus, from an industrial perspective, all sectors and interested parties should make the transition, taking advantage of the momentum created. On a macroeconomic scale, a comprehensive reevaluation of the current production and supply chains would also promote a more sustainable existence on Earth.

Funding: This work is funded by national funds through the FCT—Fundação para a Ciência e a Tecnologia, I.P. and, when eligible, by COMPETE 2020 FEDER funds, under the Scientific Employment Stimulus—Individual Call (CEEC Individual)—2021.03036.CEECIND/CP1680/CT0003.

Acknowledgments: The editor thanks all the authors, reviewers and editorial team for the support in bringing this Special Issue to life.

Conflicts of Interest: The author declares no conflicts of interest.

List of Contributions:

1. Ramos, A. Editorial for the Special Issue "Circular Economy Strategies for Sustainable Development: Applications and Impacts. *Sustainability* **2022**, *14*, 12831.
2. Nakhate, P.; van der Meer, Y. A Systematic Review on Seaweed Functionality: A Sustainable Bio-Based Material. *Sustainability* **2022**, *13*, 6174.

3.   Oliveira, M.; Ramos, A.; Monteiro, E.; Rouboa, A. Improvement of the Crude Glycerol Purification Process Derived from Biodiesel Production Waste Sources through Computational Modeling. *Sustainability* **2022**, *14*, 1747.

4.   Roy, B.; Kleine-Möllhoff, P.; Dalibard, A. Superheated Steam Torrefaction of Biomass Residues with Valorisation of Platform Chemicals—Part 1: Ecological Assessment. *Sustainability* **2022**, *14*, 1212.

5.   Roy, B.; Kleine-Möllhoff, P.; Dalibard, A. Superheated Steam Torrefaction of Biomass Residues with Valorisation of Platform Chemicals Part 2: Economic Assessment and Commercialisation Opportunities. *Sustainability* **2022**, *14*, 2338.

6.   Lama, V.; Righi, S.; Quandt, B.M.; Hischier, R.; Desing, H. Resource Pressure of Carpets: Guiding Their Circular Design. *Sustainability* **2022**, *14*, 2530.

7.   Kuzior, A.; Arefieva, O.; Poberezhna, Z.; Ihumentsev, O. The Mechanism of Forming the Strategic Potential of an Enterprise in a Circular Economy. *Sustainability* **2022**, *14*, 3258.

8.   Scheel, C.; Bello, B. Transforming Linear Production Chains into Circular Value Extended Systems. *Sustainability* **2022**, *14*, 3726.

9.   Mansilla-Obando, K.; Jeldes-Delgado, F.; Guiñez-Cabrera, N. Circular Economy Strategies with Social Implications: Findings from a Case Study. *Sustainability* **2022**, *14*, 13658.

10.   Kayaçetin, N.C.; Piccardo, C.; Versele, A. Social Impact Assessment of Circular Construction: Case of Living Lab Ghent. *Sustainability* **2023**, *15*, 721.

11.   Sun, H.; Fang, Y.; Yin, M.; Shi, F. Research on the Restrictive Factors of Vigorous Promotion of Prefabricated Buildings in Yancheng under the Background of "Double Car-bon". *Sustainability* **2023**, *15*, 1737.

12.   Chen, H.; Shen, M. Do Central Inspections of Environmental Protection Affect the Efficiency of the Green Economy? Evidence from China's Yangtze River Delta." *Sustainability* **2023**, *15*, 747.

References

1.   Foundation, E.M. What is a Circular Economy? 2020. Available online: https://ellenmacarthurfoundation.org/topics/circular-economy-introduction/overview (accessed on 13 July 2022).

2.   Foundation, E.M. *Universal Circular Economy Policy Goals*; Ellen MacArthur Foundation: Cowes, UK, 2021; p. 66.

3.   Potting, J.; Hekkert, M.P.; Worrell, E.; Hanemaaijer, A. *Circular Economy: Measuring Innovation in the Product Chain*; PBL Publishers: Utrecht, The Netherlands, 2017.

4.   Guinée, J. Life Cycle Sustainability Assessment: What Is It and What Are Its Challenges? In *Taking Stock of Industrial Ecology*; Clift, R., Druckman, A., Eds.; Springer International Publishing: Cham, Switzerland, 2016; pp. 45–68.

Disclaimer/Publisher's Note: The statements, opinions and data contained in all publications are solely those of the individual author(s) and contributor(s) and not of MDPI and/or the editor(s). MDPI and/or the editor(s) disclaim responsibility for any injury to people or property resulting from any ideas, methods, instructions or products referred to in the content.

MDPI AG

Grosspeteranlage 5

4052 Basel

Switzerland

Tel.: +41 61 683 77 34

Sustainability Editorial Office

E-mail: sustainability@mdpi.com

www.mdpi.com/journal/sustainability

Disclaimer/Publisher's Note: The title and front matter of this reprint are at the discretion of the . The publisher is not responsible for their content or any associated concerns. The statements, opinions and data contained in all individual articles are solely those of the individual Editor and contributors and not of MDPI. MDPI disclaims responsibility for any injury to people or property resulting from any ideas, methods, instructions or products referred to in the content.

www.ingramcontent.com/pod-product-compliance
Lightning Source LLC
LaVergne TN
LVHW071020170726
843515LV00006B/1164